Environmental Science
AIR and EARTH

D1434416

Block **2**

Cover image View of sandstone cliffs, Southern Utah, USA.

The Open University, Walton Hall, Milton Keynes, MK7 6AA

First published 2002. Reprinted 2006.

Copyright © 2002, 2006 The Open University

All rights reserved. No part of this publication may be reproduced, stored in a retrieval system, transmitted or utilized in any form or by any means, electronic, mechanical, photocopying, recording or otherwise, without written permission from the publisher or a licence from the Copyright Licensing Agency Ltd. Details of such licences (for reprographic reproduction) may be obtained from the Copyright Licensing Agency Ltd of 90 Tottenham Court Road, London W1T 4LP.

Edited, designed and typeset by The Open University.

Printed and bound in the United Kingdom at the University Press, Cambridge

ISBN 0 7492 6988X

This publication forms part of an Open University course, S216 *Environmental Science*. Details of this and other Open University courses can be obtained from the Student Registration and Enquiry Service, The Open University, PO Box 197, Milton Keynes MK7 6BJ, United Kingdom: tel. +44 (0)1908 653231, e-mail ces-gen@open.ac.uk

Alternatively, you may visit the Open University website at http://www.open.ac.uk where you can learn more about the wide range of courses and packs offered at all levels by The Open University.

To purchase a selection of Open University course materials visit http://www.ouw.co.uk, or, contact Open University Worldwide, Michael Young Building, Walton Hall, Milton Keynes MK7 6AA, United Kingdom for a brochure. tel. +44 (0)1908 858785; fax +44 (0)1908 858787; e-mail ouwenq@open.ac.uk; website http://www.ouw.co.uk

2.1

PART 1

AIR

Andrew Conway and Ross Reynolds

What is the atmosphere?

The Earth's atmosphere is a tiny fraction of the environment in terms of mass. At the same time, it is the most dynamic aspect of the environment. Our everyday experience tells us that the weather can change dramatically from day to day, hour to hour, and even minute to minute. On time-scales from years to millions of years, we can also see changes in the overall patterns of weather—the **climate**. Some of these changes can be induced by human activity, but many are completely natural.

Despite its relatively tiny mass, the atmosphere has a profound impact on the conditions on Earth. When cloudy, it can prevent a large fraction of sunlight from reaching the Earth's surface. It is even more effective in preventing the surface's heat (absorbed from sunlight) from escaping back into space. Both these interactions depend on the amount of water in the atmosphere: water in the form of clouds blocks out sunlight, and water vapour traps the heat. Clearly, the amount of water in the atmosphere depends on how much is evaporated from the oceans. Less well known is the fact that water is transferred to the atmosphere by evaporation from the leaves of plants, which draw water up from the soil through their roots. This one example—and there are many others—demonstrates that the atmosphere plays an integral part in the complex interactions that take place within the environment of the Earth.

Block 2, Part I is an introduction to the basics of the atmosphere. Sections 1 and 2 are concerned primarily with *what* the atmosphere is and *what* phenomena take place within it. Sections 3 and 4 delve more deeply into the *why* and *how* questions of the atmosphere.

1.1 Ascending through the atmosphere

This subsection describes the balloon ascent of Col. Kittinger, which you can view as video on the DVD.

One complete circuit of the globe requires travelling a distance of about 40 000 km. To reach space is only an upward journey of somewhere between 20 km and 100 km (depending on what you mean by space). Before describing the atmosphere in factual terms, we will describe one human's pioneering journey through the atmosphere, that of US Air Force Col. Joe Kittinger (Figure 1.1).

In 1960, three years after the first satellite entered space, and a year before Yuri Gagarin's first space flight, a balloon carried Col. Joe Kittinger to heights greater than ever achieved before that time. Before take off, the balloon is a partially inflated, saggy sack of helium. Helium is a gas that is less dense than air, meaning that a given volume of helium is lighter, i.e. has a smaller mass, than the same volume of air. Being less dense, helium is buoyant and rises up through the atmosphere, carrying its gondola and passenger with it. Another way to achieve this effect is to use hot air instead of helium, because hot air is less dense than cold air. Understanding the science behind this last statement, which we'll get to later, is crucial to understanding the workings of the atmosphere and weather.

Figure 1.1 Col. Joe Kittinger—arguably the first human to visit space.

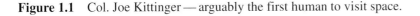

Let's now follow Kittinger's epic journey. The straps are cut and the balloon begins to rise. As it ascends, the balloon begins to inflate; a barometer carried on the balloon indicates that pressure is dropping. The pressure falls rapidly but smoothly as the balloon ascends, and continues to do so through all layers of the atmosphere. The air around the balloon also gets colder, but in a more complex way. By 1 km the temperature has dropped from the surface reading of 16 °C to 10 °C, at 2 km the temperature is 4 °C, and at 3 km it is −2 °C. This trend continues until about 11 km (about 33 000 feet), where the temperature ceases falling, and remains almost constant at about −60 °C.

○ Sketch a graph of air temperature outside the balloon against height. A common convention is to have temperature increasing to the right on the page, and height increasing up the page. There is no need to accurately plot many points; the aim is to draw the rough shape of the temperature profile.

● Your graph should look similar to the lower part of Figure 1.10 (on p. 17).

The balloon is now in a different realm of the atmosphere. All weather as we know it is below, and the sky above appears to be empty and calm. The view from Kittinger's balloon is much like that from a passenger jet. However, this is 1960 and only a privileged few have seen this view, and *no one* has seen the view from the heights where Col. Kittinger is heading.

The balloon is now well above the height of Mount Everest, the highest mountain on the Earth. As on high mountains, the air is so thin (in other words, there are so few air particles per unit volume) that breathing apparatus is required. At 15 km (50 000 feet) Col. Kittinger faces a problem not faced by mountaineers. He is now just on the verge of entering what some scientists would call space, where he must rely on his pressure suit to protect him. It is at this crucial point that Col. Kittinger notices that one glove has sprung a leak. The low pressure is causing too much blood to flow into his hand. Being more soldier that scientist, he vows to press on regardless. After another 5 km or so, the thermometer in the gondola shows us something surprising — the temperature begins to increase. This increase persists throughout the remainder of the ascent, but the temperature is still well below 0 °C. In any case, the thinness of the air here means that the usual relationship between temperature and what we perceive as heat has changed.

It is not long before the blue of the atmosphere becomes just a thin haze over the surface of the planet, and the sky gives way to the blackness of space (Figure 1.2). The air is now so very thin, and free of dust and other particles, that it no longer scatters the light. Either the light from the Sun, Earth, Moon and stars makes it straight to your eye, or you just see blackness.

After one and a half hours, at a height of 31.4 km (103 000 feet), Col. Kittinger, who is touching space and setting the record for the highest ascent made by a human, leaps out of the balloon's gondola. The world around him is a deadly environment — as deadly as being surrounded by a poisonous gas. Yet, just over 30 km below him is a safe, fertile environment supporting humans and a myriad of other kinds of life.

Figure 1.2 The blackness of space and the blue haze of the atmosphere. (The satellite in the photo is the Skylab Space Station.)

At first, apart from feeling weightless, there is no sensation of falling. The air is so thin that it cannot make the fabric of his suit rustle, or permit sound waves to travel from it. In fact he is falling at the speed of sound. In just four minutes of free fall he enters the blue haze of the lower atmosphere, and quickly the blackness of space transforms into familiar blue sky. Now he can feel the air as he rushes through it — it rustles the suit fabric and 'wooshes' as it speeds past his ears. As he approaches the Earth, the air becomes dense enough to make using a parachute feasible. The air hitting the underside of the parachute slows him down, counteracting the downward force of gravity until he slows to a speed where he can safely land upon the hard surface of the Earth. Now all around him, supported by the air, the rocks, the water and the soil, is a life-rich environment.

1.2 Constituents of the atmosphere

The gas we call **air** is a mixture of many individual gases, but can roughly be described as four-fifths nitrogen and one-fifth oxygen. Nitrogen gas is composed of a great many nitrogen molecules, each molecule being a pair of nitrogen (N) atoms that are strongly bonded together. For this reason, nitrogen is represented by the shorthand for its molecule, N_2. Oxygen is required for respiration. It is composed of O_2 molecules, in which the oxygen (O) atoms are bonded together, but not as strongly as the N atoms in N_2 molecules. Many forms of life on Earth (all multicellular organisms and most single-celled ones too) can use oxygen because they are able to break the oxygen to oxygen bond in the O_2 molecule. In contrast, only a very few species can cleave the strong bond that binds the N atoms in the N_2 molecule (e.g. certain specialized bacteria that can use atmospheric nitrogen in protein synthesis).

1.2.1 The nature of a gas

To understand the atmosphere, which is mostly gas, it is essential to understand the nature of a gas.

○ How would you define a gas, a liquid and a solid? *Hint*: imagine you have a certain volume of each in a sealed container.

● In everyday terms, it is easy to define the difference between a solid and a liquid. A brick is solid because it keeps the same shape without any need for a container. Any liquid, such as water, always settles to the bottom of a container. A gas is more difficult to imagine because, unlike solids and liquids, most gases are invisible. If there is an amount of gas in a sealed container, it will fill the container.

Before going any further, some clarification of terminology is needed. The term **particle** can be used to mean almost anything from a tiny sub-component of an atom, to a grain of dust which contains a huge number of atoms. In this block we will use 'particle' to mean an individual particle within a gas — either a molecule (as in oxygen or nitrogen) or an atom (as in argon).

The properties of a gas (or a solid or liquid) are best understood in terms of the particle model. A gas, a liquid and a solid are all made up of many individual particles, which can be either atoms or molecules. In a gas the particles are quite far apart, on average, and do not hold on to each other in any way, though they do collide with each other. This can explain two of the properties of a gas mentioned earlier. Firstly, the particles are free to move, so they can move freely around inside a container to fill it up. Secondly, because the particles are far apart, it follows that there are fewer particles in a given volume of a gas, as compared with the same volume of liquid or solid. In other words, a gas has a lower density than a liquid or a solid. So, in our everyday experience, many gases are invisible because light can pass through a gas without colliding with many gas particles (actually this is only part of the explanation). In contrast to gases, particles in liquids are closer together, and hold on to one another to some extent. Particles in solids tend to be even closer, and hold on to each other so strongly that they are almost fixed in place.

There is one more important property possessed by a gas: it is compressible. Unlike a solid or a liquid, an amount of gas (meaning either a mass, or a number of particles) can be squeezed into smaller volumes, or allowed to expand into larger volumes. When you pump up a tyre on a car or a bicycle, you are putting more air into the tyre, and if you are doing it correctly, no air should be escaping. When air is pumped into a flat tyre it causes the rubber to stretch, and the tyre expands. Once the rubber is stretched to near its limit, when it no longer looks flat, you can still put air in, but now you are compressing the air. In other words, you are putting more and more particles into the same volume. The number of particles per unit of volume is called the **number density**, and is usually denoted by N/V, where N is the number of particles and V is the volume that they occupy. In SI units, 1 m^3 is used as the standard volume, so number density has units of (number of particles) m^{-3}.

In general, if an object has a large number density, it also has a large **mass density**, which is defined as the mass per unit volume. For example, lead has a much higher density than air, in both number and mass terms, because its particles (lead atoms) are much closer together than air particles (mostly nitrogen and oxygen molecules). Although it is also true that lead atoms have a greater mass than nitrogen and oxygen molecules, the difference in mass is nowhere near enough to explain the enormous difference in the mass densities of air and lead. Finally, when the word 'density' is used by itself, in most cases it is referring to mass density. If there is only one type of gas particle present, and you know the mass of the gas particle, m, you can define the mass density, ρ, as the number density, N/V, multiplied by the particle mass:

$$\rho = \frac{\text{mass}}{\text{volume}} = \frac{Nm}{V} \tag{1.1}$$

The SI units of mass density are kg m^{-3}. If, like air, there is a mixture of gases, all with different masses, then the mass density is still defined as mass divided by volume, but the mass must take account of the different proportions and particle masses of the constituent gases.

1.2.2 The nature of air

Table 1.1 lists the amounts of the two most common gases in air (oxygen and nitrogen), together with the next most common (argon) and two important trace gases. A **trace** gas is one that makes up only a small proportion of a sample. Argon is an example of gas that does not readily form molecules, so its fundamental gas particles are just argon atoms. Higher layers of the atmosphere have similar proportions of the two main gases, oxygen and nitrogen, but can have different proportions of the trace gases. The amounts of each gas can be expressed as **mixing ratios**, which are explained in Box 1.1. Because the amount of water vapour in the air is very variable, scientists usually deal with this constituent separately, and refer to the other constituents as **dry air**.

The exact proportion of each gas can vary with location and time. A famous example is ozone. Its mixing ratio is greatest in the stratosphere, and the amount of stratospheric ozone over Antarctica is reduced by a large amount each winter due to the presence of certain synthetic chemicals. There are many smaller-scale examples as well. The air expelled in your breath has a larger mixing ratio of water vapour and carbon dioxide (CO_2) and a smaller mixing ratio of oxygen than the surrounding air. Fortunately, the changes you make to the air are slight in comparison to the total amount that surrounds you in a normal environment, and continual motions in the atmosphere ensure that 'fresh' air is brought to you.

Table 1.1 Some constituents of dry air.

Gas	Symbol	Mixing ratio
nitrogen	N_2	0.781
oxygen	O_2	0.209
argon	Ar	0.0093
carbon dioxide	CO_2	365 ppmv*
ozone†	O_3	0–200 ppbv* (troposphere)
		0.1–8 ppmv* (stratosphere)

* These are explained in Box 1.1. † Ozone's mixing ratio
varies considerably throughout the atmosphere, as described later.

Box 1.1   Mixing ratios

At first sight, the first few entries in Table 1.1 can be interpreted easily enough. Air is roughly 78% nitrogen, 21% oxygen and less than 1% trace (all other) gases. However, we must be precise and scientific. If we were to analyse a sample of air, would we find that 78% of the gas particles in the sample are nitrogen molecules, or that 78% of the sample's mass is nitrogen molecules? There is a difference because nitrogen and oxygen molecules have different masses. The answer is that the table is dealing with numbers of particles; the definition of the mixing ratio is

$$\text{mixing ratio} = \frac{\text{number of atoms or molecules of the gas}}{\text{total number of all atoms and molecules of all gases present}} \quad (1.2)$$

Lower down the table, ppmv is used, which stands for parts per million by volume. Sometimes this unit is further abbreviated to ppm. Carbon dioxide has a mixing ratio of 365 ppmv, meaning that there are 365 carbon dioxide molecules per million particles in the gas. This is a shorthand for 0.000 365. There is also ppbv, where 'b' stands for billion (i.e. 10^9 = 1 000 000 000) and ppt, where 't' stands for thousand. Be aware that there are other types of mixing ratio, e.g. the mass mixing ratio (parts per million by mass) which uses the abbreviation ppmm.

1.2.3   The dynamic atmosphere

Gravity holds the atmosphere close to the Earth's surface. Solar System bodies smaller than the Earth (e.g. Mercury, the Moon and asteroids) have long since lost their atmospheres because of their lesser mass, and therefore lower gravity. You might then wonder why the atmosphere doesn't just sink to the surface like the water in oceans and lakes. The explanation comes from the fact that air particles are in continual motion. To understand this effect we will use an analogy in which gravity acts on something more familiar. Consider dropping a ball. When the ball hits the ground, it doesn't just stop; it bounces up again. Also, the lighter the ball, the more it is affected by the presence of air, which can slow it down or blow it sideways in a gust of wind. Likewise, the gas particles (be they atoms or molecules) in the atmosphere do not just fall down and hit the ground and stop; they bounce off the ground and each other. Considering that gas particles have a very tiny mass and move at a typical speed of over 1 km s^{-1} at 20 °C, it is not too surprising that they can apparently defy gravity by continually bouncing around. However, there is of course a difference between the ball analogy and the gas particle's reality. The ball loses energy with every bounce and eventually comes to rest on the ground after just a few bounces. Unlike the ball, gas particles' energies are replenished by the influx of sunlight. Some of this energy goes straight into the gas — it is absorbed by the gas — and this keeps the particles moving. Also, sunlight heats the ground, so that the gas particles can actually gain some energy every time they hit the warmer ground.

An analogy that illustrates why the atmosphere extends above the ground is popcorn in a pan. When cold it just sits at the base of the pan. When warmed the popcorn pops and starts to bounce around the pan — in a sense, filling it up. In short, the Earth's atmosphere is held above the surface, defying gravity, by the continual input of energy from the Sun. Exactly how this energy is distributed dictates the structure of the atmosphere and can explain a great deal about weather phenomena. We will explore this further in Section 3.

You might be wondering how the atmosphere's make-up came to be as it is today, and if it will change in the future. Most crucially for us, is the amount of oxygen decreasing because it is being used by us and other life forms? The answer is no — certainly not in your lifetime or in the foreseeable future. In fact, the oxygen is here today *because* of life. Oxygen that is lost from the atmosphere because we have breathed it in, is replaced by other living organisms, i.e. plants, which produce oxygen by photosynthesis (described in Block 3). Water too cycles through the atmosphere: water vapour is lost from the atmosphere when it condenses and falls as rain and is replaced by evaporation from the land and sea. From these examples, we can see that atmospheric constituents, such as oxygen and water vapour, can vary from place to place and from time to time. But is the atmosphere in an *overall* balance so that the amount of each gas remains the same? In other words, is the atmosphere in what is called a **steady state**, in terms of its constituents?

Before answering this question, you should realize that the term 'steady state' has a precise scientific meaning, which is best explained using a simple example. Consider water flowing from a tap into a sink. The water falls from the tap, splashes around and flows down the plug-hole. The sink is a system in steady state because the rate at which water enters is matched by the rate at which water leaves. If this were not true, because for example the sink's plug-hole became blocked, then it would fill up with water. Once the water reached the level of the overflow, water would drain out again and a steady state would again be achieved. Note that this steady state would have a different water level (up to the overflow) than the initial steady state (almost empty). In general terms, a system is in a steady state if the inputs and outputs of a particular quantity are equal, and the amount of that quantity remains constant.

We now return to the question: is the Earth's entire atmosphere in a steady state in terms of the amounts of each of its constituents gases? The answer is: almost. There are changes in the amounts of certain gases, such as the increase in carbon dioxide, and the decrease in ozone in the stratosphere over the poles. A glance at Table 1.1 might give the naïve impression that this is not a worry; after all, both gases make up only a small proportion of the atmosphere. Size, or in this case amount, is not everything though. The role of both of these gases is crucial for very different reasons that will be described later. Consequently, the potential impacts of changes in the amounts of these gases have become important worldwide issues in the last few decades. In order to assess these impacts and initiate action to reverse the changes, we must understand the precise role of these gases in the whole environment of the Earth. We now illustrate the important roles played by three atmospheric gases: water vapour, carbon dioxide and ozone. Many of the points raised here will be explored later on in this block and also in other blocks.

Water: H₂O

Water has many different roles in the environment, but we will only be concerned here with its role whilst in the atmosphere. One of water's most familiar guises is as clouds, which reflect solar radiation back into space, preventing a good deal of it from reaching the lower atmosphere. As we all know, a cloudy day is often a cool day. Less obvious but equally important effects arise from water in its gaseous form: **water vapour**. Be careful not to confuse water vapour with mist, steam or clouds, which are composed of liquid water, made up of many, tiny droplets, or in the case of some clouds many tiny ice crystals. Each droplet or ice crystal is composed of an enormous number of water molecules (depicted in Figure 1.3). In contrast, water vapour is an invisible gas that is composed of many individual, freely moving water molecules. In liquid water the molecules are much closer to each other than in water vapour. Being closer, these molecules attract each other, though the attraction is not strong enough to hold the molecules together in a rigid structure, as would be the case in ice (solid water). In order to evaporate liquid water, energy must be given to overcome the attraction of the water molecules to each other and thus separate them. When water vapour condenses to form liquid water, the opposite is true, i.e. energy is released. For example, sweating cools us down because energy from our skin is transferred to the water in sweat, making it evaporate. In the same way, water evaporating from the Earth's surface takes energy from the land and transfers it to the atmosphere. Thus water acts as an important storer and transporter of energy in the atmosphere. We will consider the role of water and the energy involved in its evaporation and condensation in more detail in Section 3.

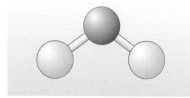

Figure 1.3   A ball-and-stick model of the H₂O (water) molecule. The atoms, represented here by coloured balls, are joined together by covalent bonds. Here a single oxygen atom (red) is bonded to two hydrogen atoms (white).

Water molecules can absorb electromagnetic radiation (see Box 1.2). In the visible part of the spectrum, i.e. the part that our eyes can see, water vapour appears to be transparent, letting most light pass through unhindered. However, it is particularly good at absorbing infrared radiation. The bulk of this radiation does not come directly from the Sun but is emitted by the Earth's surface and the atmosphere as a result of being heated by the Sun. So in addition to storing and transporting energy, water plays an important role in 'catching' energy flowing through the Earth's atmosphere as radiation.

Box 1.2   The spectrum of electromagnetic radiation

You can think of **electromagnetic radiation** as waves travelling through electric and magnetic fields which surround us. You may not be able to see these fields themselves (or even happily digest the concept just now!) but your eyes are designed to detect these waves and relay the information to the brain. A radio performs a similar function in a different part of the electromagnetic spectrum, receiving radio waves and relaying them to a speaker to make sound.

Figure 1.4 includes these two subranges, along with all of the others. The ordering from left to right in the figure is in accord with a property of electromagnetic radiation that increases in value continuously, with no abrupt changes as we go from one subrange to the next. This property is called the wavelength of the radiation.

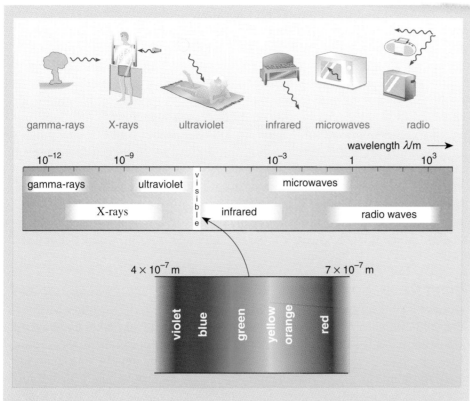

Figure 1.4 Electromagnetic radiation subdivided by wavelength. Note that the wavelength increases from left to right across the figure, so that (for example) infrared waves have longer wavelengths than visible radiation.

Carbon dioxide: CO_2

Atmospheric carbon dioxide, like water, absorbs infrared radiation. A CO_2 molecule is shown in Figure 1.5. Unlike water, carbon dioxide is always a gas at the Earth's temperatures and so does not form clouds or transfer heat through evaporation. Water vapour and carbon dioxide are the two most important greenhouse gases. We'll give a brief account of the greenhouse effect here, and then will discuss it again in more detail in Section 3 once a few key concepts have been established.

Figure 1.5 A ball-and-stick model of the CO_2 (carbon dioxide) molecule. Here, a single carbon atom (black) is bonded to two oxygen atoms (red).

The Earth's atmosphere is transparent to most of the solar radiation, which is mainly in the visible part of the electromagnetic spectrum (i.e. light). However, a fraction of the solar radiation is absorbed by the atmosphere and surface, causing them to heat up and emit infrared radiation. Greenhouse gases in the atmosphere, such as carbon dioxide and water vapour, absorb this infrared radiation and prevent most of it from escaping directly into space. Because solar radiation can enter but infrared radiation cannot easily escape, energy is trapped in the atmosphere. This means that the surface of the Earth is at a higher temperature than it would have been if there were either no atmosphere or an atmosphere devoid of greenhouse gases. The situation is likened to that of a greenhouse or a car on a sunny day because solar radiation can easily enter and warm up the interior, but the energy is trapped inside because air cannot easily flow in and out.

Carbon dioxide is also responsible for the natural acidity of rain. Acidity is measured in units of pH. A solution with a pH of 1 is extremely acidic and one with a pH of 14 is extremely alkaline. Pure water, which is neither acidic nor alkaline, i.e. neutral, has a pH value of 7. (You will learn more about the pH scale later in the course.) Rainwater is not neutral because carbon dioxide in the atmosphere dissolves in water droplets to form carbonic acid. For this reason, the pH of unpolluted rainwater is typically 5.6, i.e. rainwater is a slightly acidic solution. This is sometimes referred to as **natural acid rain**. Less commonly, certain natural processes can reduce this pH still further, e.g. volcanic emissions, biological decomposition or even ocean spray. In industrial areas, all these factors can be completely swamped by the human pollution of the atmosphere and the rain typically has a pH much lower than 5.6 and is commonly referred to as **acid rain** or **acid precipitation**. In order to avoid confusing natural acidity with that caused by pollution, some scientists reserve the term 'acid rain' for rain with a pH of 5 or lower.

Ozone: O_3

Ozone is a gas that consists of molecules made of three oxygen atoms, as depicted in Figure 1.6. The oxygen that we breathe, and that makes up about one-fifth of air, is made of two oxygen atoms. Despite this apparently slight difference, ozone is poisonous to us. Fortunately, the amount of ozone in air is small. At ground level, its mixing ratio is typically 5–30 ppbv (parts per billion by volume). Its largest mixing ratio occurs at heights of around 20–30 km (within the stratosphere), where it can reach 6–8 ppmv (parts per million by volume).

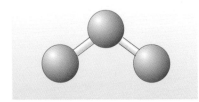

Figure 1.6 A ball-and-stick model of the O_3 (ozone) molecule. Here, three oxygen atoms are bonded together.

○ Given the above ranges, what is the largest ratio of ozone's maximum mixing ratio to its ground level value?

● A mixing ratio of 6–8 ppmv is equivalent to a mixing ratio of 6000–8000 ppbv. The largest factor by which ozone's mixing ratio is increased from that at ground level to that above ground level is obtained by dividing the largest value by the smallest value, i.e. 8000 ppbv/5 ppbv = 1600.

The ozone molecule absorbs ultraviolet (uv) solar radiation. Life on Earth can be easily damaged by too much uv radiation. It can cause genetic mutations (which are needed in moderation for life to evolve) and is infamous as a cause of skin cancer in humans. A seasonal depletion of stratospheric ozone over Antarctica — the so-called ozone hole — appears to be a direct result of certain artificial chemicals being released into the air.

In addition to gases in the atmosphere, there are also aerosols. **Aerosol** is a blanket term for many types of particles, such as dust grains, smoke particles or droplets of water, or other liquids. For example, an aerosol spray from a can is made up of many small droplets of liquid. Unlike the fundamental particles of a gas, which are atoms or molecules, aerosol particles are clumps of many different kinds of atoms and molecules and so tend to be much larger, with a much wider range of particle sizes. Aerosols can absorb radiation, as a gas can, but they also scatter light, which is why we can often easily see an aerosol spray. It is important to be clear about the distinction between **absorption** and **scattering**. If radiation is absorbed, then some or all of the energy of the radiation is given to the absorbing substance. If radiation is scattered, then no energy is exchanged, but incoming rays of light change direction.

1.3 The structure of the atmosphere

You have already seen from the description of Col. Kittinger's balloon flight that the nature of the atmosphere changes with height. Kittinger was only too aware that high in the atmosphere the air thins, the pressure drops dangerously low and the familiar blue sky fades to black space. We mentioned that the air surrounding your body, and for that matter your body itself, is supporting hundreds of tonnes of air above you. We said that water, carbon dioxide and ozone are important substances in the atmosphere, but that they are not evenly distributed through it. We will now look into the vertical structure of the atmosphere and the science that explains this structure.

Make sure you understand the concepts of pressure, density and temperature, as summarized by the perfect gas law (Box 1.3). In Sections 1.3.1 and 1.3.2 we investigate how these atmospheric properties vary with height.

Box 1.3 The perfect gas law

A gas of any kind can be described in terms of three key quantities: pressure, density and temperature. **Pressure** is usually denoted by p, and is measured in units called pascals, abbreviated to Pa. The pressure of a gas is a measure of how it pushes, or presses, on objects that come in contact with it. Since you live in a gas, namely air, you are constantly being pushed by it from all sides, though you don't notice this happening. If you push into a balloon, you can feel the air inside it push back. Pressure measures the amount by which it pushes back. The meaning of pressure and the unit of the pascal can be interpreted in a surprising way. Imagine an empty, cube-shaped box of side 1 m at normal atmospheric pressure at sea-level, which is about 100 000 Pa. The top of the box has a surface area of 1 m^2. The pressure tells us that the air *above* the box is equivalent to a weight of about 10 000 kg, or 10 metric tonnes, on top of the box. Why isn't the box crushed? Because the air inside the box pushes upwards by the exact same amount, balancing the pressure from above. (We will define pressure properly, in terms of force, in Box 1.6.)

We have already met number density, but how is it related to pressure? The particles in a gas are constantly moving, and if an object is placed in the gas, the gas particles bounce off it, giving the object a tiny kick each time. The pressure of gas can be thought of as the total effect of a great many of these particles bouncing off the box. It is not surprising then that if you double the number of particles in the gas, you also double its pressure. For this reason, we can say that pressure is proportional to number density, N/V:

$$p = C \frac{N}{V}$$

What is C? If you think about it, in saying the gas particles are constantly bombarding an object, we are saying nothing about how *fast* they are moving. If particles are moving faster, they not only hit the object more often, but they also give it a greater 'kick' with each hit. The standard measure of how fast particles in a gas are moving is **temperature**, T. In fact, when you touch something hot, it is those very many, extremely fast particles hitting you that can damage your skin, resulting in a burn (in a solid the particles actually vibrate extremely rapidly rather than move around like gas particles). It turns out that we can relate pressure, p, number density, N/V, and temperature, T, by an equation called the **perfect** or **ideal gas law**:

$$p = \frac{N}{V} kT \tag{1.3}$$

where k is a constant called the Boltzmann constant. So our C above is kT. Equation 1.3 must be used with the correct units: pressure in pascals, number density in units of number of particles per m^3 and temperature in kelvin. The kelvin temperature scale is named after the great physicist Lord Kelvin and is just the Celsius scale shifted up by 273°. So 0 °C is 273 K and 100 °C is 373 K. The general rule is:

temperature in K = temperature in °C + 273.15

The value of the Boltzmann constant, k, is 1.38×10^{-23} J K^{-1}, where J is a unit of energy, the joule. (Energy is defined properly later.) A fundamental law of nature is that *nothing* can be at a temperature lower that 0 K.

1.3.1 Temperature structure

Long before Col. Kittinger's balloon flight, more Earth-bound pioneers had established that the temperature of the Earth's atmosphere decreases with height (Box 1.4). Snow can be found on certain high peaks all year long, e.g. on Mount Everest (8850 m), shown in Figure 1.7. Even a day's walk up a smaller mountain, e.g. Ben Nevis in Scotland (1340 m, Figure 1.8), can convince you that it's noticeably colder the higher up you go. This is not just because it is more exposed and windy, but because the atmosphere is getting colder with height. As a rough rule of thumb, the temperature drops by 6 °C for every 1 km ascended, up to about 10 km. The conventional way to say this is that the **environmental lapse rate (ELR)** is roughly 6 °C km^{-1}.

Figure 1.7   Mount Everest.

Figure 1.8   Ben Nevis.

Box 1.4   Ascending heights, descending temperatures

Scientists are often laughed at because of their stereotype as obsessive individuals who become somewhat detached from reality. While most of them are simply hard-working members of an educated profession, some go above and beyond the call of duty in the name of science. One such individual was Clement Wragge (Figure 1.9). During the years 1881–1883, in order to discover how conditions of the atmosphere varied with height, Wragge ascended Britain's highest mountain, Ben Nevis, daily for 5 months of the year. He was meticulous in his routine and his methods. Observations were made at 4.40 a.m. at sea-level, at 6.30 half-way up, at 9.00, 9.30 and at 10.00 a.m. at the summit, with further observations on his descent, his final one being at sea-level at 3.30 p.m. In addition to making regular temperature and pressure measurements of the air, he also took temperature readings of streams, ponds and lakes, noted cloud cover and even drew pictures of plants. In fact, he noted anything that could have any relevance to meteorological conditions, with meticulous accuracy. Whilst he ascended the mountain, his wife would take readings at ground level at their home in Fort William. He was also careful to record occasions when he was unsure of the accuracy of any of his measurements, which in some instances was because his fingers were too numb with cold to manipulate his instruments.

It was due to Wragge's tireless efforts that a meteorological observatory was eventually established at the summit of Ben Nevis (now a ruin). Unfortunately, the powers-that-be did not put him in charge of this observatory, so he left for Australia where he became a Government Meteorologist and began his own journal, modestly entitled *Wragge — For God, King, Empire and People*. He was one of the first to attempt pseudo-scientific schemes to induce rain. His method, which did not work, involved firing air from several large 'guns' into the atmosphere. One of his more eccentric habits was naming low-pressure regions, likely to give rise to nasty weather, after individuals he despised. So his main claim to fame these days is in inspiring the system of naming hurricanes with popular forenames, e.g. Hurricane Andrew or Hurricane Mitch.

Figure 1.9   Clement Wragge (1852–1922) was born in England, but spent much of his adult life in Australia. He worked for the Surveyor-General's Department, South Australia from 1876 until 1878 when he returned to England. During the 1880s he went back to Australia and established a number of observatories. Wragge was Queensland Government Meteorologist from 1887–1902.

○ If the temperature at sea-level is 20.0 °C, what would you expect the temperature to be at the top of Ben Nevis? What would be the temperature at the top of Mount Everest?

● Ben Nevis is 1340 m high and the ELR is 6 °C km^{-1}, so the temperature drop can be estimated as:

$$\Delta T = \frac{1340 \text{ m}}{1000 \text{ m}} \times 6\,°\text{C} = 1.340 \times 6\,°\text{C} = 8.04\,°\text{C}$$

We should regard this value as 8 °C, because although the height and the temperature are given to 3 significant figures, the rough value for the ELR is given to only 1 significant figure. (If you are unsure what is meant by 'significant figures', refer to *The Sciences Good Study Guide*.) The Greek letter 'Δ' (capital delta) is used in science to indicate a change in the quantity following it. So here ΔT (spoken 'delta-tee') represents the change in temperature T. The temperature at the top of Ben Nevis is therefore 20 °C – 8 °C = 12 °C. The calculation for Mount Everest gives ΔT = 53 °C, so the temperature at the top should be 20 °C – 53 °C = –33 °C.

The ELR is given only an approximate value here for several reasons. For one thing, the weather can have a profound effect. For example, ground level could be below cloud and so hidden from the warming sunlight, whereas the mountain top could be above the clouds. It is also important to realize that the figure of 6 °C km^{-1} only applies to the lower part of the atmosphere. Above 10 km, at about passenger-jet cruising height, the temperature stops decreasing and starts to increase with height. The 'lower part' of the atmosphere, the region below about 10 km, is called the **troposphere**. The troposphere is where we live, and it is where almost all weather phenomena occur. It contains about 80% of the atmosphere's mass. There are no mountains that rise above the troposphere. In short, it is the atmosphere of our environment. For this reason, we will spend most time exploring this part of the atmosphere. It would be wrong, however, to say that we can ignore the rest of the atmosphere. The higher layers can have a dramatic effect on what goes on down below, such as the protection that ozone gives life against uv radiation. The typical temperature structure of the atmosphere is shown in Figure 1.10, along with the names for the various regions.

The only time most people visit another region of the atmosphere is on jet aircraft, which cruise in the lower **stratosphere**. The boundary between the troposphere and the stratosphere is called the **tropopause**. The boundary at the top of the stratosphere is likewise called the **stratopause**, above which is the **mesosphere** (meaning middle sphere), separated from the thermosphere above it by the **mesopause**. You can see from Figure 1.10 that the 'spheres' are layers of the atmosphere separated by the 'pauses'. It is at the pauses that the temperature changes from increasing to decreasing with height, or vice versa. In fact, if you went on a direct ascent of the atmosphere, measuring the temperature as you went, you'd see the temperature reading 'pause' between increasing and decreasing at each of the pauses. The question of why temperature varies in the way it does with height is a difficult one to answer, though we aim to lead you to the answer by the end of Section 3.

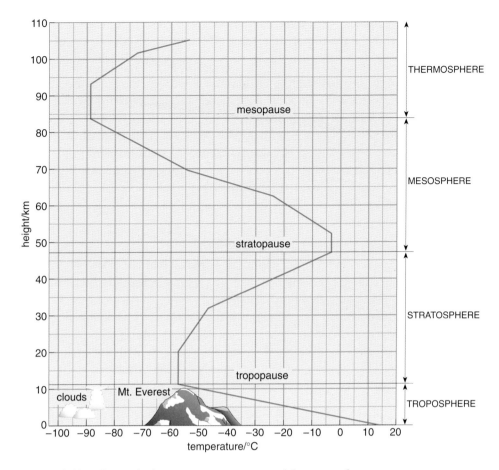

Figure 1.10 The typical temperature structure of the atmosphere.

1.3.2 Pressure and density structure

Temperature is the most familiar of all atmospheric properties because we, like most forms of life, are very sensitive to variations in temperature. We are less familiar with pressure because we do not usually notice its variations in day-to-day life. You probably know that low-pressure days are often associated with bad weather and high-pressure days are generally associated with good weather. However, these changes in pressure are too small for us to notice and we have to use a **barometer** to measure them, as will be described in Section 2.

Changes in pressure can be noticeable when ascending through the atmosphere. Imagine that you are climbing a large mountain such as Mount Everest. As an experiment, you partially inflate a small, perfectly airtight balloon at base camp, and take it with you as you climb. You notice that the balloon becomes progressively larger. Because the balloon is airtight, no air can enter or leave it. The only other explanation for the expansion, the correct one, is that the air in the balloon takes up more room at higher altitude, i.e. it expands to occupy a greater volume. This means that the same number of air particles are spread over a greater volume, which in turn means that the density of the air in the balloon is less. But why does this happen, and what has it to do with pressure? We know from the perfect gas law (Equation 1.3) that pressure $p = kTN/V$, where V is the volume of the balloon, which increases with height. This fact, together with the fact that the temperature (T) decreases with height, allows us to conclude that the pressure decreases with height. We can experience this effect, using a very sensitive pressure-measuring instrument that we all possess, as described in Box 1.5.

As the pressure of the air surrounding the balloon decreases, the air in the balloon pushes outwards more than the air outside pushes inwards. The result is that the balloon expands and causes the pressure in the balloon to drop. It keeps expanding until the pressures inside and out are equalized. The general, scientific name for a situation where such a balance is achieved is **equilibrium**. In practice, as you ascend, the balloon is continually adjusting its pressure down to the pressure of the surrounding air. This kind of continually adapting equilibrium is called **dynamic equilibrium**. The concept of equilibrium is a recurring theme throughout the environment.

Consistent with our conclusion that pressure falls with height is the fact that with increasing altitude, it becomes more and more difficult to breathe, and eventually a source of oxygen from a mask connected to an oxygen tank might be required. This breathing problem at high altitudes suggests that insufficient oxygen is inhaled in each breath, which in turn suggests that oxygen is less abundant. If you measured the composition of the air, you would find similar proportions to those at sea-level, i.e. about 78% nitrogen and 21% oxygen, with trace amounts of everything else. This means that the *total* number density of the air must decrease with height.

○ By considering the perfect gas law (Equation 1.3), can you think of a way to make a perfectly airtight balloon expand or contract without changing the density of the air around the balloon, or blowing more air into the balloon?

● To make a balloon expand or contract requires the pressure outside the balloon to be different from that inside the balloon. The pressure of a gas depends on both its density and its temperature, so if we cannot change the density, we must change the temperature. A very large difference in temperature would be needed to produce a noticeable change in the balloon's volume.

A dramatic demonstration of the effect of a change in temperature on a balloon's size is to dip part of a long balloon in liquid nitrogen, as shown in Figure 1.11. The liquid nitrogen, which is at a temperature of about 77 K (−196 °C), cools the air inside the balloon, thus reducing its pressure. The part of the balloon that is still in contact with the air shrinks because the surrounding air is at a higher pressure than the air within the balloon. When removed from the liquid nitrogen, the air in the balloon returns to room temperature, and so the balloon expands to its original size, in just a few minutes. An example of this process in reverse is the hot air balloon. Heating the air in the balloon makes it expand, and causes it to be less dense than the surrounding air. This makes the balloon buoyant and so it moves upwards.

In addition to the SI unit of pressure, the pascal, there is another unit, called the **bar**, where

$$1 \text{ bar} = 10^5 \text{ Pa} = 100\,000 \text{ Pa}$$

(a)

(b)

Figure 1.11 A balloon (a) before and (b) after it is dipped in liquid nitrogen.

Box 1.5   Ears: your own personal pressure meter

There are some special, though familiar, circumstances when you can detect pressure changes using your ears. As a plane ascends, the air surrounding it decreases in density and pressure. During the first part of the ascent the pressure, and therefore density, of the cabin air also decreases. The air behind your eardrums is like the air inside a balloon in that it is pushing outwards because it is at a higher pressure than the cabin pressure. This can cause some discomfort until your ears 'pop', releasing some of the pressurized air (into the back of the throat).

Your ears are in fact highly sensitive pressure sensors, but they are only sensitive to sudden changes in pressure. For example, when hands are clapped together, the air between them is suddenly squeezed, i.e. increased in pressure, and forced to move out from between the hands. A pulse of slightly compressed air then travels through the air, like a wave travelling in water. When this pulse reaches your eardrum, the tiny changes in pressure are detected and relayed to the brain, which interprets them as a sound.

Since the average, or mean, sea-level pressure is just slightly over 1 bar, this unit is more convenient for measuring air pressures on the Earth. In dealing with weather, quite small pressure changes can be important, so the unit of the millibar or **mbar** is used, where

$$1 \, \text{mbar} = 10^{-3} \, \text{bar} = 0.001 \, \text{bar} = 100 \, \text{Pa} = 1 \, \text{hPa}$$

The **hPa** is another commonly used unit called the hectopascal (this was introduced by meteorologists, who were accustomed to using the unit of the millibar, but who were 'forced' to move over to the more modern SI unit system).

The mass density of air is quoted in (SI) units of kilograms per metre-cubed (kg m^{-3}). The mass density is used in preference to the number density because there is such a huge number of particles in $1 \, \text{m}^3$ of air, whereas the average mass density at sea-level is conveniently close to $1 \, \text{kg m}^{-3}$. Table 1.2 gives the typical pressure and (mass) density values at selected sites of interest.

Table 1.2   Air density and pressure at various heights.

Site	Height/m	Density/kg m^{-3}	Pressure/hPa
mean sea-level	0	1.23	1013
Telecom Tower, London, UK	189	1.20	990
Sky Tower Auckland, New Zealand	328	1.18	970
Empire State Building, New York, USA	448	1.17	960
Petronas Towers, Kuala Lumpur, Malaysia	452	1.17	959
Ben Nevis, UK	1340	1.07	860
Mont Blanc, France	4809	0.75	550
Mount McKinley, Alaska, USA	6195	0.65	460
Mount Everest, Nepal	8850	0.48	315
cruising Boeing 747	11 000	0.35	230
cruising Concorde	18 000	0.13	70

You should now read Box 1.6, which explains the relationship between force and pressure.

Box 1.6 Balancing forces

We argued above that internal and external pressures are being balanced as a balloon ascends. If the pressure is different at two neighbouring points in a gas, the gas will flow from high pressure to low pressure until the pressures are the same. (This process isn't necessarily apparent from weather maps, because there are many other effects to consider as well.) It is important to understand the more basic concept of **force** to really grasp what pressure means, and understand why pressures must balance. Sir Isaac Newton introduced force to explain how objects move when they are pushed or pulled; the force simply measures the amount by which an object is pushed or pulled. Newton stated three laws about forces:

1   If no force is applied, then the speed and direction of an object's motion does not change.

2   If a force of magnitude F is applied, the object changes speed at a rate that is proportional to F.

   The rate at which the object changes speed, i.e. its acceleration, is smaller for more massive objects. For a mass m, the acceleration a is

   $a = F/m$

   which can also be written as

   $$F = ma \tag{1.4}$$

3   For every action (force), there is an equal and opposite reaction (opposing force).

Force is measured in the SI unit of the newton, which is denoted by 'N'. We are not going to be particularly concerned with force when discussing the atmosphere, but we need to understand the basic concept in order to understand pressure. To this end, it's useful to consider how you might measure a force on an object in everyday life. All you need to know is the mass of the object and its acceleration. For example, a sports car of mass $1000\,kg$ can typically accelerate to $100\,km\,h^{-1}$ (about 60 miles per hour) in $6\,s$. To apply the force law, we need to work in SI units of metres and seconds:

$$100\ \mathrm{km\ h^{-1}} = 100\,000\ \mathrm{m\ h^{-1}}$$

$$= \frac{100\,000\ \mathrm{m}}{60^2\ \mathrm{s}} = \frac{100\,000\ \mathrm{m}}{3600\ \mathrm{s}} = 28\ \mathrm{m\ s^{-1}}$$

So the sports car reaches a speed of $28\,m\,s^{-1}$ in $6\,s$. To do this it must on average increase in speed by $28/6\,m\,s^{-1} = 4.7\,m\,s^{-1}$ every second. The acceleration is therefore $4.7\,m\,s^{-1}$ per second, i.e. $4.7\,m\,s^{-2}$. The force is therefore

$$F = ma = 1000\ \mathrm{kg} \times 4.7\ \mathrm{m\ s^{-2}} = 4700\ \mathrm{N}.$$

We are now in a position to define pressure properly. Pressure is defined as the force per unit area, that is

$$p = F/A \tag{1.5}$$

The SI unit for pressure, the pascal, is equal to $1\,N\,m^{-2}$ (one newton per square metre). In more practical terms, one can interpret a gas pressure of $100\,000\,Pa$

(roughly air pressure at sea-level) as the force exerted on a $1\,m^2$ area, say a piece of card, that is suspended in the gas. So, can we measure this pressure by knowing the mass of the card and measuring how fast the card starts to move? No, because the pressure on both sides of the card is the same (assuming there is no wind). The force of $100\,000\,N$ in one direction exactly balances the force of $100\,000\,N$ in the opposite direction, so the total, or net, force is zero, as shown in Figure 1.12a. A good job too, because $100\,000\,N$ could accelerate an object as heavy as the sports car to a very high speed in just a fraction of a second! Instruments for measuring pressure will be described in Section 2.

To finish off this introduction to forces, let's consider what happens if forces are *not* balanced. What would happen to the piece of card if the pressure of the gas on either side were slightly different, say $100\,000\,Pa$ on the left, and $99\,000\,Pa$ on the right, as shown in Figure 1.12b? The forces would be $100\,000\,N$ to the right and $99\,000\,N$ to the left, so the card would experience a net force of $1000\,N$ to the right, and so would start to accelerate to the right. The card would move from high pressure towards low pressure. If there were no card, then the air where the card was would experience the net force and move to lower pressure. In this way, a difference in pressures can cause a flow of air, i.e. a wind.

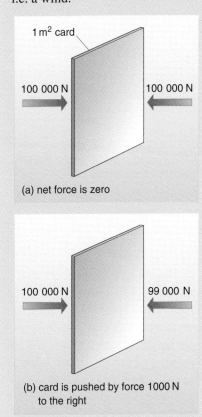

(a) net force is zero

(b) card is pushed by force $1000\,N$ to the right

Figure 1.12   A $1\,m^2$ card is suspended in air: (a) the pressure is the same on both sides; (b) the pressure is lower on the right.

○ Referring to Box 1.6, express one newton in the more familiar units of mass, length and time.

● Using Newton's second law (Equation 1.4), we see that 1 N is equivalent to the units of 1 kg multiplied by the unit of acceleration. Acceleration is change in speed per unit time, i.e. $\text{m s}^{-1}\,\text{s}^{-1} = \text{m s}^{-2}$. Therefore $1\,\text{N} = 1\,\text{kg m s}^{-2}$.

Figure 1.13 shows a typical plot of pressure against height in the atmosphere. In reality, the pressure value at a particular height varies from place to place around the world, because different layers of the atmosphere are heated differently. It is clear that pressure has a much simpler variation with height than does temperature: pressure simply decreases with height. Correspondingly, the explanation for the pressure structure is much simpler: the air lower down has to support a greater mass of air above it, and so is more compressed.

Figure 1.13 Pressure plotted against height in the atmosphere.

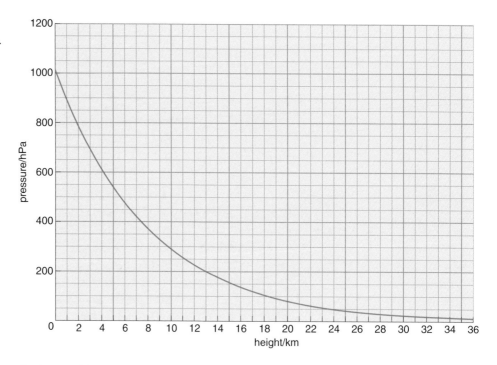

Let's consider the situation shown in Figure 1.14, where a sealed pocket of air is encased by a fixed floor and fours walls, but has a movable 'ceiling' of area $1\,\text{m}^2$, on top of which is a mass in a sealed vacuum. We will consider the forces on the ceiling, and so predict how it will move. First we need to calculate the force due to gravity that is exerted by the mass m. It turns out that all objects are accelerated towards the Earth by gravity at the same acceleration, $9.81\,\text{m s}^{-2}$, usually denoted by g. This means that according to Newton's second law (Box 1.6, Equation 1.4), the force exerted by the mass due to gravity is mg, in the downwards direction. This force is what is scientifically termed the **weight** of an object. What is the force of the gas at pressure p from beneath the ceiling? It is simply $p \times 1\,\text{m}^2$ in the upward direction. So the net force in the upwards direction is $p \times 1\,\text{m}^2 - mg$. If $p = 100\,000\,\text{Pa}$, which is roughly atmospheric pressure at sea-level, and $m = 1\,\text{kg}$, e.g. a bag of sugar or a brick, then the net force in the upward direction is $(100\,000 \times 1)\,\text{N} - (1 \times 10)\,\text{N} = 99\,990\,\text{N}$. (Here we approximate g as $10\,\text{m s}^{-2}$.)

The air would expand extremely rapidly, and accelerate the brick upwards so that it would reach 100 km s^{-1} in a second! Obviously, with such a huge force, it would be very difficult to set this experiment up in the first place. So, what mass would be needed to balance the force due to the atmospheric pressure, i.e. for mg to equal $p \times 1 \text{ m}^2$? The answer is given by rearranging $mg = p \times 1 \text{ m}^2$ to give

$$m = \frac{p \times 1 \text{ m}^2}{g} = \frac{100\,000 \text{ N}}{10 \text{ m s}^{-2}} = 10\,000 \text{ kg}$$

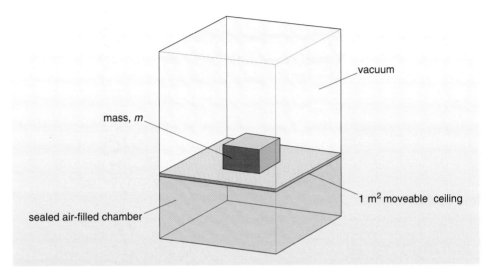

Figure 1.14   A thought experiment to show how the force due to air pressure in a sealed room can be balanced by the weight of a mass on the moveable ceiling.

So it would take a mass of 10 000 kg on top of each square metre to compress a gas to the typical sea-level atmospheric pressure of 100 000 Pa. In other words, the mass of gas in the atmosphere above a square of side 1 m on the ground is about 10 000 kg. It is the weight of all this gas (remember weight is a force) that compresses the air at sea-level to a pressure of about 100 000 Pa. At greater heights in the atmosphere, there is less mass in the atmosphere above, so the pressure is lower. Half the atmospheric mass is above a height of about 5500 m, so it follows that the air at 5500 m only has to support half the mass that air at sea-level does, so the pressure is half the sea-level value, i.e. about 50 000 Pa.

Activity 1.1   Relating pressure, density and temperature in the atmosphere

Using data from the graph in Figure 1.13, construct a table of atmospheric pressure at intervals of 5500 m. How is the value in one entry related to the next one? Also, read off from Figure 1.10 the temperature at each height and convert the values to kelvin before inserting them in your table. Then use the perfect gas law (Equation 1.3) to calculate the mass density of the air at each height, assuming the mass of the average air particle to be 5.0×10^{-26} kg. Work to 3 significant figures, bearing in mind that information obtained from the graph is probably accurate to only 2 significant figures. Include in your table, for comparison, the first (mean sea-level) and the last five entries in Table 1.2. How do the density values in successive entries compare?

For every 5500 m ascended, the pressure drops by factor of about 2. This sort of decrease is called **exponential** and is very common in science. The density has a nearly exponential fall-off, with deviations occurring because of the changing temperature. If the temperature were constant with height, then density would have the same profile as the pressure with height.

A system in which the pressure of a fluid (liquid or gas) balances the gravitational force is said to be in **hydrostatic equilibrium**. The 'hydro' literally means water, but in this context applies to any fluid. The 'equilibrium' refers to a balance of forces. As well as the atmosphere, there are other systems that are in hydrostatic equilibrium, including the oceans (liquid water) and the Sun and other stars (hydrogen and helium gas). If the pressure or density structure of the atmosphere in a particular place is suddenly altered, by the movement of air (e.g. winds) or because of heating (e.g. from the Sun rising), then the atmosphere can move away from hydrostatic equilibrium. The atmosphere always tries to return to hydrostatic equilibrium, possibly releasing energy in the process. Remembering the huge forces that result from even small changes in pressure (e.g. with the mass and the moving ceiling), it is not hard to imagine that a very large amount of energy can be involved in even small changes in the atmospheric structure. In general, the atmosphere is always very close to hydrostatic equilibrium, so this situation is often assumed in computer modelling of atmospheric processes. Nevertheless, the associated release of energy in restructuring the atmosphere (i.e. returning it to hydrostatic equilibrium) is in fact what drives the whole 'weather machine'. In the next section, we will describe all the different facets of weather, created by the continual exchange of energy. After that, in Section 3, we will discuss how energy is moved around the atmosphere. We will finish Part 1 of Block 2 by bringing together what has been learned in the first three sections and applying it to a study of weather on a global scale in Section 4.

1.4 Summary of Section 1

1   Air is mainly composed of the gases nitrogen and oxygen in a ratio of almost 4 : 1. All other gases are present in only trace amounts. All other particles in air are called aerosols.

2   Visible light is just one wavelength range of electromagnetic radiation. Others include infrared, ultraviolet and radio.

3   Although trace gases are present in small amounts, they can have a significant impact on the atmosphere, particularly in absorbing different kinds of radiation.

4   The temperature, density and pressure of a gas are all related by the perfect gas law. Pressure is force per unit area. The mass and number density are the mass and number of particles per unit volume, respectively. The temperature of a gas is a measure of how fast, on average, the gas particles are moving.

5   The concept of force, necessary to understanding pressure (i.e. force per unit area), is encapsulated in Newton's three laws.

6   The atmosphere is divided into several layers according to its temperature structure, specifically according to where the temperature switches from increasing to decreasing, or vice versa. The lowest layer, the troposphere, extends up to about 10 km and contains all mountains. The troposphere is where all weather takes place and the temperature within this layer generally decreases with height. The layer immediately above the troposphere is the stratosphere.

7   The pressure steadily decreases with height throughout the atmosphere, because the pressure at a given height is dictated by the weight of the atmosphere above that height.

Question 1.1

The density profile of the atmosphere closely follows the pressure profile, and does not show the switches between increasing and decreasing that are apparent in the temperature profile. Use the perfect gas law (Equation 1.3) to explain this density profile in terms of the ranges of pressure and temperature from sea-level to the top of the stratosphere. (Assume that the pressure at the top of the stratosphere is 1 hPa. For the height range of interest, work out both the ratio of maximum to minimum temperature and the ratio of maximum to minimum pressure.)

Question 1.2

A mass of 10 kg is spread over a horizontal platform of area of 20 m². (a) What is the force on the platform? (b) What is the pressure on the platform? (c) If the area of the platform was reduced to 10 m², what would the force and pressure be?

Question 1.3

Suppose that ozone has a mixing ratio of 5 ppbv at sea-level and 6 ppmv at Concorde's cruising height of 18 000 m. According to the data in Table 1.2, does a given volume of air at ground level contain more ozone molecules than an equal volume of air at 18 000 m? (Assume the mass of the average air particle to be 5.0×10^{-26} kg in converting from mass to number density.)

Question 1.4

Without consulting a graph or table, how many times smaller is the pressure at 16 500 m compared to that at sea-level?

Question 1.5

Consider a column of air of cross-sectional (i.e. horizontal) area 1 m², which stretches from the ground (at mean sea-level) to the tropopause (at 11 km), and assume that the pressure changes with height as in Figure 1.13. (a) What is the force experienced at the *base* of the column due to the pressure on this area? (b) What is the *total* mass of air above the base of the column? (c) What is the mass of air above the top of the column (i.e. above the tropopause)? (d) What is the mass of air in the column itself, and what percentage is this of the total mass?

Learning outcomes for Section 1

After working through this section you should be able to:

1.1 Describe the two main constituents of air.

1.2 List three trace constituents (water, carbon dioxide and ozone) and explain their importance.

1.3 Define and use the concept of mixing ratio. (*Question 1.3*)

1.4 Name at least three electromagnetic wavelength ranges (visible light, infrared and ultraviolet).

1.5 State and explain the perfect gas law in terms of how it relates the concepts of pressure, density and temperature. (*Question 1.1*)

1.6 State the definition of force and use it to define pressure. (*Question 1.2*)

1.7 Sketch the structure of the lower atmosphere (troposphere and stratosphere) in graphs of temperature and pressure versus height. (*Activity 1.1*)

1.8 Explain why pressure falls off steadily with height, and calculate the mass of air above a given height from the pressure. (*Questions 1.4 and 1.5*)

Weather watch

2

2.1 Introduction

The processes of observation and measurement are both fundamental to science. They form the basis of how we go about understanding and explaining a host of natural phenomena, including the weather. Indeed, the weather and its prediction are probably the best-known, most 'exposed' facets of environmental science on a day-to-day basis.

In order to predict the future state of the atmosphere, it is necessary to map all the current conditions at exactly the same time, as widely as is possible, for as many weather measures as are necessary, with as much precision as is realistic. The essential message in weather prediction is that the weather over a week or two into the future, on a global scale, is dependent on what's happening *now*. So, to the operational meteorologist, observing and measuring elements of the weather are the lifeblood of the forecasting trade.

In addition to this use of observations, there is of course the basic requirement of accruing knowledge about the weather variations that occur at a particular site. Such information forms the basis of defining the climate of a location by statistical analysis, which can provide information about, say, mean values of temperature and precipitation, and the rarity of certain extreme daily precipitation totals or of very high daily maximum temperatures.

The notion of **weather** is of course related to the details of the day-to-day changes that occur world-wide; it is essentially the short-term variation and variability of the atmosphere that we think of. In contrast, the term **climate** refers to the longer-term changes of weather patterns which are expressed in seasonality around the world. Climate is concerned with, for example, both the 'expected' monthly and seasonal average patterns/values of temperature, precipitation and sunshine, and the anomalies (departures from the long-term average) that occur on these time-scales. Predicting tomorrow's weather is a quite different problem from that of predicting next season's or next century's!

○ What do you consider are the main weather measures that meteorologists need to be concerned with, and why?

● They are those measures that express something useful about the day-to-day changes in the weather. They include temperature, humidity, wind speed and direction, barometric pressure, visibility, precipitation and cloud type and amount. These are important for mapping the current state of the weather and for supplying weather forecast models.

This section focuses mainly on surface, upper-air and remote-sensing observations. The last of these comes from instruments such as radars and sensors on weather satellites, which do not measure environmental properties *directly* (as thermometers do) but map rainfall or temperature, for example, from a distance. They are used in the daily routine of operational weather services. Also discussed are the international collaboration in weather monitoring and forecasting and the way in which the observations are plotted to an agreed global convention.

2.2 Surface observations

The oldest type of weather observations are those made at the Earth's surface, dating back to the second half of the 17th century for temperature in some parts of the UK.

2.2.1 Temperature

Perhaps the first ingredient of the weather we think of is how warm, cold or mild it's going to be. We need to have a measure of the heat content of the air per unit mass or volume: this can be obtained simply by measuring temperature.

Drybulb temperature

The drybulb temperature is what we usually think of as temperature. In this text, and all others, an unqualified use of the word 'temperature' means drybulb temperature. Currently, the international standard means of measuring air temperature is the mercury-in-glass thermometer exposed within a weather screen, commonly known as a Stevenson screen (Figure 2.1). This instrument has remained essentially the same for many years and is unlikely to change in the foreseeable future for routinely staffed weather stations. The principle is that liquid mercury expands or contracts evenly with respect to changes in air temperature and does so over a wide range of temperatures (Box 2.1). The vertically mounted **drybulb thermometer** used in weather screens senses the air temperature within. It is read, logged and reported to the nearest 0.1 °C every hour, on the hour, at every station, including major airports. A reservoir of mercury resides in the bulb of the thermometer and the fluid moves up and down the narrow bore within the instrument as the air temperature in the screen increases or decreases. Box 2.2 provides an illustration of how drybulb temperature data can be used to investigate the variation of temperature with height.

Figure 2.1 A Stevenson screen.

Box 2.1   How does a thermometer work?

Figure 2.2 shows a typical liquid-filled thermometer. You have probably used one at some time in the past, but how does it work? Why is the scale marked off in evenly spaced divisions?

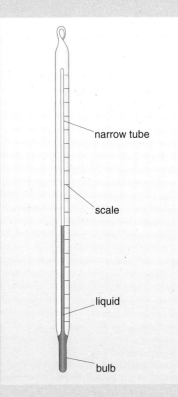

Figure 2.2   A liquid-filled thermometer.

The basis of a liquid-filled thermometer is the fact that virtually all substances expand when they are heated and contract when they cool. The amount of expansion depends on two things: the size of the temperature increase, and the substance used. Measuring the amount of expansion (or contraction) for a given substance therefore tells us the size of the temperature change.

The most convenient type of substance to use is a liquid. Water can be used but will not work at temperatures below its freezing temperature. To overcome this problem, mercury can be used (it freezes at $-38.9\,°C$), but in the very cold climates of Siberia and northwestern Canada, for example, this too is useless, so alcohol-filled thermometers are used. Ordinary alcohol (or ethanol, as found in alcoholic drinks) freezes at $-114.4\,°C$.

To observe the amount of expansion or contraction, the liquid is held in a narrow transparent tube with a bulb at the base. When the bulb is heated, the volume of liquid in the bulb increases and the excess liquid volume has to pass into the narrow tube. Because the tube is so narrow, a small change in liquid volume becomes a large change in the length of the liquid in the tube. When the bulb is cooled, the opposite happens: liquid in the tube retreats into the bulb.

The larger the temperature change, the larger is the change in length of the liquid. In fact, the connection between the changes in temperature and length of liquid is very regular. For instance, an increase of $2\,°C$ causes an increase in length that is twice that caused by a $1\,°C$ increase; an increase of $10\,°C$ causes an expansion that is ten times that of a $1\,°C$ increase, and so on. This is why the scale marks on a thermometer are evenly spaced; the change in length is proportional to the change in temperature. For every $1\,°C$ increase in temperature, water expands in volume by about 0.021%, mercury by about 0.018% and alcohol by about 0.11%.

○   What is the principal reason for putting a thermometer inside a weather screen?

●   It is because the fluid inside the thermometer should not be exposed to the heating associated with direct sunshine. Inside the screen, it is measuring the real air temperature.

Thermometers are used for spot measurement, at one instant. In contrast, the **thermograph** is a standard screen instrument that is still used operationally to provide a continuous record of temperature variation, normally over one week. It consists of a clock-driven drum around which is wrapped a weekly **thermogram** strip (Figure 2.3). A pen fixed to the end of a long arm traces the thermal changes by being connected to a bimetallic strip, made of two different metals fastened together lengthwise. Different solid metals expand and contract at different rates.

The idea behind the bimetallic strip is that two different metals with different coefficients of expansion (the rate at which they expand when the temperature increases) will, if fused as one strip, distort/curve gently in a way that can be measured. The degree of such 'curving' is an expression of how one metal is expanding or contracting at a different rate from the other and, by a mechanical connection, expresses variations in the air temperature.

Figure 2.3 A thermogram for a four-day period in May 1989.

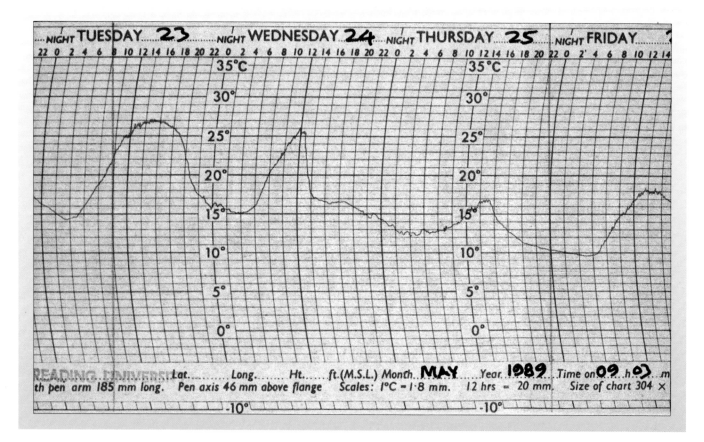

Box 2.2 Going up! Lower temperatures

We are aware that temperature usually decreases upwards in the troposphere. The reasons for this temperature gradient will be outlined in Sections 3 and 4. Let's just for now consider how average drybulb temperatures change within the Teign Valley. In the following we will assume that the temperature variations are due solely to differences in height, though in practice they can be due to other effects, which will not concern us here.

The annual mean drybulb temperature at the coastal resort of Teignmouth, where the screen is at a height of 3 m above mean sea-level (amsl), was 11.30 °C (1990–1998). The Yarner Wood site at 198 m amsl, higher up on the right bank of the River Bovey, had a corresponding value of 10.20 °C (1990–1998). These two points indicate that the mean 'lapse rate' of temperature (the rate at which it changes with increasing height) was some 0.56 °C per 100 m.

Slightly outside the upper reaches of the Teign Valley, some 5 km from its extreme upper headwaters, there is a temperature screen near Okehampton. This particular screen is at 415 m amsl, recording an annual mean temperature of 8.70 °C (1992–1996). Comparing this mean temperature with that for Teignmouth reveals a mean lapse rate of 0.63 °C per 100 m. Both lapse rates calculated for the Teign Valley are examples of environmental lapse rates (ELR), introduced in Section 1.3.1. Meteorologists usually quote a 'typical' value of 6 °C km^{-1} for the average ELR, but it can and does vary a lot from time to time and place to place.

Maximum temperature

The maximum temperature is normally defined as the highest drybulb value that occurs over a 24-hour period within the screen. It is measured using a specially designed, horizontally mounted, mercury-in-glass thermometer (Figure 2.4). The principle is the same as for the clinical thermometer, in which there is a narrowing of the bore of the thread within the instrument. This constriction permits mercury to expand out of the bulb and along the bore as the air temperature increases, which it continues to do until the highest temperature of the day. When the temperature falls, however, the constriction is such that the mercury cannot contract back into the reservoir. This means that the maximum value is 'preserved' for the observer to read. A gardener's maximum/minimum thermometer (Figure 2.5) indicates both extreme values by two shifting steel markers, or indices, that are pushed by the mercury within the bore. It is reset by using a magnet to move each index back into contact with the mercury.

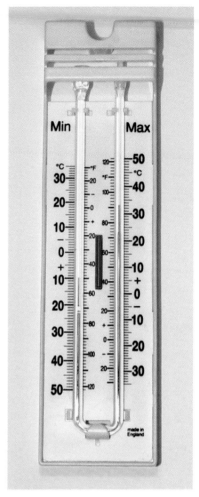

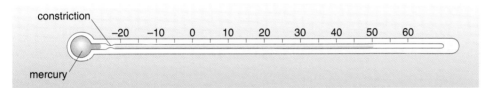

Figure 2.4   A mercury-in-glass thermometer for measuring maximum temperature.

The daily maximum drybulb temperature is read to the nearest 0.1 °C at 09.00 **UTC** (Universal Time Coordinated, the same as Greenwich Mean Time, GMT) across the UK and the Irish Republic. Around the world, it is usually read at this time *locally*, i.e. in the morning. After taking the reading, the observer must carefully re-set the instrument in the same way as a clinical one. This is done by holding it firmly, bulb end furthest from the hand, and bringing it down rapidly a few times to ensure that the mercury has moved back along towards the reservoir. It is then replaced on its mounting within the screen. Refer to Box 2.3 for an example of what maximum temperatures can tell you in practice.

Figure 2.5   A maximum/minimum thermometer.

Box 2.3   *Temperatures in the Teign Valley*

How does the average maximum temperature change with height up the Teign Valley, and how does it change between summer and winter? It is perhaps surprising that Teignmouth's highest mean monthly maximum of 21.0 °C, in August, is equalled by Yarner Wood's. Indeed, in May, June and July, Teignmouth's average maximum temperature falls short of Yarner Wood's by up to 0.50 °C. Even at Okehampton, the corresponding value in August is 19.3 °C, which is only just over 1.50 °C cooler than at the coast. The situation is reversed in the winter, when Teignmouth's lowest monthly mean maximum is 9.10 °C in January; it is 7.60 °C at Yarner Wood and 6.70 °C at Okehampton, both also in January.

Here, we are seeing a small-scale example of the role played by continents in climate (see later), which is made a little complicated by elevation changes. Teignmouth displays a smaller annual range of monthly *mean maximum* temperature than the middle of the Teign Valley, for the difference between the *maximum* in the warmest and coolest months is 11.9 °C at Teignmouth and 13.4 °C at Yarner Wood. This pattern is mirrored by the difference in *monthly mean* temperatures of 10.5 °C at the coast and 11.4 °C higher up. Although sunnier, the coast is cooled by sea breezes in the summer, making the coastal strip cooler than some of the higher regions of the catchment. In the winter, the relatively warm sea keeps temperatures slightly but significantly higher at the coast.

The mean monthly minimum temperature of 4.2 °C in the coolest month of January at Teignmouth is significantly higher than at both Yarner Wood (2.6 °C) and Okehampton (0.7 °C). Why is this difference perhaps more significant than those during the warmer months? The higher value at the coast is related to quite frequent moist southwesterly flow over a reasonably warm sea *and* to the fact that it's breezier there than inland. This is because inland the surface is rougher than the sea, aerodynamically that is. Although the sea surface does of course become choppy and disturbed, it is not as rough (from the point of view of slowing the wind down by friction) as the land. All this means that windy weather keeps the air well mixed, so that low minima and frosts are much less common on the coast.

There is a tendency on clear and calm nights for the air right at the land surface to become quickly chilly, as the ground loses heat to space by radiation. This condition can produce a ground frost. If it's windy on a clear night, however, the chilly surface air doesn't have a chance to develop because of the mixing down of warmer air from not far aloft.

The highest maxima ever recorded in the UK, Europe and globally are:

England	37.1 °C at Cheltenham on 3 August 1990
Scotland	32.8 °C at Dumfries on 2 July 1908
Wales	35.2 °C at Hawarden Bridge, Clwyd on 2 August 1990
N. Ireland:	30.8 °C at Knockarevon, County Fermanagh on 30 June 1976
Spain:	50.0 °C at Seville on 4 August 1881
Libya:	57.8 °C at El Aziziya on 13 September 1922.

Question 2.1

Referring to Box 2.3 and from your everyday experience, briefly describe the factors that might influence the maximum daytime temperature.

Activity 2.1 Temperature measurement at your own weather station

One way to measure temperature is to use a 'combined' maximum/minimum mercury thermometer, which can be purchased for just a few pounds. Electronic sensors that have a digital display are also available. The problem is how to expose the instrument as best you can. This means buying a kit-form or pre-constructed weather screen or placing the instrument on or near a north-facing wall, away from any heat sources, so that it is permanently shaded.

There are some interesting things you can do with such an instrument:

1   Maintain a daily log of the maximum and minimum temperatures in degrees Celsius (°C). If you can, you should take the reading to the nearest 0.5 °C, because a low-cost thermometer is unlikely to be more accurate than that. Just like the professionals, you should aim to take both values at, or very near to, the same time every day. Ideally, this should be in the morning sometime after sunrise, at 08.00 or 09.00 for example. Meteorologists in the UK observe these values at 09.00 UTC which is 09.00 local time in the winter but 10.00 when on British Summer Time.

Note: it is important to always 'throw back' the maximum temperature you read in the morning. This is because, much more often than not, it will have occurred during mid-afternoon on the *previous* day. In contrast, the minimum is allotted to the day on which you read the thermometer because it in most likely to have occurred between midnight and the time you take your observation.

2   Study a one-week period in depth by collecting the North Atlantic and Europe weather maps, UK weather data and regional forecasts published daily in the broadsheet newspapers. Keep a log of your data and this other information.

3   To help you try to understand why there might be day-to-day variations in the maximum and minimum temperatures, you should also keep a log of wind direction and cloud type and amount at the morning observation time *and* at a few times during daylight hours — at lunchtime and in the late afternoon/evening, for example.

4   Compare your daily values of the two temperatures with those from around Britain (using newspaper data). How do yours compare? Where do you lie within the range of values on a particular day? Do you live in a rural valley, in a large city, near the coast, in a hilly district?

5   Think about the probable lack of proper exposure of your instrument. This will affect your comparison, apart from any geographical considerations. You should also assess the quality of the published newspaper forecast for your region in terms of maximum and minimum temperatures each day. Think about how these values would vary within the large region for which you have the forecast. A forecast of the overnight low, for example, will be mapped as one number to the nearest whole degree Celsius, and will be relevant for an area of two or three counties or so.

6   You may be able to invest in a second maximum/minimum thermometer to enable you to look at different locations within your garden (if you have one). Microclimates can be markedly different even on the back-garden scale: you could investigate the impact of shaded and sunny locations and/or the difference between a level and sloping site.

7   You could use two instruments to also investigate the role of their exposure on recordings. Locate one on a building north wall and another on the north side of a tree or fence, for example.

Minimum temperature

Minimum temperature is normally defined as the lowest drybulb temperature that occurs over a 24-hour period within the screen. It is sensed by a horizontally mounted, alcohol-in-glass thermometer in order to make measurements in the cold climates of the world when mercury would freeze (see Box 2.1). The design principle employed in this case involves the use of an index (a short, very thin bar) inside the bore (Figure 2.6).

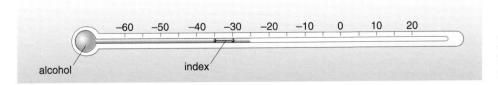

Figure 2.6 An alcohol-filled thermometer for measuring minimum temperature.

The alcohol expands out of the bulb and along the bore as the day warms up; the index does *not* move 'forwards' with the meniscus (the curved surface) of the alcohol as this happens. When the air cools, the alcohol contracts back towards the reservoir and its meniscus drags the index back with it. This 'retreat' continues until the minimum temperature occurs, after which the alcohol expands again with any warming. The index, however, is abandoned by the expanding alcohol at the point of the minimum value, which is read to the nearest 0.1 °C at the end of the index furthest from the bulb.

The value is normally logged once every 24 hours at the same time as the maximum temperature. The thermometer must be reset after the reading is taken, by gently inclining it so that the bulb is slightly higher, to enable the index to slide slowly along to contact the meniscus. This means that any subsequent cooling is mirrored by the movement of the bar back towards the bulb. Refer to Box 2.3 for an example of what minimum temperatures can tell you in practice.

The lowest recorded minima in the UK, Europe and globally are:

England	−26.1 °C at Newport, Shropshire on 10 January 1982
Scotland	−27.2 °C at Braemar on 10 January 1982 and 11 February 1995
Wales	−23.3 °C at Rhayader, Powys on 1 January 1940
N. Ireland:	−17.5 °C at Magherally, County Down on 1 January 1979
Siberia:	−68.0 °C at Oymyakon in 1933 (date unknown)
Antarctica:	−89.6 °C at Vostok on 21 July 1983

A **ground frost** occurs when the temperature measured over a short-grass surface falls to 0.0 °C; an **air frost** occurs when the temperature within the weather screen falls to 0.0 °C. Ground frosts are more common than air ones because the cooling that leads to them occurs most strongly right at the surface. This is because land, which is heated by solar radiation during the day, also loses its heat by emitting infrared radiation. At night, with no Sun to continue the heating, land cools quickly and in turn chills the air at and near the surface. This explains the curious fact that gardeners living inland in the UK remain wary of ground frosts until early June!

Referring back to the Teign Valley in Devon, UK, the annual average number of days with air frosts at Teignmouth was 13, at Yarner Wood it was 24 and up at Okehampton, 44. This threefold increase from sea-level to around 400 m is partly related to the change in elevation and thus in the air temperature. It is also related to the local setting (exposure) of each site.

Sites in upland valleys tend to experience most frosts, when on clear, calm nights the chilling air drains down from the colder, higher ground into valley bottoms. Across hilly and mountainous areas, the cooling air flows downhill to become pooled in hollows or in valleys. The lowest minima on record in Britain come from the Scottish glens (the deep valleys in the highlands), not from the peaks. At Okehampton, on about one day in three during an average December and January, there was an air frost, whereas in Teignmouth the frequency of air frosts was roughly one day in nine. The mean minimum temperature in January at Okehampton was 0.7 °C while at Teignmouth it was 4.2 °C. There is also of course a higher frequency of snow and sleet higher up the valley.

The last of the thermometers usually found in a Stevenson screen, the wetbulb thermometer, is described later, in Section 2.2.4.

2.2.2 Barometric pressure

As well as temperature, another important ingredient of the weather that springs to mind is the **barometric pressure**, which is measured using one of two types of barometer. The **aneroid barometer** was developed at roughly the same time as the thermometer. The other type, the **mercury barometer**, was invented in 1643 by the Italian, Evangelista Torricelli.

Mercury barometer

The significance of the height of a column of mercury in a glass tube closed at its upper end and immersed at its lower end in a vessel of mercury was quickly realized. Above the mercury column, at the upper end of the tube, a vacuum is created; the downward pressure of the outside air on the surface of the mercury in the reservoir balances the weight of the mercury in the column (Figure 2.7).

The height of the mercury (Hg) column produced by a typical barometric pressure value at or near sea-level turns out to be around 76 cm (it is often quoted as 760 mmHg). Mercury is 13.6 times denser than water, so a water barometer would have to be some 10 m high! Neither fluid is perfect (because their volumes are temperature-dependent for example) but mercury does have the advantage of providing an instrument of convenient dimensions. Such a barometer for official use has to be certificated, with corrections required due to the effects of temperature and very small, local variations in the value of the gravitational acceleration.

The mercury barometer is read every hour, to the nearest 0.1 hPa, at major reporting stations. The reading represents a downward force that is exerted per unit area of horizontal surface. A conceptually easier way of appreciating the magnitude of the downward force is to think of the *height* of the mercury column that is supported by the weight of air overlying the barometer. Since 76 cm of mercury corresponds to the typical mean sea-level atmospheric pressure of 1013 hPa, a change in height of 1 cm of mercury is equivalent to a change in pressure of $(1013/76)\,\text{hPa} = 13.3\,\text{hPa}$.

Another important barometric pressure measure to report is that of its **tendency**. This normally consists of the magnitude and the sign of the *change* in barometric pressure in the three hours leading up to an observation time. If tendencies are mapped over a large region, it is possible to see areas where pressure is, for example, falling fast, holding more-or-less steady, or rising rapidly. The magnitude of the tendency is related to the movement of regions of high and low pressure across the surface *and* to their development — that is, whether a low is deepening (pressure falling) or filling (pressure rising) or whether a high is intensifying (pressure rising) or weakening (pressure falling). So a low that is approaching a location as well as deepening is expressed by a marked tendency of falling pressure.

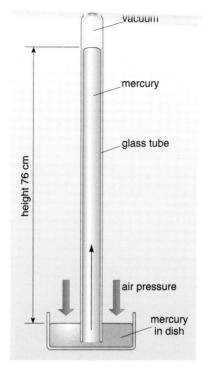

Figure 2.7   A mercury column barometer.

Aneroid barometer

An aneroid barometer is the type that is used as an ornament in many homes (Figure 2.8). The term 'aneroid' derives from the Greek and means 'not wet'; it refers to the fact that a liquid, such as mercury, is not used. Instead there is an evacuated metal capsule (the aneroid cell) in this type of barometer, which is 'squashed' by the pressure of the air surrounding the instrument. The amount of 'squashing' increases when the outside air pressure increases. Conversely, as pressure falls, the capsule 'relaxes' to a subtly different shape. These changes of shape are magnified by an indicator arm attached to the capsule. As they occur, the arm moves around the graduated dial on the aneroid barometer.

Figure 2.8 An aneroid barometer.

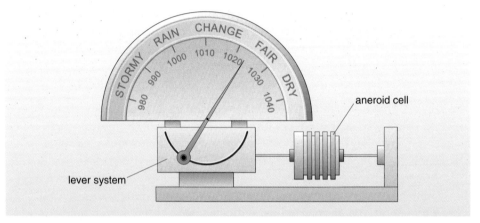

Domestic barometers of this type are not as accurate as mercury ones or the expensive precision aneroid instruments used in weather stations. The weather descriptions written around the edge of the face of domestic instruments provide very broad indications of the weather related to the actual pressure reading and to the likely change associated with rising or falling pressure.

Before any pressure value is transmitted to aid weather services, it has to be corrected to mean sea-level. This means that the precise elevation of the instrument must be known so that a correction can be applied. Apart from a very few of the world's weather stations that lie below mean sea-level, such as Schipol Airport in Amsterdam, an additional pressure must be added to that measured by the weather station, to represent the column of air between the barometer and sea-level.

Activity 2.2   Pressure measurement at your home weather station

For those who either already possess, or who wish to buy a barometer.

You may have on occasion heard the TV weather forecaster say that a particular day was good for adjusting or correcting your home barometer to mean sea-level. As stated above, the instrument you may have at home reads the station pressure, meaning the value at your particular height above mean sea-level. One simple way to adjust the value approximately is to add 1 hPa (1 mbar) for every 10 m your home is above mean sea-level. (This is an approximation and only works well up to heights of 1000 m. Strictly, the profile of pressure is exponential, i.e. the pressure falls off by a constant factor for a constant increase in height — see Section 1.3.2).

Another strategy is to wait for a day when a high-pressure area (anticyclone) is over your region, because on such days mean sea-level pressure does not vary by very much within a region. On such a day, you should look at the value for your home region, either from the TV map or from the forecast map issued in the broadsheets (making sure that, if possible, you adjust your reading at the validity time of the official information). Once you've done this, you turn the pressure indicator on the barometer to the sea-level value. You don't need to adjust it again after taking this action.

There is a crucially important reason for adjusting values to the sea-level values. It is that pressure decreases so rapidly in the vertical direction compared to the horizontal. As stated earlier, pressure is halved for every 5.5 km ascended, whereas across the entire Earth's surface the typical pressure variations are only 5–10%. The all-important, but relatively subtle changes of pressure that are related to **highs** (**anticyclones**) and **lows** (**cyclones** or **depressions**) across the surface would be 'swamped' if the non-modified values were plotted. Doing so would lead to a contour map of pressure that would be very similar to a contour map of height. The all-important weather disturbances would be 'buried' somewhere in this pattern.

Once simultaneous values of mean sea-level pressure are plotted onto a base map, an automatic or human-produced 'analysis' is carried out in which contours of constant mean sea-level pressure or **isobars** are analysed. This is done because the resultant pattern will indicate the location and intensity of highs, lows and other weather-related features, which all move and evolve incessantly.

○ Study the surface analysis map for an early autumn day (Figure 2.9). What and where are the highest pressure value, and the lowest?

● The highest is 1035 hPa and centred over southern Finland, while the lowest is 985 hPa, just southwest of Iceland.

The weather systems in Figure 2.9 — areas where contours encircle centres of high or low pressure — are typically about a few thousand kilometres across. You can assess the scale of features on the map by realizing that 10° of latitude (angular distance between circle arcs shown in Figure 2.9) is equivalent to 1110 km along any longitude line (the straight lines crossing the circle arcs). The Scandinavian high dominates a huge area of northern Europe while the low is part of a larger feature that influenced much of the North-East Atlantic. Such areas of influence are quite typical.

Intuitively, one can imagine that air in a high-pressure zone is squeezed out into low-pressure zones. Indeed, whenever there is a pressure difference between two neighbouring places there is a force pushing air in the direction of high to low pressure. How strong the force is depends on the magnitude of the horizontal **pressure gradient**, which is the rate at which barometric pressure changes with distance:

$$\text{horizontal pressure gradient} = \frac{\text{difference in pressure between two points}}{\text{distance between two points}} \quad (2.1)$$

Figure 2.9 The mean sea-level surface analysis map for an early autumn day in part of the Northern Hemisphere. Red lines and symbols denote warm fronts, blue lines and symbols denote cold fronts, and purple lines and symbols denote occluded fronts — which you will learn about in Section 4. The numbers are pressure values in hPa.

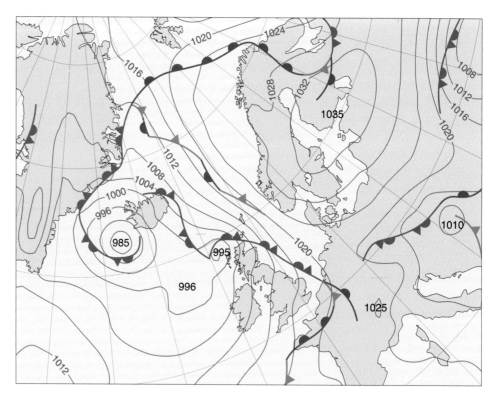

A steep pressure gradient, i.e. a rapid change of pressure with distance, is linked to strong winds while a weak gradient is associated with light winds. In practice, the pressure gradient can be measured between a higher contour and a lower one, over a direction that is at right angles to the contours (isobars). This is exactly the same as the way you estimate the slope of a hill from a contour map, hence the use of the word 'gradient'. A television weather forecaster might point to where the isobars are tightly packed as a region at risk of gales.

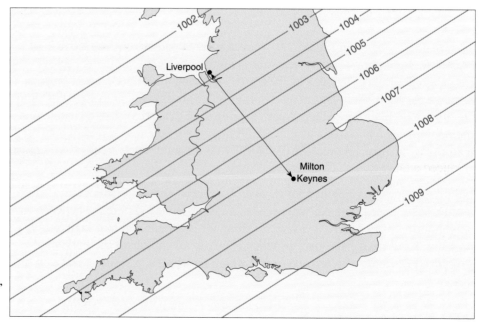

Figure 2.10 Contours of mean sea-level pressure (hPa), or isobars, in part of the UK at noon on a particular day.

○ The mean sea-level pressure in Milton Keynes (UK) one day at noon was 1007.4 hPa while at exactly the same time it was 1002.9 hPa in Liverpool (UK) to the northwest (Figure 2.10). The distance between the two weather stations is 210 km along a line which on this particular day was exactly at right angles to the isobars. What was the horizontal pressure gradient?

● The horizontal pressure gradient is calculated by dividing the difference in the two pressure values by the distance between the stations, strictly at right angles to the isobars:

$$\frac{1007.4 \text{ hPa} - 1002.9 \text{ hPa}}{210 \text{ km}} = 2.1 \times 10^{-5} \text{ hPa m}^{-1}$$

2.2.3 Wind speed and direction

Wind speed and direction combined define a vector for which direction (from true north) and speed are logged. A **vector** is a quantity that has both magnitude (e.g. speed, in metres per second) *and* direction (e.g. northwesterly). Vectors are usually represented on maps using arrows. Wind speeds are quoted in knots (abbreviated to kn), or nautical miles per hour, by the vast majority of nations. One **knot** is 1.15 statute miles per hour or 0.51 m s^{-1}.

By international agreement, wind speed and direction are measured at a height of 10 m above the ground and are not instantaneous values because the direction and speed fluctuate a lot over very short time periods. Each observation represents an average over a few minutes.

Wind speed

Wind speed is measured by an **anemometer**, comprising a rotating vertical axis on which are mounted three equally spaced hemispherical cups (Figure 2.11). These structures are very responsive to even small wind speeds; their rate of rotation is converted mechanically or electrically into a wind speed value, which is taken to be that of the horizontal airflow. Wind speed can also be *estimated* by considering how different wind strengths affect either the sea surface or features of the land surface. This is the basis of the well-known **Beaufort scale** (Table 2.1) developed in 1805 by a Royal Navy officer, Francis Beaufort, who later attained the rank of Admiral. Beaufort's efforts were used very widely, to relate how disturbed the sea surface is to the speed of the wind blowing across it. He produced a table to link Beaufort force as a number to these increasingly disturbed states. The table has since been adapted to link the same numbers to land-based indicators of surface wind speed.

Figure 2.11　An anemometer. The instrument on the right counts the number of rotations.

Table 2.1 The Beaufort scale of wind force.

Beaufort scale number	Typical wind speed*		Limits of wind speed*		Descriptive terms	Sea criterion†	Probable maximum height of waves/m†
	kn	m s⁻¹	kn	m s⁻¹			
0	0	0.0	less than 1	0.0–0.4	calm	Sea like mirror.	–
1	2	1.0	1–3	0.4–2.2	light air	Ripples with the appearance of scales are formed but without foam crests.	0.1
2	5	2.6	3–6	2.2–3.6	light breeze	Small wavelets, still short but more pronounced; crests have a glassy appearance and do not break.	0.3
3	9	4.6	6–10	3.6–5.7	gentle breeze	Large wavelets. Crests begin to break. Foam of glassy appearance. Perhaps scattered white horses.	1.0
4	13	6.6	10–16	5.7–8.7	moderate breeze	Small waves, becoming longer; fairly frequent white horses.	1.5
5	19	9.7	16–21	8.7–11.3	fresh breeze	Moderate waves, taking a more pronounced long form; many white horses are formed. (Chance of some spray.)	2.5
6	24	12.2	21–27	11.3–14.4	strong breeze	Large waves begin to form; the white foam crests are more extensive everywhere. (Probably some spray.)	4.0
7	30	15.3	27–33	14.4–17.5	near gale	Sea heaps up and white foam from breaking waves begins to be blown in streaks along the direction of the wind.	5.5
8	37	18.9	33–40	17.5–21.2	gale	Moderately high waves of greater length. The foam is blown in well-marked streaks along the direction of the wind.	7.5
9	44	22.4	40–47	21.2–24.7	strong gale	High waves. Dense streaks of foam along the direction of the winds. Crests of waves begin to topple, tumble and roll over. Spray may affect visibility.	10.0
10	52	26.5	47–55	24.7–28.8	storm	Very high waves with long overhanging crests. The resulting foam in great patches is blown in dense white streaks along the direction of the wind. On the whole, the surface of the sea takes on a white appearance. The tumbling of the sea becomes heavy and shock-like. Visibility affected.	12.5

11	60	30.6	56–63	28.8–32.9	violent storm	Exceptionally high waves. (Small and medium-sized ships might be for a time lost to view behind the waves.) The sea is completely covered with long white patches of foam lying along the direction of the wind. Everywhere the edges of the wave crests are blown into froth. Visibility affected.	16.0
12	–	–	63 and over	32.9 and over	hurricane	The air is filled with foam and spray. Sea completely white with driving spray; visibility very seriously affected.	–

* Measured at a height of 10 m above sea-level.

† These columns are added as a guide to show roughly what may be expected in the open sea, remote from land. In enclosed waters, or when near land with an offshore wind, wave heights will be smaller and the waves steeper. *Note*: (a) It is difficult at night to estimate wind force by the sea criterion; (b) The lag effect between the wind getting up and the sea increasing should be borne in mind; (c) Depth, swell, heavy rain and tide effects should be considered when estimating the wind from the appearance of the sea.

The record gusts recorded in the UK and globally are:

England:	103 kn at Gwennap Head, Cornwall on 15 December 1979
Scotland:	123 kn at Fraserburgh, Aberdeenshire on 13 February 1989; 150 kn at Cairngorm Automatic Weather Station (near the boundary between Highland and Moray, a high-level site) on 20 March 1986
Wales:	108 kn at Rhoose, South Glamorgan on 28 October 1989
N. Ireland:	108 kn at Kilkeel, County Down on 12 January 1974
USA:	276 kn near Oklahoma City (tornado) on 3 May 1999.

Wind direction

The direction of the air's flow is reported to the nearest 10° from true north in such a way that values go round the 360° of a circle in a clockwise sense. This means, for example, that a wind report of '090' is an 'easterly'; meteorologists have to be unambiguous about direction nomenclature. An easterly wind is one *from* the east, a southerly is from the south (reported as 180°), and so on. So a westerly is reported as 270° and a northerly as 360° (Figure 2.12). If there is a calm, then the direction is recorded as 000°.

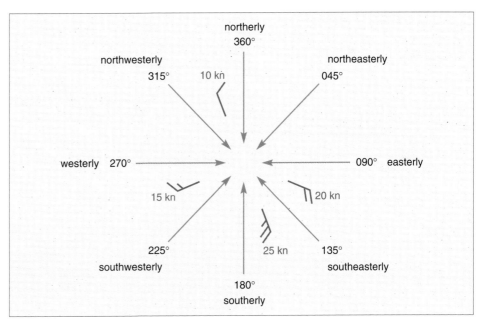

Figure 2.12 Meteorological definition of wind direction, including four examples of wind direction and speed. The orientation of the stem of the wind symbol shows direction and the branches denote speed: each long branch = 10 kn and each short branch = 5 kn.

Wind direction is indicated by a weather vane (Figure 2.13), designed aerodynamically so that it always points into the wind. Operational vanes consist of a horizontal arm that has a pointer at one end and an aerofoil plate at the other. Ornamental types are not as responsive or as accurate as this pattern.

Figure 2.13   A professional standard weather vane (right) with anemometer (left).

The term **veering** means that the wind direction changes with time at one place in a clockwise sense. For example, if at 12.00 UTC at Milton Keynes it was reported to be 270° but at 13.00 UTC it had changed to 300°, and the observer knows that it changed clockwise through 30° to do so (it is feasible but unlikely in the extreme that it would shift in the other sense, through 330°), then the wind has veered. Conversely, it would be called **backing** if it had changed over an hour at one place from, say, 120° to 090°, i.e. in an anticlockwise sense. Such shifts are often very significant, and are related to the passage of features like fronts, of which we shall say more in Section 4.

In addition, veering and backing are terms used for changes in wind direction *at one time* but at different heights above a place, normally an upper-air station. For example, a balloon might be released and return the wind data shown in Table 2.2 at three successive points of its ascent. These data indicate that the wind veers with height between 1000 and 950 hPa, then it backs with height between 950 and 900 hPa — such changes are not unusual.

You can see this for yourself occasionally when there are two or more layers of cloud, if you watch them carefully. It is most often the case that the cloud layers are drifting in different directions, allowing you to deduce whether the wind is veering or backing with height.

Table 2.2   Typical weather balloon data.

Height/hPa*	Wind direction
1000	350°
950	020°
900	340°

* Because pressure drops steadily with height, it is used as a measure of the height ascended by a weather balloon.

2.2.4   Humidity

Humidity, the water vapour content of air, is quantified according to several different definitions. Some (e.g. relative humidity) are easier to measure directly, but are not of immediate use, and others are useful (e.g. absolute humidity), but are more difficult to measure.

Absolute humidity

The easiest humidity measure to understand is **absolute humidity**, which is simply the humidity mass mixing ratio. (Mixing ratios were described in general terms in Box 1.1.) Absolute humidity is usually quoted in units of $g\,kg^{-1}$, i.e. grams of water vapour per kilogram of air. Air at sea-level at a temperature of $10\,°C$ can hold up to $7.6\,g\,kg^{-1}$, when it is said to be **saturated**. The mixing ratio of air at saturation increases with increasing temperature — warm air can carry more water vapour than cold air. The advantage of the mixing ratio as a measure of humidity is that it does not depend on other factors such as pressure or temperature. Its disadvantage is that there is no straightforward method for directly measuring it. The absolute humidity is of immediate use if you want to calculate the amount of rain likely to fall on a showery day. It is essential to know how much water is being supplied in air that is going to form rain clouds. Indeed, weather forecast models require humidity to be made available as an absolute measure, e.g. in $g\,kg^{-1}$.

Dewpoint temperature

A more practically-based quantity is the dewpoint temperature, often just called the **dewpoint**. It can be measured directly by cooling a sample of air until a temperature is reached at which water begins to condense out as very tiny droplets. This method works because as the air cools, its capacity to hold water is reduced until it reaches a point — the dewpoint — at which it is saturated, only just able to hold the water it started with. A further reduction in temperature causes water to condense out, forming dew on surfaces in contact with the air sample — hence the name. The dewpoint is therefore, by definition, always less than or equal to the actual drybulb temperature of an air sample. You can think of this method of measuring humidity, which is not routinely used, as like squeezing water out of a sponge, where the sponge is an air sample and the squeezing is a reduction in temperature. A disadvantage of dewpoint is that it depends on pressure, so that it is not just a measure of the water content of air.

Relative humidity

The relative humidity can be defined as the ratio (expressed as a percentage) of the air's *actual* humidity mass mixing ratio (the absolute humidity) to the *saturation* value of the humidity mass mixing ratio. In plainer words, the relative humidity is a percentage value, expressing how close the air is to being saturated *at the current temperature*.

Relative humidity is measured *directly* inside a Stevenson screen by a **hair hygrometer**, which provides a weekly trace of its fluctuations on a strip chart, wrapped around a mechanically or electronically driven drum. The pen used is connected via a long arm to a bar across which a sheaf of horse hair is stretched; the hair changes length in response to changes in relative humidity. Temperature is measured concurrently and printed out on the chart in the same way. So the record, called a thermohygrogram, comprises continuous traces of both temperature and relative humidity (Figure 2.14).

To meteorologists, the relative humidity is probably the least useful index of humidity, due to its dependence on temperature. In fact, the relative humidity of an air sample can change without the actual amount of water vapour in the

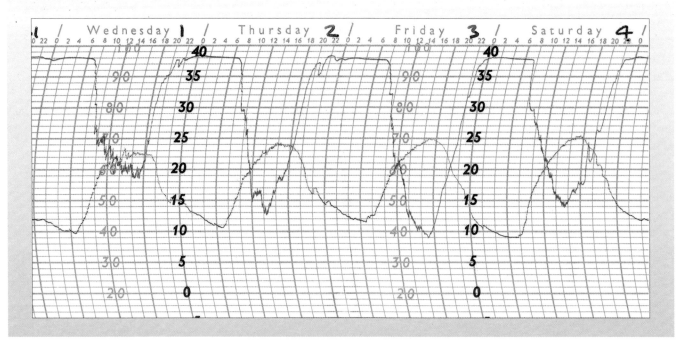

Figure 2.14   Part of a thermohygrogram (data for 4 days). The red trace and scale show the relative humidity; the black trace and scale show the temperature.

sample changing! For example, relative humidity decreases if the temperature increases because the air becomes able to hold more water; the absolute humidity will remain constant. This means that relative humidity is frequently at a maximum near dawn, when the minimum temperature often occurs, and is at a minimum in the early to mid-afternoon, when the drybulb value is highest (Figure 2.14). The relative humidity is a quantity that is often useful in modelling certain atmospheric processes, e.g. the formation of clouds.

One observation we're all familiar with is the occurrence of dews and frosts on some mornings. One sign of the end of summer is a noticeable dew on lawns — and an indicator of progress into autumn is the first ground frost. These are both a consequence of moist air at low levels being chilled to its dewpoint by prolonged cooling during gradually longer nights. As saturation (100% relative humidity) is reached, further cooling leads to a deposit of liquid (dew) or solid water (frost) on the surface.

Question 2.2

Why are ground frosts more common than air frosts (i.e. those measured in a temperature screen)?

Vapour pressure

Water vapour is just one of many gases that can be present in a sample of air. As such, water vapour contributes to the total pressure of air. Clearly, the (water) vapour pressure, p_v, is much less than the total pressure, p, of a sample of air. Also, for a given temperature the vapour pressure cannot rise above a value called the **saturated vapour pressure**, denoted by p_s, which occurs, as the name suggests, when the air is saturated. A useful measure of humidity is the difference $p_s - p_v$, called the **vapour pressure deficit** or the **saturation deficit**.

Also, it turns out that the ratio p_v/p_s is equal to the **relative humidity**. (Note that definitions of saturated vapour pressure do differ in different branches of science.)

Wetbulb temperature

In addition to the drybulb thermometer, there is a second vertically mounted mercury-in-glass thermometer housed in the Stevenson screen. This type of thermometer differs from the drybulb one in that its bulb is covered snugly by a muslin wick which is stretched over it. The wick leads to a bottle of distilled water to ensure that the thermometer bulb is always wet. Not surprisingly, this instrument is known as the 'wetbulb' thermometer. The temperature it measures depends on the rate of evaporation of water from the soaked muslin surrounding its bulb. As we will see in Section 3, as water evaporates from the muslin, it takes heat with it (in the same way that evaporation of sweat from skin removes heat from the body) so cooling it down. If the air within the screen is very damp, the evaporation rate is low and the consequent cooling of the bulb and mercury inside it is small. In contrast, if the air is dry, there is a lot of evaporative cooling of the mercury, so the temperature drops substantially. The wetbulb temperature is thus a measure of the air's humidity; it is read and logged to the nearest 0.1 °C and so can be used to accurately calculate the other humidity measures. Box 2.4 compares and contrasts the humidity measurements at two different sites, both in Europe.

Box 2.4   Examples of humidity

The data needed to calculate the relative humidity are the drybulb temperature (to establish the saturation value) and either the wetbulb or the dewpoint temperature (to establish the *actual* concentration of water vapour). When using data from routinely plotted surface weather maps, drybulb and dewpoint temperatures are the values used.

On an early autumn day in mid-afternoon, the drybulb temperature/dewpoint temperature at London's Heathrow Airport and at Rome's Leonardo da Vinci Airport were 16 °C/4 °C, and 28 °C/16 °C respectively.

These figures mean that, assuming both sites reported barometric pressures of 1000 hPa, the actual and saturation values and the relative humidities were as shown in Table 2.3.

Table 2.3   Humidity measurements at two major airports.

Airport	Absolute humidity/g kg^{-1}		Relative humidity/%
	Actual	Saturation	
London Heathrow	5.1	11.5	44
Rome Leonardo da Vinci	11.5	24.3	47

The point about this variety of values is that although Heathrow and Leonardo da Vinci airports experienced just about the same relative humidity, they clearly were seeing really very different drybulb temperatures, dewpoint temperatures and thus absolute humidities (humidity mass mixing ratios). Rome Airport was much warmer, at 28 °C, and had an absolute humidity that was more than twice that at Heathrow.

○ A 1 kg sample of air at 15 °C is kept at sea-level. It is found to contain 7.6 g of water vapour. (a) What is the humidity mass mixing ratio of this sample? (b) If the temperature is lowered to the dewpoint, keeping the pressure constant, what are the sample's relative humidity and mixing ratio?

● (a) The mixing ratio is 7.6 g kg^{-1}. (b) According to the definition of dewpoint, the air must be saturated, so the relative humidity must be 100%. Although on the verge of condensing, the water is still all water vapour, so the mixing ratio is the same as in (a).

The psychrometric chart

The psychrometric (*not* psychometric!) chart enables you to quickly convert between different humidity measures (Figure 2.15). Most commonly, you will have a reading of wet- and drybulb temperatures that you will want to convert to relative humidity or vapour pressure. The chart lets you calculate any humidity measure, if you already know two of them, by crossing two lines or curves on the chart to define a point. The following examples illustrate how it works.

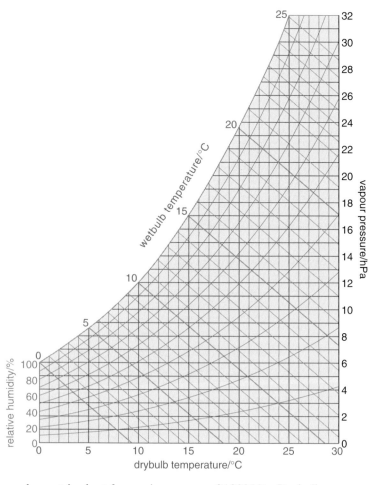

Figure 2.15   The psychrometric chart for an air pressure of 1000 hPa. Drybulb temperature is constant along vertical lines; wetbulb temperature is constant along diagonal lines; vapour pressure is constant along horizontal lines; and relative humidity is constant along curves.

Example 1

Drybulb temperature = 21 °C; wet bulb temperature = 16 °C.

First locate the vertical line that meets the drybulb axis (horizontal one at the bottom of the chart) at a value of 21 °C. Placing the edge of a transparent ruler along the vertical line will help keep your eye on track. Now locate the intersection point, where the diagonal line from 16 °C on the wetbulb temperature axis (the uppermost curve on the chart) meets the vertical line at 21 °C. If you trace the horizontal line through this point to the vapour pressure axis (right side of the chart), you can read off a vapour pressure of 15 hPa. One of the curves also conveniently passes through the intersection point — we've picked the dry- and wetbulb values to make this first example a bit simpler! If you follow the curve back to the relative humidity axis (left side of the chart), you can read off a value of 60%.

Example 2

Relative humidity = 40%; vapour pressure = 10 hPa.

This time place your ruler along the 10 hPa line, and follow the curve for 40% humidity. The intersection point conveniently lies on one of the drybulb temperature lines, that of 21 °C. The wetbulb temperature is a bit more tricky. Place your ruler so that one edge is parallel to the diagonal lines, making sure that it still passes through the intersection point. If you follow the ruler up to the wetbulb axis, you'll see that it crosses it somewhere between 13 °C and 14 °C. In fact, it is about one-fifth of the way along, so our best estimate for the wetbulb temperature is 13.2 °C.

○ Referring to Box 2.4 for drybulb temperatures and relative humidities, what would the wetbulb temperature and vapour pressures be, according to Figure 2.15, for (a) London Heathrow and (b) Rome Leonardo da Vinci?

● (a) London has a drybulb temperature of 16 °C and a relative humidity of 44%. Although there isn't a relative humidity curve for 44%, we can look roughly half-way between the 40% and 50% ones. The intersection point would therefore lead us to a wetbulb temperature of 10 °C and a vapour pressure of 8 hPa.

(b) For Rome we have a drybulb temperature of 28 °C and a relative humidity of 47%. Again there isn't a curve for 47%, but if we go three-quarters of the way between the 40% and 50% curves, we find that the wetbulb temperature is about 20 °C, and the vapour pressure is about 18 hPa.

2.2.5 Cloud type and amount

Clouds are formed when moist air is cooled by some means, so that it reaches saturation, and is then cooled further. This further cooling causes vast numbers of minute water droplets (or more rarely, ice crystals) to condense out in the form of a cloud. Box 2.5 provides a calculation of the mass of water in a cloud.

The most common cooling mechanism is that of ascent, when the air cools adiabatically (as explained in Box 2.6). This produces both cumulus cloud on quite a small scale and also the vast areas of layer cloud that characterize the very large-scale ascent within a depression (low-pressure region). In addition, air that has to ascend over hills or mountains often cools enough to produce an **orographic cloud**. 'Orographic' means related to or caused by surface features such as mountain ranges. **Fog** is in essence cloud at the ground, formed when air is chilled at the surface, either because the ground has cooled radiatively or because air is blown across a much colder surface, in which case it is called advection fog. (You will meet the term fog again in Section 2.2.7, on visibility.)

Box 2.5 Weighing a cloud

Clouds are of course composed of water droplets and/or ice crystals so they have mass. We can estimate how massive a cloud is from its dimensions and composition.

Let's assume we see a fair-weather cumulus, the kind that drifts across the sky on a pleasant sunny day. Let's also assume it is hemispherical with a diameter of 400 metres. The cloud's volume is then $\frac{2}{3}\pi r^3$ (half the volume of a sphere):

$$\text{cloud volume} = 0.667 \times 3.14 \times (200\,\text{m})^3 = 1.68 \times 10^7\,\text{m}^3$$

Next, we need to know the size and the number of cloud droplets contained in a cubic metre of such a cloud. Suppose there are some 2×10^8 droplets in a cubic metre of cloud, each one having a typical radius of 6×10^{-6} m, i.e. $6\,\mu$m.

So, the volume of one droplet, assuming it is sphere of radius r, is $\frac{4}{3}\pi r^3$:

$$\text{droplet volume} = 1.33 \times 3.14 \times (6 \times 10^{-6}\,\text{m})^3 = 9.02 \times 10^{-16}\,\text{m}^3$$

So, in 1 m³ of cloud the volume of water is

$$2 \times 10^8 \times 9.02 \times 10^{-16}\,\text{m}^3 = 1.80 \times 10^{-7}\,\text{m}^3$$

The entire cloud therefore contains

$$1.80 \times 10^{-7} \times 1.68 \times 10^7\,\text{m}^3 = 3.02\,\text{m}^3 \text{ of water.}$$

The density of water is about $1000\,\text{kg m}^{-3}$, so such a cloud weighs about 3000 kg, or 3 tonnes!

Box 2.6 The adiabatic concept

Adiabatic means 'without heat exchange'. So, a parcel of air that cools or warms adiabatically is one that exchanges no heat with the surrounding air. Basic physics tells us that an air parcel that is cooling adiabatically must be expanding. Conversely, a parcel that is expanding in adiabatic conditions must be cooling. Likewise, warming is associated with contraction under adiabatic conditions. In discussing the balloon in Section 1.3.2, we assumed the opposite was true, that because the balloon was being carried up at walking pace, it had plenty of time to exchange heat with its surroundings. A rising air parcel usually does not have time to exchange heat in this way, and so it is reasonable to assume that the parcel is under adiabatic conditions. Being adiabatic, we then know that a rising air parcel's decrease in temperature is solely because it is expanding, or conversely, that a sinking air parcel's increase in temperature is because it is being compressed.

You might wonder why adiabatic conditions relate a change in temperature to a change in volume. You can understand this by considering what you have to do to compress a gas under adiabatic conditions: you have to put energy in to work against its pressure. The energy you put into compressing the gas cannot disappear because energy is always conserved, nor can it be given to the surroundings because we have supposed the conditions are adiabatic. So the energy you put in must turn into heat energy, raising the temperature of the gas. You can check this for yourself by putting your finger over the end of a hand pump for a bike, and feeling the finger get hot as you compress the air inside by pumping.

Unlike the measures of weather taken at the Earth's surface described earlier, cloud type is not a quantitative assessment but simply a visual recognition of whether the cloud or clouds present are low, middle or high and whether they are stratiform (layered) or cumuliform (heaped), for example. Layered cloud has the appearance of an extensive continuous sheet, while heaped clouds are often as tall as they are wide at the base, and have a lumpy top surface where the currents within the cloud are bubbling up.

Cloud type

The broadest definition of cloud type rests on the height of their base above the surface, whether it be at sea-level or in a mountain region. What constitutes low, middle or high cloud varies with latitude; values are highest in the tropics and lowest in the higher latitudes. The British Isles are in middle latitudes, and the approximate heights are:

low cloud	surface to 2000 m
middle cloud	2000–5000 m
high cloud	above 5000 m

One way of estimating the height of a cloud's base is to use a searchlight pointing vertically upwards to illuminate the cloud from underneath. This instrument is at a known distance (d) from the observer who measures the angle of elevation (θ) of the bright spot on the cloud's base from their vantage point. This means that the height of the cloud base is $d \tan \theta$ (Figure 2.16).

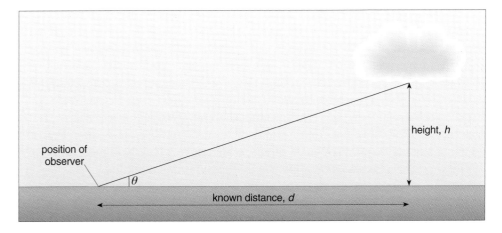

Figure 2.16   Estimating the height of a cloud base.

Don't be put off by the tan θ. Think of it as a way to relate an angle, which you can measure, and a distance, which you know, to the height of distant object, which you want to find out. On your calculator you can work out the 'tan' or 'tangent' of an angle θ by typing in the angle and then pressing the 'tan' button (making sure you are in 'degree mode', if the angle is in degrees). Some models of calculator might require you to do it the other way around, i.e. press 'tan', then type in the angle, then press the '=' or 'enter' button. Try the following question if you are unsure.

○ If the searchlight is 3.00 km away, and you measure an angle of 18°, what is the height of the cloud base?

● Use the 'tan' button on your calculator to obtain the value of tan 18°. This is 0.325 to 3 significant figures. Now you must multiply tan θ by d, which is 3.00. This gives 0.975. The cloud base is therefore 975 m above the ground.

Cloud genera

Overall, ten basic cloud types, or *genera*, are now recognized. It was Luke Howard, a London pharmacist, who in 1803 proposed a simple classification which forms the basis of the one used today. He termed a sheet of cloud, **stratus** (Latin for a layer), a heap cloud, **cumulus** (a pile or heap), a wispy, streak cloud, **cirrus** (a hair) and a rain cloud, 'nimbus'. This latter name is used today only in combination with two other component names, i.e. **cumulonimbus** and **nimbostratus**.

By international agreement, the genera are:

low cloud   stratocumulus (Sc), stratus (St), cumulus (Cu), cumulonimbus (Cb)

middle cloud   altostratus (As), altocumulus (Ac) and nimbostratus (Ns)

high cloud   cirrus (Ci), cirrostratus (Cs), cirrocumulus (Cc)

The Field Studies Council/Royal Meteorological Society basic cloud identification card, *The Cloud Name Trail* (included with the course materials) can be used to aid the task of keeping the daily weather log as an extension to Activity 2.1.

Observers need of course to be trained to be able to identify these ten basic cloud genera, and in fact to be able to refine the report by adding the *species*, which depends upon the detail of cloud shape and structure. (This terminology of 'genus and species' has been borrowed from the way in which biologists have classified plants and animals over the last two centuries or so.) There are 14 cloud species but we won't go into detail here! A couple of examples should suffice. A shallow, flattened-top cumulus cloud is termed cumulus humilis, whereas a very deep 'cauliflower', bubbling-top cumulus may be termed cumulus congestus. 'Fractus' are clouds in the form of irregular shreds, which have a clearly ragged appearance; this classification applies only to stratus and cumulus.

Cloud amount

The amount of cloud is quoted numerically as eighths (or oktas, from the Greek for 'eight') of the sky covered both by individual types of cloud and as the total cover of all cloud types. Eighths are used because the transmitted report of cloud amount is allowed just one digit from 0 to 9; 0 oktas clearly means no cloud at all, 4 oktas means half the sky covered with cloud, 8 oktas means overcast, but 9 oktas (i.e. nine-eighths) may seem strange. It means that the sky is obscured by fog or blowing snow for example, so that nothing can be reported. These values are assessed by eye but can be aided at airports by planes landing and taking-off to enrich the obviously limited ground-based view.

Sunshine

The total sunshine hours recorded at a site is of course inversely related to daytime total cloud amount. At most places the hours of 'bright sunshine' measured in a day still come from the Campbell–Stokes sunshine recorder. This instrument comprises a glass sphere which concentrates the Sun's rays onto a sensitized, curved card that burns for the daily duration of 'bright sunshine'.

The mean annual totals of sunshine hours during 1990–1998 were 1785 h and 1610 h at Teignmouth and Yarner Wood respectively. At Teignmouth the sunniest month was June (7 h 50 min per day) while at Yarner Wood it was July (7 h 2 min per day). In the dullest month, Teignmouth managed 1 h 40 min per day in January while at Yarner Wood it averaged out at 1 h 27 min daily in the same month.

Why are sunshine totals smaller at the higher elevation? Sunshine totals are obviously greater when cloud amounts are smaller. Sea breezes (which blow from sea to land, and are explained in Section 3) mean that the coastal strip is less cloudy, because they are associated with the landward penetration of cloud-free sea air. In addition, the ascent of damp air leads to cloud inland and upslope more commonly than at the coast.

Record monthly maximum sunshine totals for the UK countries are:

England:	383.9 h at Eastbourne, East Sussex in July 1911
Scotland:	329.0 h at Tiree, Argyll and Bute in May 1946 and May 1975
Wales:	354.3 h at Dale Fort, Dyfed in July 1955
N. Ireland:	298.0 h at Mt. Stewart, County Down in June 1940.

Record monthly minimum sunshine totals for the UK countries are:

England:	0.0 h at Westminster, London in December 1890
Scotland:	0.6 h at Cape Wrath, Highlands in January 1983
Wales:	2.7 h at Llwynon, Powys in January 1962
N. Ireland:	8.3 h at Silent Valley, County Down in January 1996.

2.2.6 Precipitation

Precipitation is defined as any liquid or solid aqueous particle that is deposited from the atmosphere, and so includes rain, drizzle, sleet, snow and hail, for example. The first reasonably prolonged set of rainfall observations kept in the British Isles was taken by a Lancashire squire by the name of Richard Towneley. He was the master of Towneley Hall near Burnley, on the west flank of the Pennines. He designed his own raingauge to provide a precipitation record from 1677 to 1704.

Later in this period, William Derham, an Essex clergyman in Upminster, set the fashion that was to persist in that profession for some decades. He kept rainfall records from 1697 to 1716. Much later in the 18th century, in 1799 in fact, the famed John Dalton was able to put together a collection of rainfall records from a variety of sites.

Rainfall

Rainfall is defined as the total liquid product of precipitation and condensation from the atmosphere, as received and measured in a raingauge (Figure 2.17). This means that snow, sleet and hail are counted in addition to rain and drizzle, as are contributions from dew and frost deposited in the gauge's funnel.

Figure 2.17   A raingauge.

In the British Isles the standard instrument has a 12.7 cm (five inch) diameter funnel placed into an upright cylindrical sheet-copper can, in which the funnel's brass bevelled lip or outer edge is 30.5 cm (12 inches) above the ground. The funnel guides precipitation into a narrow-bore tube at its base, to minimize evaporative losses from the glass measuring bottle into which the precipitation flows. Most gauges are used to record a daily total of precipitation in millimetres, representing the depth the precipitation would accumulate to over 24 hours if it didn't soak into the ground, evaporate, etc. To gain this value, the observer carefully decants any water from the bottle in the raingauge into a tapered measuring vessel that enables amounts as small as 0.05 mm to be recorded.

Question 2.3

On a very wet day, 20 mm of rain can fall per m². What is the mass of rainwater that falls into each square metre? (Take the mass density of liquid water to be 1000 kg m⁻³.)

○ Consider the Teign Valley. The average annual precipitation is 899 mm for Teignmouth, 1384 mm for Yarner Wood and 1645 mm for Okehampton — almost a doubling in amount for an increase in elevation of some 400 m. The average number of wet days (those on which 1.0 mm or more of rain fell) increased from 124 at Teignmouth to 150 at Yarner Wood to 185 near Okehampton, i.e. roughly 1 in 3 days at the coast to 1 in 2 on the high moors (wet days are most frequent at *all* stations from November to March). Calculate the mean precipitation rate (in mm per day) for wet days, based on the annual total divided by the number of wet days, to show that it increases from Teignmouth to Yarner Wood.

● The average precipitation rates are: Teignmouth, 7.25 mm d⁻¹; Yarner Wood, 9.23 mm d⁻¹.

This difference is related essentially to the impact of orographic enhancement (described in Box 2.7), which can lead to more than a twofold increase in rainfall rate between a coastal site and one in adjacent uplands. The increase between these two places due to orographic enhancement would in fact be larger if we were able to focus only on those weather situations where the seeder–feeder mechanism (see below) was operating. The difference estimated here also probably includes some showers which wouldn't affect the coast as much as they might inland in the summer.

The record daily falls for the UK and globally are:

England:	279.0 mm at Martinstown, Dorset on 18 July 1955
Scotland:	238.0 mm at Sloy Main Adit, Loch Lomond on 17 January 1974
Wales:	211.0 mm at Rhondda, Cynon Taff on 11 November 1929
N. Ireland:	158.9 mm at Tollymore, County Down on 31 October 1968
La Réunion:	1880.0 mm at Cilaos on 15–16 March 1952.

Box 2.7 *Wetter upwards*

We all know that it gets wetter the higher we go up into hills. Quite why this should be is not obvious. It is not just because air flowing across hills or mountains ascends, cools adiabatically and sometimes produces cloud, that upland areas are wetter.

High-lying regions are *cloudier* from this 'orographic' lifting of moist air, but the significantly *wetter* conditions occur within very particular weather systems that cross them. In fact, precipitation totals actually decrease with height in some mountainous areas, like subtropical islands such as Hawaii. The largest mountains, like the Himalayas, are so high that in their upper reaches the air is so cold, and therefore so dry, that substantial precipitation falls occur very rarely.

Consider the wettest parts of the UK, such as the Scottish Highlands and the Lake District of Northern England. The reason why these hilly areas receive so much precipitation is related to what is called the **seeder–feeder** mechanism. The first necessary ingredient is **orographic enhancement** of rainfall, which requires a pre-existent layer of precipitating cloud (at say 2 km up). As this layer crosses parts of the UK, it is very often accompanied by a fast-flowing, lower-level, moist airstream that crosses the coast and ascends hills.

Any rain on the coast is simply from the cloud 2 km above. The ascending moist airstream, however, produces persistent and water-rich orographic cloud that caps the hills. If the low-level flow is strong enough to keep replenishing this cloud with 'new' water from upwind, the rain from the 2 km high (seeder) cloud falls through it and washes out significant amounts of extra rainfall from the lower (feeder) one, which does not however vanish but persists as long as supplies are forthcoming (Figure 2.18). This is the prime reason for wetter hills. It doesn't happen every day and is linked to a quite specific type of weather pattern.

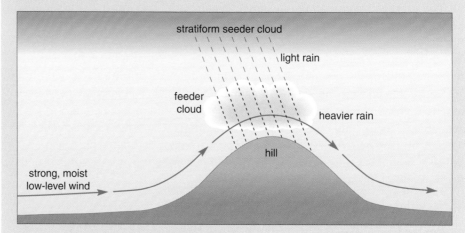

Figure 2.18   Orographic enhancement of precipitation by the seeder–feeder mechanism.

2.2.7   Visibility

Visibility is assessed by the naked eye at most surface reporting stations, although there are automatic means of doing so too. The observer must quantify the poorest horizontal visibility if it varies spatially, which is done using a set of objects that are situated at known distances from the observation point. It is naturally an easier task in daylight than at night, when lights at known distances have to be used, and easier on land than at sea. Indeed, maritime observations of visibility are of necessity quoted on a much coarser scale than those on land.

Visibility is obviously very important at airports and for transport in general. The observation is finely scaled for poor visibilities and more coarsely for moderate to good ones.

The summary or synoptic code for visibility has one hundred possible values from 00 to 99 and is used for official visibility reports. Numbers 00 to 50 report generally poor to moderate visibility, to the nearest 100 m in steps up to and including 5 km. For example, a coded report of 32 means a visibility of 32 × 100 m = 3.2 km. As the visibility improves, there is less need for precision, so the observer reports in 1 km steps from 5 km to 30 km. Beyond this visibility, each unit on the visibility scale represents an increases of 5 km. There is a special code that is used if the visibility falls below 100 m, which is clearly very dangerously low. In the British Isles, such a value is invariably associated with the presence of fog. Fog is defined internationally as a visibility of 1000 m or less due to the presence of suspended liquid water droplets at and just above the surface. Visibility of 100 m or lower is officially thick fog.

2.2.8 Present weather

It may seem that all the surface observations and measurements described in Sections 2.2.1–2.2.7 define present weather. However, to the meteorologist the term **present weather** has a special meaning. Is it raining, snowing or drizzling, for example? If so, is the precipitation intermittent or continuous and is it light, moderate or heavy? If none of these occurs at the time of the observation, is it foggy, showery, thundery, or has there simply been no significant change in cloud type and amount? Again, the observer selects a single two-digit code number from 00 to 99 to record the present weather. The code numbers are ranked in such a way that the most innocuous weather has the low code values, up to 10 for instance. The code number increases as the significance of the weather reported increases, so numbers 40–49 represent fog (including whether it's thickening, thinning, freezing, etc.), 50–59 represent drizzle, 60–69 rain and 70–79 snow (including the details of whether they are light, moderate or heavy rates and whether they are intermittent or continuous). It is possible to organize your own observations using low-cost instruments and taking visual data relating to cloud type, cloud amount and visibility.

Activity 2.3 Do try this at home!

This activity suggests some straightforward weather observations that you can perform in your local environment.

Temperature

Temperature can be sensed using a simple maximum/minimum thermometer (as described in Section 2.2.1). Not only are the two temperature extremes measured once each day (ideally), but you can also read the actual drybulb temperature at one or more 'routine' times each day from the same instrument. Values taken each morning at a regular time, and perhaps in the early evening too, can be combined with the maximum and minimum values to provide four a day. Whatever instrument you use should of course be exposed out-of-doors in a permanently or mainly shaded spot if at all possible. Siting thermometers away from buildings is also advisable, if you can do it.

Pressure

You can read pressure from an aneroid barometer at the same morning and early evening times as your temperature measurements. Each time note the tendency, i.e. the difference from the previous value (Section 2.2.2), albeit over more than the usual 3 h time interval.

Visibility

You can estimate visibility simply, but only if you live or travel to work in an area where you experience an extensive panorama, ideally in all directions. If you do, then it's a small job to note from an Ordnance Survey map the location of places that are, say, 1, 2, 3, 4, 5, 10 and 20 km away. It doesn't matter that your mapping won't be as precise as that used for official visibility recordings!

Cloud type and amount

This is a good point at which to look at the cloud images supplied on the DVD in order to familiarize yourself with the range of cloud types.

Choose several days that are obviously different with respect to cloud type and use *The Cloud Name Trail* card to identify each of the cloud types you observe.

Estimate total cloud (here it is not necessary to distinguish the different cloud types present) as eighths of the sky covered, or oktas (as described in Section 2.2.5).

Question 2.4

How would you convert daily maximum and minimum temperatures to: (a) daily mean temperature, (b) weekly or monthly mean temperature and (c) daily range of temperature? (d) What are the limitations of using the same averaging procedures for daily rainfall totals?

2.2.9 Plotting the observations: surface weather maps

It is crucial that the surface observations are taken simultaneously around the world, and that they are transmitted as rapidly as possible to processing centres such as the UK Meteorological Office in Bracknell (in Exeter from 2003) and the European Centre for Medium-Range Weather Forecasts (ECMWF) in Reading, both in Berkshire. Once they have undergone and passed a sophisticated quality control check, the values are plotted on a base map using an international convention which means that any meteorologist anywhere in the world can read the report. A map that brings together all the station reports in this way is termed a **synoptic chart**. The plotted data are then used to perform an analysis by manually or automatically drawing mean sea-level isobars (contours of the same mean sea-level pressure value) for instance. In the UK, these are produced at intervals of 4 hPa for charts on the scale of much of the North Atlantic and Europe (Figure 2.19).

This mean sea-level pressure analysis map is in fact just a fraction of a global map that defines where the synoptic-scale highs, lows, ridges and troughs are at one instant — most commonly one of the main **synoptic hours** of 00.00, 06.00, 12.00 and 18.00 UTC. These are the times that weather services have as priority ones if they cannot support hourly observations.

Highs and lows are extensive regions of *relatively* high and *relatively* low mean sea-level pressure, respectively. There is no 'magic' value of pressure that defines a high or a low; it simply depends on the range of values on each day. One day a centre of 1015 hPa could be a high while some time later the same

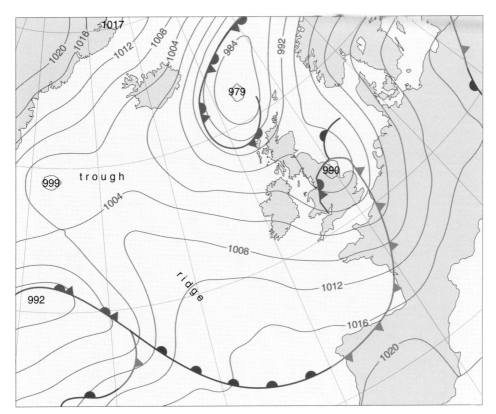

Figure 2.19 A mean sea-level pressure surface analysis map for part of Europe and the North Atlantic. The numbers on the isobars denote pressures in hPa. Notice the ridge of relatively high pressure and the trough of relatively low pressure.

Figure 2.19 A mean sea-level pressure surface analysis map for part of Europe and the North Atlantic. The numbers on the isobars denote pressures in hPa. Notice the ridge of relatively high pressure and the trough of relatively low pressure.

value (not the same feature) might represent a low centre. Lows are also termed cyclones or depressions while highs are also termed anticyclones (Section 2.2.2). **Ridges** and **troughs** are pressure features that are reminiscent of ridges and troughs (valleys) on a land topography map. Ridges are linked to highs while troughs are linked to lows (Figure 2.19).

Figure 2.20 shows the wind speed and direction for the same time, in a subregion of the surface analysis map in Figure 2.19. It illustrates the fact that the wind direction is more-or-less parallel to the isobars, which is somewhat surprising, given the fact that a pressure gradient is supposed to cause a force *across* the isobars, as we said at the end of Section 2.2.2. The explanation for this involves the rotation of the Earth, and is given in Section 4. Reassuringly though, as we also said before, the wind strength is inversely related to the spacing between the isobars. In other words, in areas where there is a large horizontal pressure gradient, the air is driven to flow more quickly than with a smaller gradient.

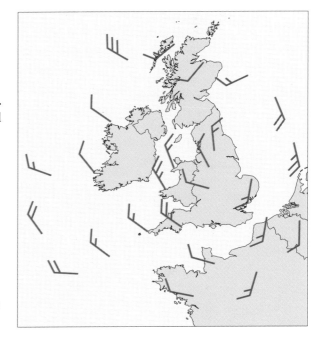

Figure 2.20 Surface wind speed and direction over part of the region shown in Figure 2.19. The meaning of the wind symbols was explained with Figure 2.12.

2.3 Upper-air observations

It was not until after World War II that the global upper-air weather-mapping network began to take shape. It is certainly true that surface observations are of great importance in the analysis and prediction of weather, but the details of by how much and where temperature, humidity and wind change with height above the Earth's surface are absolutely critical in these procedures. Today there are some 600 upper-air stations around the world, which usually report up to four times every day. These are sites from which weather balloons are released routinely (Figure 2.21).

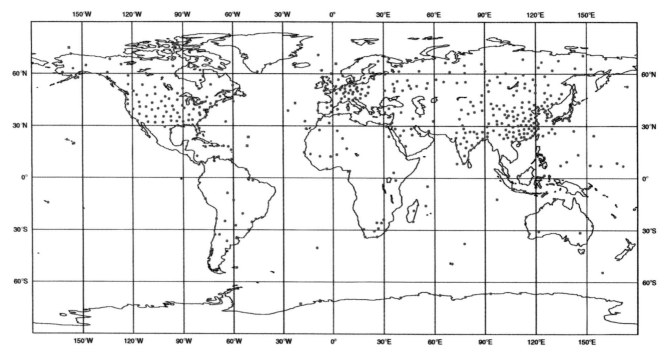

Figure 2.21 Global upper-air reports received at the European Centre for Medium-Range Weather Forecasts in one day. Notice the sparseness of reports in the Southern Hemisphere, which hampers the ability to forecast weather there.

A great pioneer of high-altitude ballooning was James Glaisher (Figure 2.22), appointed in 1840 as Superintendent of Meteorological Observations at the Royal Greenwich Observatory.

Although the observatory was founded in 1697, it was Glaisher who began routine weather records in 1841. Glaisher was concerned to investigate how weather measures changed with height in the atmosphere, so in September 1862, accompanied by a Mr Coxwell, he ascended from a site near Wolverhampton.

In under one hour they had reached some 8 km where the air temperature was recorded to be below 0 °C; the temperature at launch was 15 °C. They continued to ascend, and at an altitude of some 9 km they were suffering ill-effects. Some time after the event, James Glaisher recounted that

> I seemed to have no limbs. I struggled and shook my body, but could not move my arms. Getting my head upright for an instant, it fell on my right shoulder; then I fell backward, my body resting against the side of the car, and my head on the edge. I dimly saw Mr Coxwell, who was up in the ring of the balloon, and endeavoured to speak, but could not. … I cannot tell anything of the sense of hearing, as no sound reaches the ear to break the perfect stillness and silence of the regions between 6 and 7 miles above the Earth.

Their ascent reached about 11 km before Mr Coxwell managed to pull the valve line with his teeth to release air and so effect the balloon's descent. A number of such ascents confirmed that there is an average decline (or lapse) of temperature with height within the atmosphere of $6\,°C\,km^{-1}$. This value varies greatly for an individual ascent.

Much has happened of course since these daring ascents. Today, with the advent of cheap electronic instruments for measuring temperature, pressure and humidity, balloon-borne instrument packages, or **radiosondes,** are carried aloft to some 30 km, well into the stratosphere, for an hour or more until the balloon bursts and the package is parachuted back down to the surface. Radiosondes sense barometric pressure (used to express height above mean sea-level, as in Table 2.2), drybulb temperature and relative humidity. The measurements are transmitted to radiosonde stations on the ground. Radiosonde observations are made (ideally) twice daily, where the balloons are released near 00.00 and 12.00 UTC. These are the same times that the surface measurements are taken.

Operational meteorologists make great use of the vertical profiles of temperature and humidity by plotting them manually (although many weather services plot them automatically nowadays) on a chart. The vertical profile of drybulb temperature is plotted directly from the radiosonde observations while the humidity is represented on the same chart by plotting the dewpoint temperature (Figure 2.23). This is calculated from the values of drybulb temperature, relative humidity and pressure automatically, using known relationships between these variables.

Figure 2.22 James Glaisher (1809–1903).

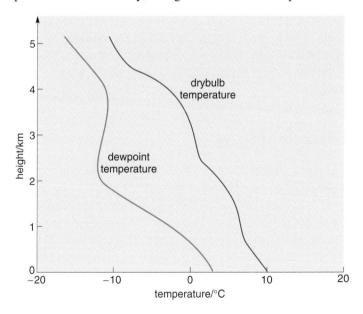

Figure 2.23 Vertical profile of drybulb and dewpoint temperature.

Plotting these two curves on an appropriate figure has great value to the trained eye; a forecaster can estimate the likelihood of showers or fog, for example, by carefully modifying the temperature and perhaps humidity of the lowest layers of the atmosphere represented on the diagram. This modification might for instance take the form of increasing the surface temperature (and possibly the humidity) to represent the situation in the middle of the day following the midnight ascent. Showers may develop in the afternoon when the temperature exceeds a critical value.

Some plots of temperature and humidity versus height are still drawn quickly by hand, but many are computer-drawn nowadays. Forecasters can even request predicted vertical profiles of temperature and humidity some hours and days ahead for a wide choice of locations.

In addition to temperature and humidity, wind direction and speed are also reported, not only at 00.00 and 12.00 but also at 06.00 and 18.00 UTC for the best-equipped stations. A target carried by the balloon is tracked by a radar from which wind speed and direction can be calculated. Some upper-air stations use the Global Positioning System (GPS) to track balloons and so deduce the wind speed and direction.

Knowledge of how the air's thermal, humidity and flow patterns evolve on the large-scale, at a variety of elevations, is crucial to predicting the surface weather. What we observe at the surface is, in the main, an expression of processes at work through great depths of the atmosphere. So we need to know how the thermal and humidity patterns are evolving within the troposphere and stratosphere. We need to know also how the wind, including its upward and downward components, is changing too. Knowledge of the upward and downward components of the atmosphere's circulation is critical because these relate to the amount of cloud present and the generation of precipitation.

Ascending air is organized on a variety of space scales from that of fair-weather (cotton wool) cumulus on a sunny summer's day to the massively extensive depressions that sweep across the British Isles most commonly in the autumn and winter. Both these cloudy features exist *because* moist air is ascending within them. If there were no rising motion, we would see no cloud and no rain. Conversely, sinking motion tends to suppress any cloud and precipitation, because the air is compressed as it subsides. This warms the air adiabatically, thus reducing its relative humidity.

2.4 Satellite observations

The launch of the Soviet Union's 'Sputnik' in October 1957 was a hugely significant step in the provision of a 'grandstand' view of the Earth and its clouds. The very first application of artificial Earth satellites was in fact in meteorology with the launch in 1960 of 'TIROS I' (the first Television and InfraRed Observation Satellite) — perhaps unfortunately on 1 April! This particular satellite provided what were essentially television pictures of the Earth in visible light only.

After early experimentation using different orbits, almost all weather satellites are now polar-orbited. Such satellites are in fact strictly near-polar orbiters (Figure 2.24) but they nevertheless provide reasonably frequent global coverage.

The early orbits were also used to test the optimum elevation (height above the Earth's surface) and orbital ellipticity (whether circular, near-circular, etc.) for providing frequent, detailed cloud imagery. After some years, the American series that has developed from the TIROS I satellite of some four decades ago has settled on a virtually circular orbit with an altitude of 850 km. This orbital height fixes the satellite's period (the time taken for it to travel once round the Earth) at around 100 minutes.

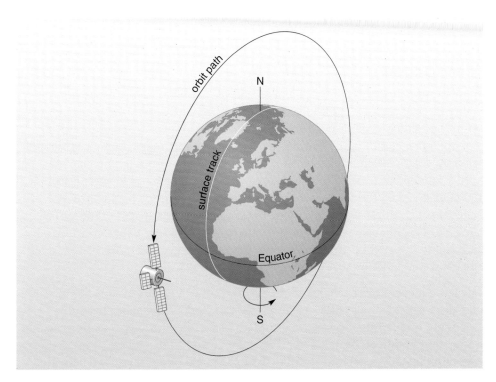

Figure 2.24 The orbit and surface track (the line of the satellite's path projected onto the Earth) of a polar-orbiting weather satellite.

The satellite sweep width is set, given the above parameters, such that images produced by successive passes just meet at the Equator but overlap more and more as the orbit progresses polewards.

So why are satellites so useful to operational meteorologists? In the early days, they provided invaluable large-scale images of the Earth's cloud patterns, both day and night, with their visible and thermal infrared sensors. Before this time, meteorologists relied on scattered surface observations to gain an overview of where weather systems were, so had particular problems over data-sparse regions such as the world's oceans and uninhabited continents. Then, during the 1960s in particular, the amazing service provided by a guaranteed weather satellite programme changed all that.

The modern American NOAA (National Oceanic and Atmospheric Administration) weather satellites are direct descendants of the TIROS series and do a lot more than just provide cloud images round-the-clock. For some years now, they have been providing thousands of vertical profiles of temperature and humidity down through the atmosphere. In fact, they produce a few thousand every day, the best-quality data coming from cloud-free regions. Such profiles are incorporated as rapidly as possible, after quality checking, into all of the operational global weather forecast models that are run in many centres. These data are an important supplement to those from the radiosonde network which, like the surface network, is limited in its coverage, i.e. there are large areas from which no radiosonde observations are available. Satellites quite literally fill in the gaps.

Another role of weather satellites nowadays is that of sensing the total column mixing ratio of ozone on a routine basis. The word 'column' is used because a satellite instrument can only easily measure properties over the range of heights below it; that is, a value is measured that is a kind of average over the air column beneath it. The satellite coverage is global, available in almost real time and includes mapping of other chemical constituents too.

Visible imagery

Weather satellites map the Earth and its clouds in visible light, although in reality the waveband is the visible and near-infrared (the part of the infrared just off the end of the visible spectrum). Sensing radiation in this spectral region — looking down at the Earth — means that the instrument maps the flux or flow of *reflected* solar radiation that has streamed to our planet from the Sun and has been scattered back out to space in varying degree. Quite how powerful the signal received by the sensors is depends on the **albedo** (the percentage of incident solar radiation that is reflected back out towards space) of the surface in question. The albedo can vary enormously from a very low value (near zero) for dark soil to a very high value (near one) for fresh snow or the tops of very deep clouds like cumulonimbus (thunderclouds).

The basic incentive for sensing in the visible part of the spectrum is therefore to map the world's clouds (Figure 2.25). This can be done because, generally, clouds are brighter than cloud-free surfaces. As a rule, the deeper (taller) a cloud is, the more it reflects from its upper surface; in addition, a cloud composed of water droplets reflects more strongly than one made of ice crystals. There are ambiguities sometimes, at least to the naked eye. A thin veil of cirrus that overlies a bright surface such as a sandy desert or an extensive ice mass may not be obvious to a human analyst. Generally, however, there are few problems with identification.

Figure 2.25 A whole-Earth disc visible image from Meteosat. Note the streamer of cloud over Spain and France, characteristic of a front, and the tropical storm in the mid-Atlantic. Meteosat is a series of meteorological satellites launched by the European Space Agency and stationed at longitude 0°.

Infrared imagery

One important drawback of visible imagery is that, by its very nature, it is available only during daylight hours. This means that for weather services in the world's middle and higher latitudes their utility in the winter months can be seriously limited.

The way round this is to use a sensor that maps radiation in the infrared part of the electromagnetic spectrum. In contrast to the 'visible', radiation measured in this waveband (usually 10.5–12.5 μm) represents a flux *emitted* by the Earth and its atmosphere (Figure 2.26). It is termed the thermal infrared because the strength of the flux at the satellite's sensors is dependent on the temperature of the emitting surface. So a hot surface, such as the daytime Sahara with a temperature of around 60 °C, emits very powerfully while a very cold one, like the top of a deep tropical cumulonimbus at say −70 °C, emits very weakly.

Figure 2.26   A whole-Earth disc thermal infrared image from Meteosat. The image is processed such that strong emission in the infrared appears black and weak emission appears white.

So a thermal infrared image is essentially a heat map. One boon for operational forecasters is that the Earth and its atmosphere radiate continuously in this waveband, to provide 24 h coverage. Another advantage is that clouds can be mapped by eye or automatically because, generally, they are colder than the cloud-free surfaces surrounding them. Clouds lower down in the troposphere, such as stratus, are relatively warm, so emit quite strongly in comparison with very cold cirrus in the upper reaches of the troposphere. These lower stratiform clouds are still colder than the adjacent cloud-free surfaces, and can thus also be mapped.

Some satellites are equipped with detectors that are specifically designed to be sensitive to infrared emission from water vapour. These are especially useful for understanding the greenhouse effect.

2.5 Radar observations

Radar (from *ra*dio *d*etection *a*nd *r*anging) is the use of high-energy radio-frequency pulses for locating objects. The technique was developed before World War II for obvious military purposes. Quite what a radar 'sees' depends on the wavelength(s) of the emitted radiation. Part of the emitted beam is scattered back to the radar antenna by the target, allowing its location to be mapped (see Figure 2.27).

A radar that emits radiation with a wavelength of 5 cm or so receives a signal that has been scattered back by precipitation-size particles — not the very tiny cloud droplets. This means that such a precipitation radar can map the location and in fact the intensity of precipitation, but it cannot see what's falling right at the surface because of the problem of 'clutter' (e.g. hills and buildings). The emitted beam must be elevated by a small angle to avoid detecting hills and other irrelevant echoes. The radar sweeps out a sequence of such circular, slightly inclined scans, stepping through a number of increasingly steeper but still shallow angles as shown in Figure 2.27. This enables the instrument to look at slightly different elevations within rain-bearing systems, thereby providing detail of their structure in the vertical.

Figure 2.27 Scan angles for an operational weather radar. By bouncing radio pulses off different parts of a cloud, it is possible to detect rainfall from it.

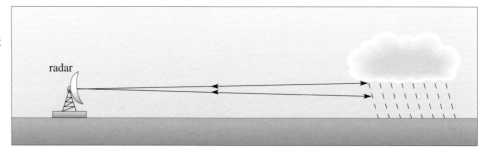

Operational radars of this type are all linked automatically to recording raingauges in their field of view, which measure the precipitation at the surface. The latter are used to modify the radar estimates so that they are more realistic. The UK network of operational radars provides a unified image of the national rainfall pattern every 15 minutes (Figure 2.28). The image consists of 1 km square pixels (boxes), each one containing an instantaneous rainfall rate observation. These data are used to monitor the movement and evolution of precipitation patterns by forecasters and by the Environment Agency in river flood prediction.

Question 2.5

Hills are regions where heavy rain often falls and so are important to map. What problems are associated with radar observations of rainfall in hilly areas?

The United States National Weather Service has established a national network of more advanced **Doppler radars**. These are capable of mapping the detailed pattern of wind speed and direction at lower and middle tropospheric levels.

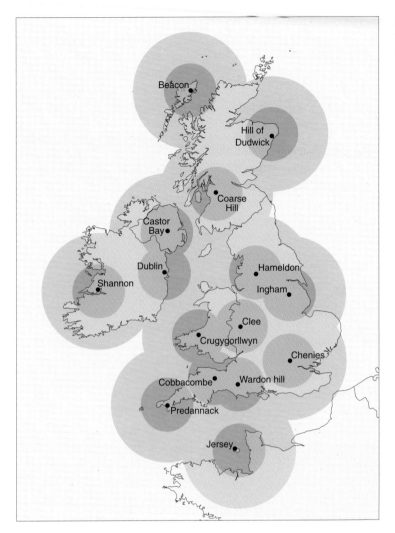

Figure 2.28 The UK and Irish weather radar network. Small, deep-coloured circles: best-quality data; large, pale-coloured circles: less reliable data.

They do this by measuring small shifts in the wavelengths of the reflected radio pulses to obtain information on how fast tracers (e.g. raindrops or small, dry particulate matter) are moving towards or away from the radar. The **Doppler effect** means that wavelengths are shorter from pulses reflected by objects moving towards the radar, and longer from those moving away — the greater the shift, the faster the movement. This information can be used to reconstruct the airflow in the area around the Doppler radar. The map of the flow component towards or away from the radar also indicates, if represented over a wide area surrounding the site, lines or zones where the air at low levels flows as a confluence. This is often a surface line along which the air flows smoothly in from two directions. Such a region of confluence may be indicative of the risk of intense thunderstorms.

2.6 The World Weather Watch

The World Meteorological Organization (WMO), which is based in Geneva, fulfils the role of an agency that promotes good observational practice in weather services world-wide and stimulates the development and execution of regional or global networks. It is a specialized agency of the United Nations.

One such global programme is the **World Weather Watch** (WWW), which is aimed at maintaining and improving the global system of surface, upper-air and satellite observations. It aims also to promote the efficient gathering and rapid transmission of observations to data-processing centres where weather forecasts are produced. Additionally, the WMO sets down criteria for the accuracy of a whole range of observations which member states (all UN member states) are duty-bound to achieve, or do their best to try to achieve.

The WWW is a unique achievement in international cooperation. There are few other fields of human endeavour in which there is unquestioning international collaboration, day-in, day-out, for the good of humankind. Indeed, the initiation of carefully planned, large-scale observation programmes within the WWW are aimed not only to advance our understanding of how the atmosphere and its weather and climate work, but also to secure the future of a hopefully cleaner planet.

The Global Observing System

An important section of the WWW is the Global Observing System (GOS). The GOS comprises facilities for making observations on land and sea, and from aircraft and satellite. The world is divided into six regions, each of which draws up a regional network of observing stations, which meets the requirements of member services. Around 75% of surface stations globally achieve the aim of reporting eight times daily (i.e. every three hours). The number of possible reports received at weather centres varies from some 50% for Region I (Africa and adjacent oceans) to about 72% for Region VI (Europe and Greenland). The number of surface stations reporting a complete set of eight observations a day was 3042 globally in 1988. By 1996 this figure had declined to 2969, although by 1998 it had crept up to 3005. The region with the lowest percentage of possible reports was South America with just 35% in 1998. The distribution of surface stations that reported on a randomly selected day is illustrated in Figure 2.29.

Figure 2.29 The distribution of surface stations that reported on a particular day. (The total number of observations was 12 683.)

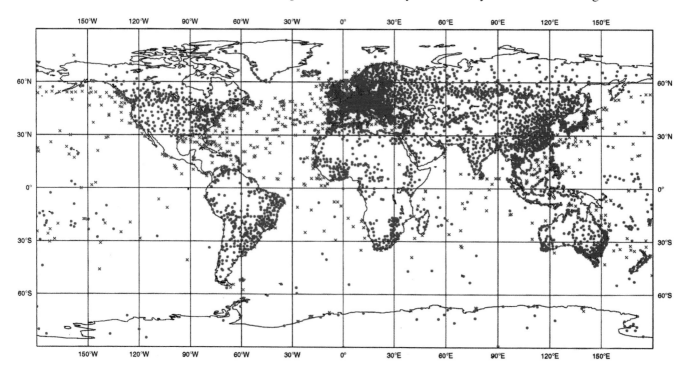

To counter the very real problems of poor quality or missing data, the WMO has established a special GOS Surface Network (GSN) consisting of some 1000 globally consistent, high-quality 'baseline' sites, at a density of one station for every 250 000 km^2.

Around 300 moored buoys and 600 fixed platforms (including gas and oil rigs) provide automatically sensed marine data (Figure 2.30). In a recent ten-year period, the number of barometric pressure observations from drifting buoys increased fourfold to around 120 000 a month. Such drifting buoys are about 1000 in number, of which 60% transmit about 100 observations.

Figure 2.30 The distribution of marine stations that reported on a particular day. (The total number of observations was 1611.)

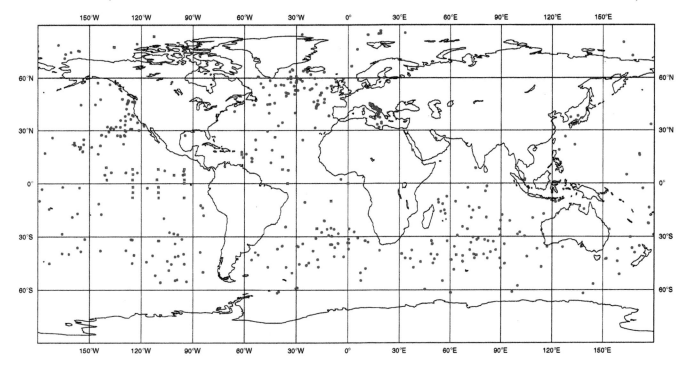

Some 69–70% of upper-air stations globally report twice a day, with a smaller percentage in the Southern Hemisphere compared to the Northern. During a three-month trial in October 1998, almost one-half of the expected upper-air reports simply did not arrive at processing centres. One important factor in this was the withdrawal of the worldwide Omega radio navigation system that was used by about 25% of the stations for deriving wind data. This system has been replaced by a Global Positioning System (GPS)-based system, but the greater expense of the latter means that many sites are cutting back from two to one ascent a day.

The WMO requested number of upper-air stations was 879 in 1998 but the operational total of twice-daily reporting sites in that year was down to 605 (from 610 in 1996). As with the surface data, South America fared worst, with 29% of planned, while Europe did best, with 89%.

Commercial aircraft supply conventional observations from the flight deck by radio link to ground. However, in the last decade or so, the automated Aircraft-to-Satellite Data Acquisition and Relay scheme (ASDAR) has increased data flow many-fold. It has been particularly useful (providing drybulb temperature, wind speed and direction) over South America and Africa where the operating upper-air network is normally poor.

One innate problem of meteorological data coverage is the 'top-heavy' nature in the distribution of surface and upper-air observations. This relates quite simply to the fact that most of the Northern Hemisphere is land while the Southern Hemisphere is mainly water. This leads to the quality of the 'initial' data fed into a forecast model being better in one hemisphere than the other. Forecasts for the Southern Hemisphere are therefore generally poorer than for the Northern. It is also the case that because the weather at the forecast outset is more poorly represented in the Southern Hemisphere, errors there can ultimately affect the quality of the prediction for the Northern Hemisphere.

The criteria in Table 2.4 are needed to meet the needs of weather prediction models on computers. It is these criteria that guide today's observational networks with regard to precision, resolution in the horizontal and in the vertical (for ascents) and in time. There is a significant shift to the automation of surface observations, which means that many places no longer provide those variables that are most easily estimated by eye. These can include cloud type and amount and present weather.

It is a fact that many weather services are developing in this way, but they are putting more money and effort into expanding and improving specific aspects of the network; for example, making better marine observations if they serve maritime users, or putting more automatic stations into remote, poorly monitored land areas. There is a constant thrust to improve both national and global coverage, which is not always successful of course.

Table 2.4 Spatial resolution, frequency and accuracy of wind, temperature and humidity measurements.

Measure	Spatial resolution/km		Time interval/h	Accuracy
	Horizontal	Vertical		
wind	100	0.1†	3	$2 \, \text{m s}^{-1}$ (troposphere)
		0.5‡		$3 \, \text{m s}^{-1}$ (stratosphere)
		2.0§		
temperature	100	0.1†	3	0.5 K (troposphere)
		0.5‡		1.0 K (stratosphere)
		2.0§		
relative humidity	100	0.1†	3	5%
		0.5*		

* up to tropopause; † up to 2 km; ‡ up to 16 km; § up to 30 km.

2.7 Summary of Section 2

1   In order to study and to predict the weather, it is essential to observe and record a range of parameters to a specified level of precision. These comprise temperature, barometric pressure, humidity, wind speed and direction, cloud type and amount, visibility and present weather.

2   Simultaneous, widespread, rapidly transmitted and high-quality weather observations are vital to the process of mapping the current situation and of forecasting its future state. The data are also invaluable for determining the detailed nature of a place's climate.

3   Operational meteorology is hallmarked by the free international exchange of all data, including upper-air, radar and satellite observations in addition to surface values. The World Meteorological Organization oversees this aspect of the science as well as promoting properly taken observations from well-kept stations. Instruments must all be correctly exposed to sense their respective atmospheric variable, and reset if necessary.

4   Weather satellites are an integral component of the international observation network. They provide visible and infrared imagery of clouds and the Earth's surface features, thousands of vertical profiles down through the atmosphere, and near real-time maps of selected chemical constituents such as stratospheric ozone.

5   Radars provide real-time or near real-time maps of the distribution and intensity of precipitation and, with Doppler radar, maps of low-to-middle tropospheric flows can be obtained.

6   Screen temperature is influenced by many factors including the local exposure/siting, distance from the sea and elevation above mean sea-level. The impact of the latter factors are quantified and explained with respect to the Teign Valley in South Devon. Related measures, such as the incidence of frost and dew, also exhibit variation due to the same factors.

7   The increase of rainfall totals with height within the Teign Valley is quantified and related to very specific events. The operating seeder–feeder mechanism influences individual days' rainfall intensities and therefore the difference in monthly, seasonal and annual catches.

8   The wind is driven by horizontal pressure gradients that are mapped on synoptic charts that depict highs, lows and isobars. Barometric pressure can be seen as a means of weighing the atmosphere above a barometer; horizontal variations in this 'weight' across a level surface are what drives the wind. The strength and direction of the wind are in general related to the steepness of the horizontal pressure gradient and the orientation of the isobars respectively.

9   There are a number of importantly different ways of defining humidity in meteorology: absolute humidity, dewpoint temperature, relative humidity, vapour pressure and wetbulb temperature. The most useful measure is absolute humidity, which expresses the absolute quantity of water vapour as a fraction of the mass of air, as this can, for example, indicate potential rainfall amounts from clouds.

Question 2.6

Imagine that you have two maximum/minimum thermometer sets placed in your back garden, which is to the south of the house and level near the house but slopes away steeply at the far end into a small basin-shaped depression. Imagine also that it's all grass except for the area just outside the back of the house, which is laid to tarmac. Assume you've placed one set in the depression at the bottom of the garden, and the other over the tarmac, a metre from the house (both inside weather screens).

(a) Describe and explain the differences you'd expect to see between the two maxima and minima over a complete day when there is no wind and the sky is completely clear.

(b) Why would these measurements be different if the day were overcast and windy?

Learning outcomes for Section 2

After working through this section you should be able to:

2.1   Explain how surface weather variables are measured, and how the instruments function. (*Activity 2.1*)

2.2   Explain why meteorologists analyse basic pressure features on weather maps and why the pattern of such features is related to wind speed. (*Activity 2.1*)

2.3   Explain why it is necessary to adjust a barometer reading to mean sea-level. (*Activity 2.2*)

2.4   Identify, with the aid of a chart, the nine basic cloud types that are the foundation of cloud observation. (*Activity 2.3*)

2.5   Calculate the mass of liquid water in a cloud, assuming a simple shape for it.

2.6   Plan and execute a programme of home weather observations, process the data and compare it with other sites' data. (*Activity 2.1–2.3*)

2.7   State typical rates of change of drybulb temperature upwards in the atmosphere, and explain what consequence this lapse rate has for the incidence of frost, snow, etc. (*Question 2.2*)

2.8   Describe how a polar-orbiting weather satellite covers the Earth and explain the nature of both visible and thermal infrared satellite images.

2.9   Explain briefly how radars sense precipitation and air motion and how radar mapping is useful.

2.10 List the challenges of making, collecting and transmitting observations on a global scale.

The ins and outs of the atmosphere

3

In Sections 1 and 2 we mainly *described* various aspects of the atmosphere. In this section we will begin to *ask why* the atmosphere is the way it is and also *ask how* the various weather phenomena occur. The key concept in explaining all this is *energy* and it provides the theme that will run through the rest of this Part of Block 2. We begin in Section 3.1 by considering how radiation relates to energy in the Earth's environment. In Section 3.2 we explore how energy is balanced, and in Section 3.3, how it is transported in the atmosphere. This part of the story provides the crucial link into how the 'weather machine' operates. To round off the whole section, we explore, by means of a DVD activity, how much solar radiation falls on different parts of the Earth, at different times of the year.

3.1 Radiation

3.1.1 Energy and radiation

Energy is present in the atmosphere in a variety of forms. It is manifest as thermal energy of warm air, as kinetic energy in the weather from the slightest breeze to the most violent thunderstorm, and in radiation. Almost all of this energy is ultimately supplied from our local star, the Sun, in the form of solar radiation. (Electromagnetic radiation, often just called radiation, was introduced in Box 1.2.) Through various cycles, some energy does come from thermal energy stored within the Earth, e.g. volcanoes, hot springs and geysers, but this represents an insignificant input into the atmosphere's total energy budget.

Box 3.1 explains the concept of energy, and the terms used to describe it.

Box 3.1 *What is energy?*

Let us consider energy in practical terms. Two identical cars, one travelling fast and one travelling more slowly, have different amounts of energy. We say that the fast car has more energy than the slow car. More precisely, we say the fast car has more kinetic energy than the slow car. **Kinetic energy** is energy of motion. If a car and a bus are travelling at the same speed, then the bus has more kinetic energy because it has a greater mass. So, in this sense, energy is a measure of motion, accounting for both mass and speed. In general, the kinetic energy, E_K, of an object of mass m and speed v is defined by the equation

$$E_K = \tfrac{1}{2}mv^2 \tag{3.1}$$

Energy can also be possessed by radiation (the subject of Section 3.1) or by a gas. We can understand the energy carried by a gas in terms of kinetic energy as follows.

Consider two identical amounts of the same gas, where one has a higher temperature than the other. The one with the higher temperature has, by definition, more **thermal energy**. We can understand this in terms of kinetic energy, in that the hotter gas has particles that are, on average, moving faster

---

than those of the cooler gas. In fact, the typical kinetic energy of an individual gas particle is about kT, where k is the Boltzmann constant from the perfect gas law (Equation 1.3). Notice also that the perfect gas law, $p = (N/V)kT$, can be interpreted as pressure equals number density of particles times the typical energy of a particle. So, pressure is a measure of the amount of energy per unit volume, or **energy density**.

There are forms of energy where no motion is involved at all. An everyday example is gravitational energy. We know that releasing a ball from a height results in it moving towards the ground. Its kinetic energy increases from zero when held, to some larger value by the time it hits the ground. It is a fundamental law of physics that *energy is conserved*: it can neither appear nor disappear. (You met this law in Box 2.6.) So where did the falling ball's energy come from? It comes from gravitational potential energy. When the ball is raised, it gains potential energy, which is then converted to kinetic energy by gravity as it falls. In mathematical terms, the gravitational potential energy E_P of an object of mass m raised a height h above the ground is:

$$E_P = mgh \tag{3.2}$$

where g is the gravitational acceleration.

Potential energy is so named because objects that possess it have the *potential* to release energy, often as kinetic energy. The gravitational potential energy stored in columns of air is particularly important in understanding the atmosphere. The other form of potential energy we will be concerned with is chemical energy. The fuel in the tank of a car is a reservoir of chemical energy that is released in a chemical reaction (i.e. burning) to give the car kinetic energy.

○ A mischievous child drops a coin from the top of the Eiffel tower, at a height of 310 m above the ground. If all the coin's potential energy is converted to kinetic energy, how fast is it travelling when it (a) hits the ground and (b) is half-way down?

● (a) Before being dropped, the coin has potential energy mgh, where $h = 310$ m. When it lands, it has a potential energy of zero. By conservation of energy we know that the energy lost, mgh, must all have become kinetic energy:

$\frac{1}{2}mv^2 = mgh$

The mass m cancels out, so the coin's speed v is

$v = \sqrt{2gh} = \sqrt{2 \times 9.81\,\mathrm{m\,s^{-2}} \times 310\,\mathrm{m}} = 78.0\,\mathrm{m\,s^{-1}}$

(b) When the coin has fallen to height $h_1 = 155$ m it has potential energy mgh_1. The potential energy lost is therefore $mg(h - h_1)$. By conservation of energy, we know that this has become kinetic energy, so

$\frac{1}{2}mv^2 = mg(h - h_1)$

which gives us

$v = \sqrt{2g(h - h_1)} = \sqrt{2 \times 9.81\,\mathrm{m\,s^{-2}} \times 155\,\mathrm{m}} = 55.1\,\mathrm{m\,s^{-1}}$

The total energy output per second from the Sun, or radiative power, amounts to a colossal 3.85×10^{26} W, where 1 watt (W) equals one joule per second (J s^{-1}). The watt is the SI unit of **power**, which is the rate of energy transfer. To attempt to put some meaning to this huge number, this amount of radiation corresponds to covering the whole surface of the Sun with light bulbs that each emit 600 000 W of power. A standard household light bulb is 100 W. Of course, since the Sun emits in all directions, we receive only a tiny fraction of this; in fact, only about 4.52×10^{-10} of emitted solar energy hits the top of the Earth's atmosphere.

○ What is the power, i.e. energy per unit time, of solar radiation reaching the top of the Earth's atmosphere?

● The Sun's total power output is 3.85×10^{26} W. Of this 3.85×10^{26} J of energy released every second, only a fraction 4.52×10^{-10} arrives at the Earth, again every second. So the power is

$$3.85 \times 10^{26} \text{ W} \times 4.52 \times 10^{-10} = 1.74 \times 10^{17} \text{ W}$$

A commonly used term is **insolation**, which is the energy arriving per unit area of the Earth's surface.

3.1.2 Absorption and scattering

An important property of radiation is that it can be absorbed. When you step out into the sunlight and feel the warmth of the Sun upon your face, you are feeling your skin being warmed by the energy absorbed from the radiation. Not all the radiation hitting your skin is absorbed, however; some of it is scattered. It is because radiation is scattered that we can see objects, as the light can arrive from any direction, and bounce off again in (almost) any direction. The amount of energy absorbed by an object depends on a number of factors. Firstly, and most obviously, it depends on the power of the radiation falling on it. Secondly, an object with a high reflectivity (you can think of reflection as a special kind of scattering), such as a mirror, absorbs little energy, whereas a black surface absorbs most of the radiant energy falling on it. Finally, the amount of radiation absorbed depends on its wavelength.

Absorption and scattering of radiation are very important in the atmosphere, and all of the factors just mentioned in the context of solid objects also apply to gases. For example, both water vapour and carbon dioxide absorb infrared radiation much more strongly than they absorb radiation in the visible part of the spectrum. Scattering of radiation can also take place in a gas because of the presence of molecules and aerosol particles; the appearance of smoke and mist are two everyday examples of such scattering. Both gas particles (atoms and molecules) and aerosol particles absorb radiation. The energy gained by gas particles from absorbed radiation causes them to move faster, which means that the temperature of the gas has risen. In this way, solar radiation can directly heat gas in the atmosphere. Similarly, the ground can absorb and scatter radiation, though particles in solids *vibrate* faster instead of moving around faster. Figure 3.1 illustrates the path of radiation experiencing absorption and scattering in the atmosphere.

Figure 3.1 A schematic diagram illustrating the processes of absorption and scattering in the atmosphere and at the Earth's surface.

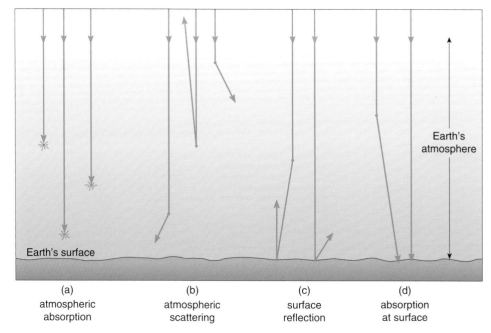

(a)
atmospheric
absorption

(b)
atmospheric
scattering

(c)
surface
reflection

(d)
absorption
at surface

Figure 3.2 is a plot of the percentage of absorption in the atmosphere against wavelength of radiation. There are two main regions of the spectrum where atmospheric absorption of radiation is weak: the visible range and part of the radio range (not shown in Figure 3.2). The fact that we have evolved to see in the visible and have developed technology in the radio is of course no coincidence; these are the only two broad bands where radiation can travel freely through the atmosphere. It is also clear from the plot that absorption can change quite dramatically with small changes in wavelength. This is because individual gases generally absorb in quite narrow ranges of wavelength. Figure 3.3 shows the spectrum of blue sky obtained by splitting the light into different colours using a prism. Superimposed on the coloured spectrum, several dark lines can be seen. These are **absorption lines**, where light of a particular wavelength is being absorbed by a particular gas (in this case by a gas in the Sun's atmosphere, not the Earth's). The wavelength of an absorption line is determined by the structure of a particular atom or molecule. So air, being a mixture of different gases, has a number of different absorption lines which are always found at the same wavelengths.

A particularly important 'window' in the atmospheric absorption spectrum is found at a wavelength of around 10×10^{-6} m (Figure 3.2). This small transparent part

Figure 3.2 The percentage of radiation absorbed by air plotted against wavelength. The shape of the spectrum is extremely complex because of the great many narrow (in terms of wavelength) absorption lines associated with the many gases present in air. Regions are indicated where absorption is dominated by water (H_2O) and carbon dioxide (CO_2), in the infrared, and ozone (O_3), in the ultraviolet.

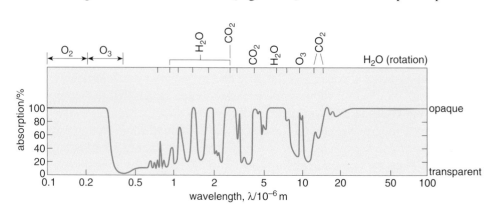

Figure 3.3 The spectrum of blue sky. Light from a clear sky is passed through a prism which bends light-rays of different wavelengths by different amounts. Although the sky appears blue, all the colours of the visible spectrum can be found in daylight, and so we see a complete visible spectrum here. The dark lines are absorption lines, where a particular wavelength of light has been absorbed by atoms and molecules in a gas. All of these lines are in fact due to absorption that occurred in the Sun's atmosphere.

of the atmospheric spectrum is what allows the land to rapidly cool on a cloudless night. This window is 'closed' if clouds are present.

In the atmosphere, and on the ground, scattering is important. On entering the atmosphere, radiation is scattered by air molecules and aerosols. Some of this scattering is back into space, but much of the radiation is scattered around inside the atmosphere. You can observe this for yourself on a cloudless day. The sky is not just bright where the Sun is, but is bright (and blue) all over. This is the result of scattering in the atmosphere. In fact, the reason the sky is blue is because the scattering at short wavelengths (blue) is much stronger than that of long wavelengths (red).

Even on a cloudy day, sunlight still makes it through the clouds because of scattering. However, it is certainly darker on a cloudy day than on a sunny day. Where is the light going on cloudy days? Have glance at Figure 3.4, which shows the Earth from space. What are the brightest regions? The clouds. On a cloudy day, much of the light is scattered back into space, and so is lost to the Earth. For this reason, clouds have a significant effect on the Earth, both locally in deciding whether its warm and bright or dark and cold, and also globally in affecting the climate. In fact, accounting for cloud cover and cloud type is one of the biggest problems in modelling the Earth's climate. In addition to light reflection by clouds, the surface of the Earth can also reflect light by different amounts. For example, snow and ice reflect much more light than forests or grassland. All of these factors contribute to the albedo (Section 2) of the Earth.

Figure 3.4 The Earth from space.

The net result of absorption and scattering is summarized in Figure 3.5, where for simplicity, 100 units is used to represent the 1.74×10^{17} W of radiation arriving at the Earth (Section 3.1.1). In this figure, the 100 units of solar radiation entering the atmosphere is divided into three parts: just under a third of the radiation is scattered back into space; about a third of what is left is absorbed by the atmosphere, leaving the surface to absorb the rest, roughly half of the total radiation incident at the top of the Earth's atmosphere. Obviously these numbers are averages, as they can vary from place to place and from time to time across the planet.

○ What is the average albedo of the Earth?

● The albedo is simply the fraction of energy reflected back into space, which, according to Figure 3.5, is 31/100 = 0.31.

Figure 3.5 Typical rates of energy entry into the Earth's environment. Note that even though arrows are depicted as stopping at particular places in the atmosphere, they represent processes that take place throughout the atmosphere. The width of the arrow indicates the rate of energy transfer, which is given more precisely by the numbers inside the arrows (100 units = the power of solar radiation incident at the top of the atmosphere, which is 1.74×10^{17} W).

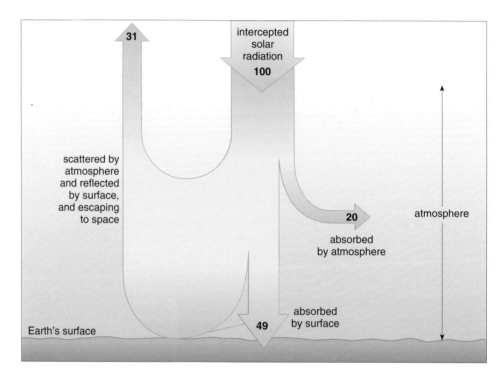

Different layers of the atmosphere absorb different wavelengths of solar radiation. Some wavelengths of solar radiation are strongly absorbed high in the atmosphere, such as ultraviolet which is absorbed by ozone in the stratosphere. Some wavelengths are absorbed nearer the ground (infrared) and some are completely absorbed by the atmosphere (X-rays). However, most of the energy carried by solar radiation, in the form of visible light, makes it through the atmosphere and is absorbed by the surface, whether it is land or sea. So, what happens to a particular wavelength of radiation — most crucially, where it deposits its energy — depends on both the structure of the atmosphere and the wavelength of the radiation.

3.1.3 The spectrum

From what we have discussed so far, you might be tempted to conclude that only visible wavelengths are important in understanding the role of radiative energy in the Earth's atmosphere and at its surface. After all, the Sun emits in the visible and air is transparent to visible light. However, non-visible wavelengths, particularly the infrared, are crucially important, precisely because the Earth's atmosphere absorbs them and because the surface emits them. For this reason it's worth investigating the spectrum of electromagnetic radiation in a little more detail.

Electromagnetic radiation is a kind of wave, and like all waves, its wavelength is defined as the distance from one crest (or trough) to the next. However, unlike waves travelling in water, which we can see as the water level rises and falls, radiation is a wave in the electric and magnetic fields, which we cannot see directly. Despite this, our eyes and brain can detect and distinguish between different wavelengths in the visible range, interpreting them as different colours. Wavelength is usually represented by the Greek letter lambda (λ) and measured in metres or nanometres (nm), where $1\,\text{nm} = 10^{-9}\,\text{m}$. Visible light has wavelengths from about 400 nm (blue–violet) to about 700 nm (red). The very shortest wavelengths of radiation are known as gamma (γ)-rays and the very longest wavelengths are known as radio waves. In between there is a whole set of names for (sometimes overlapping) ranges of wavelength. Radiation can consist of many different wavelengths all mixed together. When our eye is exposed to light that contains all wavelengths of the visible spectrum we see it as white light.

When you turn on an electric fire, or a ring on an electric cooker, before it begins to glow you can feel heat from it if you are close enough. This energy is transferred from the warming electric element to you by infrared radiation. Eventually, it increases in temperature to the point where it glows red, and then perhaps orange. As the metal in the electric element heats up, it emits radiation that is both increasing in intensity and changing in wavelength. This change in wavelength is a shifting to shorter wavelengths, which in the case of the electric element, starts beyond the visible spectrum, in the infrared, and ends at the red end of the visible spectrum. This does not just apply to electric elements; all bodies at all temperatures (above zero kelvin) emit radiation, and they emit it with a spectrum that peaks at a wavelength that depends on temperature. The Sun's spectrum, shown in Figure 3.6, is that of a hot body at a temperature of about 5800 K. The Earth's spectrum, also shown in Figure 3.6, has a similar shape, but, being at the lower temperature of about 288 K, peaks at a longer wavelength. Fundamental physics tells us that all objects at the same temperature have the same spectrum. **Wien's law** states that an object at temperature T (units of kelvin, K) has its spectrum peaked at a wavelength λ (units of metres, m) given by

$$\lambda T = 2.9 \times 10^{-3}\,\text{m K} \tag{3.3}$$

○ At what wavelength does the spectrum of solar radiation peak if the surface of the Sun is at a temperature of 5800 K? What is the peak wavelength for your body (assume body temperature is 37 °C)? Express your answers in nm. What parts of the spectrum corresponds to these wavelengths (see Box 1.2)?

● Applying Equation 3.3, we find for the Sun

$$\lambda = \frac{2.9 \times 10^{-3}\,\text{m K}}{5800\,\text{K}} = 5 \times 10^{-7}\,\text{m} = 500\,\text{nm}$$

This wavelength corresponds to the visible part of the spectrum.

Body temperature is 273 + 37 K = 310 K, so the peak is at wavelength

$$\lambda = \frac{2.9 \times 10^{-3} \text{ m K}}{310 \text{ K}} = 9.4 \times 10^{-6} \text{ m} = 9400 \text{ nm}$$

This wavelength is longer than that of red light, and is in the infrared part of the spectrum.

As an object increases in temperature, it loses energy at an increasing rate, by emitting radiation. In fact, the rate of loss of energy, or power loss, is very sensitive to temperature, being proportional to its fourth power, T^4 (=$T \times T \times T \times T$). So, if the temperature doubles, the power emitted as radiation increases by a factor of $2^4 = 16$.

○ Suppose a patch of land in the desert falls to a temperature of 288 K (about 15 °C) at night and rises to 323 K (about 50 °C) in the afternoon. Without working out the actual values, calculate the factor by which the peak wavelength and total radiated power changes as a result of this increase in temperature. State your factor as a number larger than 1, but be careful to state whether your factor is an increase or a decrease.

● Wien's law (Equation 3.3) tells us that peak wavelength is inversely proportional to temperature, i.e. is proportional to $1/T$. So the new wavelength is 288/323 = 0.892 times the old one. This is a decrease by a factor of 1/0.892 = 1.12. The radiated power is proportional to the fourth power of temperature, i.e. T^4. So the change in energy output is $(323/288)^4$ = 1.58. (The easiest way to do this calculation on a calculator is to divide 323 by 288, and then square the result twice by pressing the 'x^2' key twice.) This is an increase by a factor of 1.58.

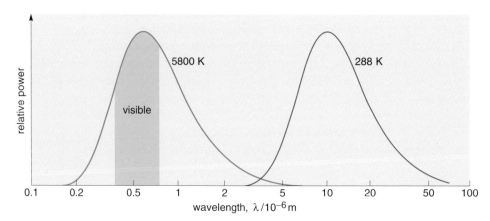

Figure 3.6   The wavelength spectrum of the Sun (left) and the Earth (right). The curves plotted indicate how the power emitted as radiation is distributed over different wavelengths. The Sun, whose surface is at a temperature of 5800 K, has an emission peak in the visible, but also emits in the infrared and ultraviolet. The Earth's surface, at the much lower temperature of 288 K, peaks and emits significantly only in the infrared. (*Note*: the spectra are scaled to have the same peak value.)

The surface of the Earth, and the atmosphere itself, are at temperatures that emit at infrared wavelengths. Some of this emission is reabsorbed again and some of it escapes into space. The radiation that escapes into space represents the outflow of energy from the Earth. In the example above, there was a 12% decrease in the wavelength of emitted light and a 58% increase of radiative power. The change in the spectrum of radiation due to this change in temperature is shown in Figure 3.7. The small change in peak wavelength is not important, because the radiation is still in the infrared range and is still absorbed in the atmosphere in much the same way. However, the 58% change in the energy emitted can be very significant, as we will now demonstrate by considering two different cases.

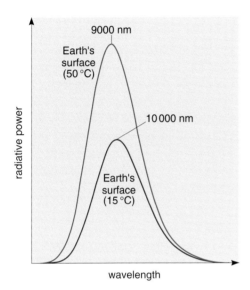

Figure 3.7 The emitted spectrum of the Earth at two different temperatures. The radiated power increases greatly with temperature, but the wavelength range varies only slightly.

3.2 Energy balance

Before we begin looking at the energy balance of the Earth, we need to explain the terms heat and thermal energy. This is done in Box 3.2 below.

Box 3.2 Heat

In science, heat has a very specific meaning: **heat** is energy that is transferred from high temperature to low temperature. Heat is not a property like temperature or thermal energy, in that it is not possessed by an object. There are three basic forms of heat:

1 Energy carried by *radiation* can be heat, which, for example, you can feel on your skin when visible solar radiation (sunlight), or infrared radiation from an electric fire that is warming up, is incident on your skin.

2 *Conduction* is the direct transfer of energy between two substances in contact; for example, when you burn your skin as it touches a hot object.

3 *Convection* is where energy is carried as thermal energy in a gas or liquid moving from high to low temperature, which you can see evidence of when heating a pan of water.

Conduction is not important in the Earth's atmosphere, whereas radiation and convection are very important (you will learn more about convection later).

The thermal energy carried by a gas (for example in convection) is often referred to as **sensible heat**. The name comes from the fact that it can be easily measured, or 'sensed', by measuring the temperature. (Unfortunately, the term 'heat' is not really appropriate in this context, according to the definition given above, but is nevertheless in common usage.)

In the atmosphere, the term latent heat is used to refer to a kind of chemical energy, which is different to that carried by a tank of fuel in a car because no chemical reaction takes place to release it. **Latent heat** is that exchanged when water (or any liquid) evaporates or condenses; for example, it is the energy taken from your body when the liquid water in your sweat evaporates to become water vapour.

Energy is always conserved. In the Earth's environment taken *as a whole*, this means that the energy that goes in must either come out or be stored inside it. The energy going in is almost all from absorbed solar radiation, and energy going out is almost all infrared radiation. Energy is stored as thermal energy of the land, sea and air and also in motions inside the atmosphere (e.g. winds). It can also be stored as potential energy in the forms of latent heat and the gravitational energy of airmasses. We will not be concerned with these until Section 3.3, because for now we are interested in the balance of energies, which, amongst many other things, brings in the well-known greenhouse effect.

3.2.1 Steady states and energy

Firstly, imagine that the outflow of energy into space is less than the inflow of energy entering as solar radiation. This excess of inflow over outflow means that energy must be accumulating. The energy goes into heating the surface and atmosphere, and raising the overall temperature. As mentioned above, the amount of energy emitted as radiation is very sensitive to temperature. So, as the temperature continues to rise, the amount of infrared emission increases significantly, though the wavelength of emission doesn't change much. This means that the total amount of energy going into space increases as the temperature increases. Clearly, there will come a point when a temperature is reached such that the emission into space balances the inflow of solar energy.

Now, imagine that the outflow of energy into space is greater than the inflow of energy entering as solar radiation. In this case, energy is being lost into space overall. This means the energy stored in the environment of the Earth decreases, resulting in a corresponding temperature decrease. As the temperature drops, the amount of infrared emission will decrease, so that the rate of energy loss into space decreases (again, the change in wavelength is not important). Eventually, a temperature is reached where the inflow and outflow of energies are equal.

The above two cases end up in a steady state, a concept we met in Section 1. The new feature we have discussed is that any change in the relative amounts of incoming and outgoing energy tends to return the Earth's atmosphere to a new steady state. A change that does this is called **negative feedback**. Of course we made a number of great simplifications. For a start, not all energy in the Earth is stored as thermal energy associated with temperature (latent heat for example). Also, a change in temperature can result in a range of effects that are difficult to describe, such as the melting of ice-caps and changes of sea-level, and present the difficult problem of determining the amount of cloud cover. These kinds of effects can only be investigated in complicated simulations on computers. These considerations are more important for studying climate change, which we will come to again in Section 4.

3.2.2 Estimating the Earth's temperature

We will now attempt to estimate the average temperature of the Earth, T_E. We will do this by making two important assumptions, whose validity we will reconsider when we obtain the result:

(1) that we can treat the Earth as a whole system, taking surface and atmosphere together;

(2) that this system is in a steady state in terms of energy.

We can therefore proceed by setting the absorbed solar radiative power equal to the outgoing power of infrared radiation. At this point we need to know the power emitted as radiation when the Earth is at a temperature T. The **Stefan–Boltzmann law** states that the rate of energy emitted, R, by an object of surface area A, at temperature T is

$$R = A\sigma T^4 \qquad\qquad (3.4)$$

where $\sigma = 5.67 \times 10^{-8}\,\text{W m}^{-2}\,\text{K}^{-4}$ and is known as Stefan's constant. If the Earth is in a steady state, R is equal to the rate of energy absorption from solar radiation.

○ If a 1 m² patch of land is at 300 K, how much energy is it emitting every second? How many household 100 W bulbs would give out this energy per second?

● Applying the Stefan–Boltzmann law (Equation 3.4), we find the rate of energy emission is

$$R = 1\,\text{m}^2 \times 5.67 \times 10^{-8}\,\text{W m}^{-2}\,\text{K}^{-4} \times (300\,\text{K})^4 = 460\,\text{W}$$

This means that 460 J of energy are given out each second, equivalent to about 4 or 5 household light bulbs.

○ Referring back to Section 3.1, what is the rate of absorption from solar radiation?

● From Section 3.1, we know that the rate at which energy reaches the Earth is $1.74 \times 10^{17}\,\text{W}$. However, we also know the albedo is 0.31 (Figure 3.5), so that only $1 - 0.31 = 0.69$ is absorbed. This means that the rate of energy absorption from solar radiation is

$$0.69 \times 1.74 \times 10^{17}\,\text{W} = 1.20 \times 10^{17}\,\text{W}$$

So, if the rate of emission is equal to the rate of energy absorption, we have

$$R = A\sigma T_E^4 = 1.20 \times 10^{17}\,\text{W}$$

Rearranging, and inserting values, where the Earth's surface area $A = 5.15 \times 10^{14}\,\text{m}^2$

$$T_E^4 = \frac{1.20 \times 10^{17}\,\text{W}}{A\sigma} = \frac{1.20 \times 10^{17}\,\text{W}}{5.15 \times 10^{14}\,\text{m}^2 \times 5.67 \times 10^{-8}\,\text{W m}^{-2}\,\text{K}^{-4}} = 4.11 \times 10^9\,\text{K}^4$$

To obtain T_E requires typing the number 4.11×10^9 into a calculator and hitting the square root button ($\sqrt{}$ or $\sqrt{x}$) twice. Doing this gives us our final result:

$$T_E = 253 \, \text{K} = -20\,°\text{C}$$

Averaging surface temperature data across the world, over the last 30 years, gives a value of 15 °C. Obviously, our estimate is much too low. The reason it is so wrong is because of the first assumption we made at the start — that we can treat the surface and atmosphere as a whole. You may think that the second assumption — that the Earth is in a steady state — is invalid too, because of reports of global warming in recent years. However, for the purposes of this calculation (but not for all aspects of the environment!) the temperature increase over the past 30 years (less than 1 K) is insignificant. We will now address the shortcomings of our first assumption by considering how energy is stored in the atmosphere.

3.2.3 Energy in the atmosphere

Let's take a short diversion and consider the thermal energy of a body considerably smaller than the Earth — your body. When you feel cold, it is because the air around you is cold. To get warm, you have two choices: either warm up the air around your body by turning on a heater, or put on an extra layer of clothing. The first works because the heater supplies energy to the air around you and so raises its temperature. This energy comes from electricity, gas, oil, or some other supply. The second may seem a little more surprising when you think of the science — how can clothing provide energy to keep you warm? It doesn't, it just makes more use of the energy that your body is constantly giving out. Heat (defined in Box 3.2) from your body is given to a thin layer of air around you. This thin layer is then mixed with the colder air surrounding you, and the heat is taken away from your body. If a wind is blowing, or there is a draught, the whole process is speeded up. This cooling effect is most noticeable if you are naked. If you put clothes on, then the thin layer of warm air around your body stays for longer because it is partly trapped between your skin and your clothes. Obviously, the more clothes and the more insulating the fabric, the better the trapping of the air, and the warmer you will be.

In some senses, the atmosphere acts like the Earth's clothing, trapping energy near its surface, though the trapping of the energy happens for different reasons. The crucial point is that the energy emitted by the surface as infrared radiation energy is then mostly reabsorbed by the atmosphere. The atmosphere in turn radiates infrared radiation itself, but since most of the mass of the atmosphere is close to the surface, most of the energy is returned to the surface, with only a smaller fraction escaping into space. In this way, energy is trapped in the lower part of the atmosphere, raising the temperature above the −20 °C we calculated in Section 3.2.2. The transfer of energy within the atmosphere is illustrated on the right-hand side of Figure 3.8.

Let's try to untangle the arrows in the figure and understand how the energy is being trapped. First, consider the radiation, both solar and **terrestrial** (the infrared radiation emitted by the surface and atmosphere), that is absorbed by the atmosphere. The latter absorbs 20 of the 100 units (remember these are units of energy per unit time, or power) of solar radiation and 102 units of the 114 units of infrared emitted by the surface. That is, the atmosphere absorbs only about 20% of the solar radiation, but almost 90% of the surface's infrared radiation.

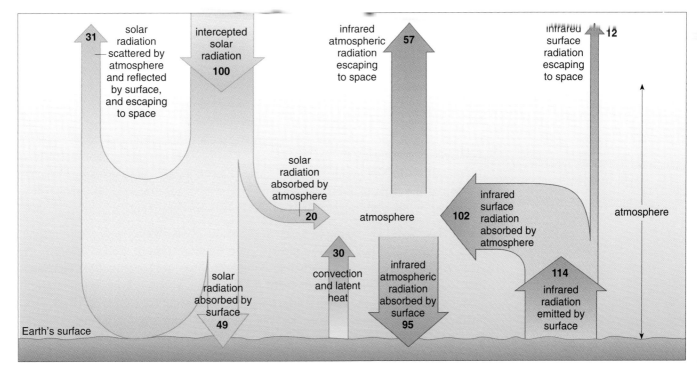

We have now established that the atmosphere is readily absorbing the infrared radiation from the ground, at a rate of 102 units. But, how can it be receiving 102 units, when only 100 units is coming in from the Sun? The answer is that the atmosphere sends most of this flow of energy — 95 units worth — back to the ground, which in turn emits it again, returning it to the atmosphere. We can now summarize how the 'trapping' of energy works: the Earth's surface and atmosphere emit infrared radiation, and because the gases in the atmosphere strongly absorb that same radiation, the energy from the Sun is effectively recycled, thus trapping energy.

○ (a) How much energy is leaving (in the units of Figure 3.8) the Earth–atmosphere system? (b) How would this figure change if the atmosphere absorbed more infrared radiation? (c) Explain briefly what the immediate effect would be on the temperature of the Earth.

● (a) Summing the numbers in all arrows pointing out of the top of Figure 3.8, we find that 31 + 57 + 12 = 100 units are leaving, exactly equal to the amount of incoming solar radiation. (b) If the atmosphere absorbed more that this, less infrared radiation would escape so that the last number in the sum would be smaller, making the total less than 100. (c) With more energy entering than leaving, the energy content of the atmosphere would increase, resulting in an increase in temperature.

○ Draw or describe a diagram similar to Figure 3.8 for a planet with no atmosphere and an albedo of 0.6.

● Of the 100 units, 60 would be reflected and 40 absorbed by the surface. The surface would emit 40 units of infrared radiation.

Figure 3.8 Typical rates of energy transfer within the Earth's environment. Note that even though arrows are depicted as stopping at particular places in the atmosphere, they represent processes that take place throughout the atmosphere. The width of the arrow indicates the rate of energy transferred, which is given more precisely by the numbers inside the arrows (100 units = the power of solar radiation incident at the top of the atmosphere, which is 1.74×10^{17} W).

The energy balance of the Earth can be disturbed by many different factors. First and foremost, any variation in the output of the Sun would be important. In terms of energy, the Sun is remarkably steady in its output, and any variations in the total power output of the Sun are small over time-scales of a few years, certainly less than 0.5%. Over longer time-scales, that of billions of years, the Sun is known to be steadily brightening, as would any star like the Sun. When the Sun was a young star, some 4.6 billion years ago, it was emitting only 70% of the energy it does now. The other way the solar radiation might change is if the Earth–Sun distance varied. However, as with the Sun's energy output, it turns out that this effect is small. If any of these external factors have a major impact on the Earth's climate, then it is almost certainly because their small changes are amplified by the internal workings of the Earth's environment.

As we have seen, greenhouses gases in the atmosphere, which are present only in trace amounts, can have a profound effect on the temperature at the Earth's surface. So, on time-scales comparable with our lives, it is reasonable to expect that small variations in the amounts of these gases can have an important effect. In our discussion of the Earth's atmosphere, considering only emission and absorption of radiation, we have left out many details in order to be clear. However, as is so often the case in science, it is easy to 'throw the baby out with the bath water'; in other words, to overlook an important effect in making assumptions about what at first sight seemed like just a detail. Imagine again that the Earth environment is not in energy balance and is receiving more energy than it is emitting. As a consequence, the temperature increases. Earlier, we said that the infrared emission would rise with temperature until the outgoing infrared emission balanced the incoming solar radiation. But imagine that the increased temperature causes more water to evaporate, increasing the amount of water vapour in the atmosphere. An increase in water vapour would result in more infrared absorption, and so more energy trapping. The importance of this is that instead of the rising temperature causing another change that helps the system to return to a steady state, it causes a change that pushes it further from a steady state. This kind of change, which reinforces itself, is called **positive feedback**. It is this kind of effect that might cause what is commonly called a **runaway greenhouse effect**. It is believed that this has happened on Venus and has led to the current balanced situation where surface temperatures can be as high 750 K, even though this planet is only about 70% of the Earth's distance from the Sun.

○   In the above runaway greenhouse effect scenario, where evaporated water increases the energy trapping in the Earth's atmosphere, can you think of another effect with a related cause that could have the opposite effect?

●   An increased amount of water in the atmosphere could result in more clouds. More clouds would mean a greater albedo, meaning that a greater fraction of the incident solar radiation is scattered back into space. This would mean less energy entering, and so oppose the water vapour's energy-increasing effect. (This scenario is by no means clear-cut however! Recent models suggest that clouds might actually contribute overall to the positive feedback because their contribution in increasing the albedo is outweighed by their role in trapping infrared radiation.)

3.3 Moving air, moving energy

So far, we have considered how energy is transported by radiation, how it enters and leaves our atmosphere, and the principle of energy balance. In doing so, we paid little attention to the small arrow pointing upwards from the ground in the centre of Figure 3.8. This innocent-looking arrow, labelled 'convection and latent heat', carries only 30 energy units from the ground and is a significant, but still minor contribution relative to the 114 emitted by infrared radiation. However, the processes represented by the convection and latent heat arrow are the key to understanding a familiar aspect of the atmosphere, or more precisely the troposphere: *weather*. In this section, we begin to explain some of the workings of the 'weather machine'. Firstly, we will consider the typical energies involved.

○ Consider the volume of air in a typical back-garden: let's say the garden is 10 m by 5 m, extending up to the height of the house, say 5 m. (a) What is the volume of this air? (b) If the air's mass density is the sea-level value (Table 1.2), what is the mass of air? (c) Calculate the kinetic energy involved in this motion if the air is moving at (i) $2\,\mathrm{m\,s^{-1}}$ (the speed of a breeze) and (ii) $50\,\mathrm{m\,s^{-1}}$ (the speed of a strong hurricane). (d) What kind of everyday electrical devices or machines would consume similar amounts of energy in a second?

● (a) The volume V is obtain by multiplying the dimensions:

$$V = 10\,\mathrm{m} \times 5\,\mathrm{m} \times 5\,\mathrm{m} = 250\,\mathrm{m^3}$$

(b) The mass m is given by multiplying the mass density $\rho = 1.23\,\mathrm{kg\,m^{-3}}$ by the volume:

$$m = \rho V = 1.23\,\mathrm{kg\,m^{-3}} \times 250\,\mathrm{m^3} = 307.5\,\mathrm{kg}$$

So the mass of air is 308 kg (rounding to the same number of significant figures given in the density).

(c) (i) Using the kinetic energy formula from Box 3.1 (Equation 3.1), the kinetic energy of the breeze, E_K, is given by

$$E_K = \tfrac{1}{2}\,mv^2 = 0.5 \times 308\,\mathrm{kg} \times (2\,\mathrm{m\,s^{-1}})^2 = 616\,\mathrm{J}$$

(ii) The energy for the wind speed of a hurricane can be calculated in exactly the same way, replacing the 2 by 50. The calculation can be simplified because $\tfrac{1}{2}$ and m are common to both calculations — only the value of v is different — so the energy for the hurricane can be obtained by multiplying 616 J by $(50/2)^2 = 625$, giving an energy of 385 000 J = 385 kJ.

(d) The breeze flowing through a typical back-garden involves an energy equal to that used by several household light bulbs in a second, or the energy consumed by a kettle in a second. The hurricane, however, has the energy consumed by several hundred light bulbs, or a very large car engine, in a second.

The above question deals with the energy of wind in a very small volume compared to the volume of the entire atmosphere. So it is clear that the 'weather machine' involves a huge amount of energy, and ultimately this energy must come from the Sun, as solar radiation. But how does energy in radiation end up as kinetic energy of air, or in producing rainfall, or contribute to any of the other phenomena collectively known as weather? The whole answer is extremely complicated, so to begin with, we will explore one example that illustrates how solar energy can become gravitational potential energy, and then kinetic energy of air.

3.3.1 Heat becoming potential energy

Consider a column of air above a certain area on the ground, as depicted in Figure 3.9a. For simplicity, let's suppose that the air can only move up or down inside this column. As described in Section 1.2, the reason that this column of air doesn't sink to the surface is that it is constantly absorbing energy. If the air in this column is not moving, then it is in a steady state. (Here we mean the gas as a whole is not moving — the individual gas particles are always moving.) The energy absorbed by the column is just enough to keep the mass 'suspended' throughout its height. As we discovered in Section 1.2, the gas near the bottom is compressed by the weight of the gas above it. The density profile of this column is also shown in Figure 3.9a.

Let's consider the effect of heating this column by exposing it to a greater amount of solar radiation. (In practice, this could be due to sunrise, or clouds parting higher in the atmosphere.) Consider material in a particular layer of the column, for example the one marked on the diagram. This layer is not defined by its height, but by the amount of material above it (or below it) in the column. We will pose the following question: what is the final steady state, once the column has readjusted to the increased heating?

As the column is heated, the temperature, T, rises in the layer, as it does throughout the column. Remember that the pressure, p, of this layer of air has to support the weight of material above it. The amount of material above this layer cannot change, by definition (i.e. that's what we meant by a 'layer'!), so the pressure required for any new steady state is the same. The perfect gas law, $p = (N/V)kT$, must be obeyed. If T rises in the layer, and p cannot change, then the number density, N/V, must decrease. This must be true whatever layer we are looking at, so the density in all layers decreases. By definition, the number of particles in the layer, N, has not changed, so the decrease in number density must be entirely due to an increase in the volume, V. In other words, the column expands when heated, and ends up in a new steady state with a profile that is indicated in Figure 3.9b.

This new steady state has more energy, stored in the kinetic energy of the gas particles, and some extra gravitational potential energy (see Box 3.1) because, on the whole, gas particles have moved up to greater heights. Now consider what happens to this new state if the absorbed energy drops back to its original level, because of sunset or increase in cloud cover, for example. As the temperature drops, the density must increase in each layer because, as we argued above, the pressure must remain constant, and the perfect gas law must be obeyed. The air becomes compressed and reverts back to the profile illustrated in Figure 3.9a.

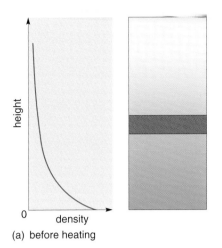

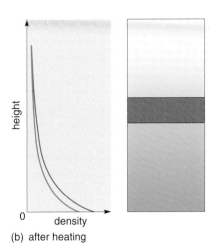

(a) before heating   (b) after heating

Figure 3.9 The expansion of a column of air on being heated. (a) The density profile along the column (red curve) and a particular layer of air in the column (red band) before heating has occurred. (b) After heating, the density profile has changed (blue curve) from the original (red curve) and the air layer identified in (a) has moved upwards and expanded, so now has a lower density.

This example is admittedly a little contrived, mainly because the troposphere is primarily heated near the ground by terrestrial radiation, not by solar radiation throughout all layers. However, even if we imagined that the column received increased heating from below, then, under certain temperature and humidity conditions, the end result would be much the same. The important point being illustrated is that heating can lead to storage of gravitational potential energy. If the temperature and humidity conditions mentioned above are not met, the heating would have resulted in quite different behaviour, which is the subject of the next subsection.

3.3.2 Heat becoming kinetic energy — convection

A significant fraction of the infrared radiation emitted by the surface is absorbed just above the ground (putting exact numbers to this statement is tricky because absorption varies greatly with the water content of the air and wavelength range of interest). This absorbed energy heats the air just above the surface, causing it to expand. If the heating is great enough, this expansion can cause a 'parcel' of air to expand and become less dense that the air around it. Being less dense, it begins to ascend into the atmosphere. In this way heat received from the ground can result in kinetic energy of moving air. In addition, the parcel carries energy up into the higher and cooler parts of troposphere. It loses energy in heating its surroundings and, having cooled, sinks back to the surface again. This process is called **convection** and is described in Box 3.3. As well as transporting energy, convection keeps the air well-mixed in the troposphere — the significance of this mixing will become clear later.

When air is warmed, it is not a foregone conclusion that it will undergo convective motion. Whether it will or not depends primarily on the temperature and humidity conditions. Specifically, it depends on how fast the temperature falls with height, quantified by the environmental lapse rate (ELR) and the appropriate lapse rate for the air that is being warmed (see Box 3.4). We will consider heating dry (unsaturated) air, so we will use the dry adiabatic lapse rate (DALR), which is about $10\,°\text{C}\,\text{km}^{-1}$ (the precise value is $9.8\,°\text{C}\,\text{km}^{-1}$). We will now explore how these lapse rates relate to the onset of convection. Bear in mind that a parcel of warm air placed in surroundings of cooler air begins to move upwards, with an acceleration (change in speed per unit time) that increases with the temperature difference.

Box 3.3   Convection

The most familiar example of convection is that which takes place in a pan of water being heated on a cooker. To see the convection, it's best to add something more visible than water—frozen peas are ideal. This is shown in Figure 3.10. In Figure 3.10a, before the heat is turned on, the water has no net movement. When the heat is applied, the water at the base of the pan becomes hotter. As it does, it expands (as the air did in the column example above) and so slightly decreases in density. The cooler water above has a higher density, so the warm water begins to rise. When the warm water reaches the surface of the pan it cools and sinks down, making way for the further hot water rising from the base. This process involves circulation, as indicated in Figure 3.10b.

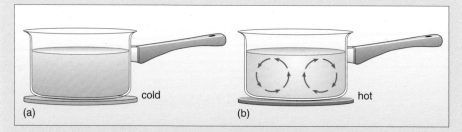

Figure 3.10   Convection in a pan of heated water.

In the atmosphere, the same process can operate, where the infrared-emitting surface plays the role of the cooker, and the air in the atmosphere plays the role of the water in the pan. Figure 3.11 illustrates this scenario. We can understand why the air expands on being heated, using the perfect gas law, just as we did in considering the air column. In contrast to the air column, which was heated throughout, the heating is restricted to air near the ground. It is the rising of hot material, coupled with the loss of energy at the top of the circulation, that makes convection a very efficient mechanism for transporting energy from the surface into the atmosphere. In the atmosphere, ascent of air occurs in confined regions, whereas descent occurs across more extended areas.

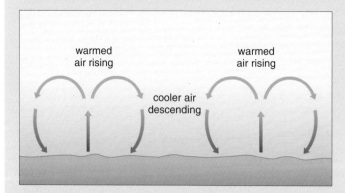

Figure 3.11   Convection in the atmosphere.

Box 3.4 Adiabatic lapse rates

The rate of change in a property, such as temperature, with height is called a **lapse rate**. We have already met the environmental lapse rate (ELR, see Section 1.3), which measured how fast temperature drops with height. We can also define lapse rates that relate to the temperature change as a parcel of air changes height. If a parcel of unsaturated air (i.e. containing less than its maximum amount of water vapour) rises through the atmosphere, then it expands and cools at what is known as the **dry adiabatic lapse rate (DALR)**, which is 9.8 °C km^{-1}. If the air contains as much water vapour as it can carry, i.e. it is saturated, then the lapse rate is less that the DALR and is called the **saturated adiabatic lapse rate (SALR)**. It is not possible to just give a single value for the SALR because the maximum amount of water vapour air can carry depends on temperature; warm air can carry more than cold air. Remember that the word 'adiabatic' (see Box 2.6) refers to the fact that the parcel is not exchanging energy with its surroundings, which is why its temperature change is not governed by the ELR.

Figure 3.12 illustrates the ascent of a warm parcel of air. Consider a situation where the ELR is 12 °C km^{-1}. Remember that an ELR of 12 °C km^{-1} means that the air in the atmosphere has a temperature drop of 12 °C for every km ascended. This is the temperature change that would be measured in practice by a thermometer on a rising balloon. Now, consider a dry air parcel, at ground level, which is heated to a temperature of 21 °C, slightly higher than that of the surrounding air which is at 20 °C. This air parcel rises, and we can make the reasonable assumption that it does not exchange heat with the air surrounding it, i.e. conditions are adiabatic. The air parcel therefore cools according to the DALR. So, having risen 100 m, the temperature change is

$$0.1 \, \text{km} \times 10 \,°\text{C km}^{-1} = 1.0 \,°\text{C}$$

The temperature is therefore $(21.0 - 1.0) \,°\text{C} = 20.0 \,°\text{C}$. In a similar way, we can work out the temperature of the atmosphere surrounding this air using the ELR. The drop in temperature is

$$0.1 \, \text{km} \times 12 \,°\text{C km}^{-1} = 1.2 \,°\text{C}$$

The surrounding air is therefore at a temperature of $(20.0 - 1.2) \,°\text{C} = 18.8 \,°\text{C}$, 1.2 °C less than the parcel. The parcel remains warmer than the air surrounding it, so it continues to rise. After rising to 500 m, the temperature difference has increased to 2 °C. When it reaches 1 km, the difference is 3.0 °C, which is just the initial difference, 1°C, plus the difference between the ELR and the DALR, i.e. $(12 - 10) \,°\text{C} = 2 \,°\text{C}$. We can conclude two things from this: (1) the parcel will continue to rise; and (2) as the temperature difference *grows* with increasing height, the parcel *accelerates* in its upward movement. All that we did to start it on this course of accelerating upward motion was to heat it slightly. When dry air in the atmosphere has a larger ELR than DALR, then we say that it is **unstable**, in that a slight change, caused by either moving or heating the air, leads to continued motion. Conversely, if such heating or movement leads to suppression of motion, the situation is said to be **stable**.

○ Draw a diagram, similar to Figure 3.12, and explain how an unsaturated parcel of air at the same temperature as it surroundings at a height of 1 km would behave if it moved downward by 100 m. Assume the same ELR and DALR values as above.

● Its temperature would rise by 1.0 °C, but the surrounding air at 900 m would be 1.2 °C warmer than the air at 1 km. Being cooler, the parcel would continue to descend, and in fact be accelerated downward until eventually it reached the ground.

○ If the ELR was 8.0 °C km⁻¹, how would an unsaturated air parcel behave if it were (a) moved upward, and (b) moved downward? (c) Are the conditions stable or unstable for dry air with this ELR?

● (a) If a parcel of air was moved upward, its temperature would decrease faster than the temperature of the surrounding air. It would therefore be at a lower temperature and so sink back down again. (b) If it were moved downward, its temperature would rise faster than that of the surrounding air, so it would move back up again. (c) This situation is clearly a stable one.

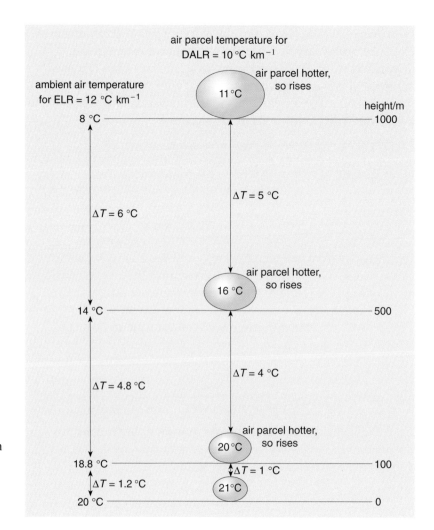

Figure 3.12 A parcel of air (shown as a blue ellipse) rises in an unstable situation. As it rises its temperature drops, but by a smaller amount than that of the air surrounding it.

Convection can start only if there is instability. An unstable situation may arise for a number of reasons, but can always be brought down to a comparison of the ELR with the DALR or SALR (saturated adiabatic lapse rate), the latter being appropriate when the air is carrying its full capacity of water vapour. Instability occurs when the ELR is larger. Such a value of ELR is said to be a **superadiabatic lapse rate**. Superadiabatic lapse rates can be caused by the surface warming the air close to it, thereby raising the temperature near the ground well above the air temperature higher up. Alternatively, instability can occur as air becomes saturated, i.e. as it reaches its maximum capacity for holding water vapour. The SALR is either less than or almost equal to the DALR. The reason for the difference is that as saturated air cools, some of its water content condenses out, releasing latent heat energy (we'll return to this again below) and this slows the temperature decrease of a rising parcel of air. Hot, dense air can carry the most water, so, for example, the SALR is significantly smaller than the DALR over the tropical forests. In contrast, the cold and much less dense air of the upper troposphere has a similar SALR and DALR because the air cannot carry much water vapour.

For convection to occur, there must be a downward motion of air to accompany the rising of the warm air. The air that descends is air that has risen, cooled and then been displaced by the continuing flow of warm air rising below it. For the rising air to turn over in this manner, it is clear that the temperature of the rising air must eventually drop to the temperature of its surroundings. This happens ultimately because the ELR drops to a smaller value at some point in the troposphere. We will now move on to consider the role of water in the atmosphere and its vertical motions.

3.3.3 Water vapour and energy

We know from everyday experience that water falls from the atmosphere onto the land and seas as various kinds of precipitation: rain, hail, snow, sleet. What is not immediately clear from our everyday experience is that roughly the same amount of water travels back up into the atmosphere.

○ Under normal circumstances, the immediate environment of your home should be in a steady state as far as water is concerned. Unfortunately, this is not always the case. What could cause a major departure from this steady state, and what are the potential consequences?

● If the rate at which rain falls exceeds the rate at which water can be drained or evaporated away, then flooding occurs. Alternatively, the amount of rain can be normal, but a blockage of drainage, such as by leaves in autumn, can cause the same effect: the draining of water is slower than the rate of rainfall. If the precipitation involves frozen water, most commonly snow, and the temperature is below freezing, then there is no drainage at all. A sudden rise in temperature could then cause a rapid melting, releasing a vast amount of water in a short time. Once again, the rate of appearance of liquid water can easily exceed the capacity of drains to carry it away.

One way or another, a large amount of precipitated water makes its way into lakes, seas and oceans. If this were the end of the story, then these bodies of water would just grow, and water in the atmosphere would eventually all precipitate down to the surface. Obviously this is not happening. Water is evaporated from the surface and re-enters the atmosphere as water vapour. So, just as water falls from the atmosphere onto the surface as various forms of precipitation, water rises upward at (roughly) the same rate, but as invisible water vapour. But remember, at a given place on the surface of the Earth, these two processes are not necessarily balanced, and movement of water on or below the Earth's surface is required to complete this **water cycle**. Also, plants and animals (e.g. humans) can have a significant effect on the local water cycle. Plants in particular transport liquid water in soil into the atmosphere by a process known as transpiration.

Here we are going to take a particular interest in evaporation. Not only is it the atmosphere's main water supply route, but it is also a very important energy transporter. Energy is stored in the form of latent heat, discussed in Box 3.5.

Box 3.5   Latent heat—water vapour carrying energy

Sweating is a physiological process that gets rid of excess thermal energy. Water emerges on top of your skin and thermal energy from the body goes into separating the liquid water molecules from each other, turning the water from liquid into gas. The water vapour then leaves the skin and is carried away in flows of air. In effect, the water vapour mixed with air leaving the body is carrying some thermal energy away. This energy is called latent heat. More than that, should this air cool, for example by coming into contact with the inside of a window dividing a warm room from a cold outside environment, then latent heat energy is released as the water condenses, turning back into a liquid. Of course, any surface, not just a living body, can release energy in this way. Here, we are particularly interested in how the Earth's surface gives up its energy to the atmosphere.

A vital fact to remember about air is that *cool air can carry less water vapour than warm air*. For example, common experience tells us that glasses, i.e. spectacles, steam up when the wearer walks into the warm indoors from the cold outside. Whilst outside, the lenses cool down, but when entering an environment with warm air, the cold lenses cause air around them to cool slightly. As this air cools, it cannot carry as much water vapour, resulting in it condensing as water on the surface of the glass lenses.

○ Before sunrise, a driver is about to set off for work in a car on a cold winter's morning. (1) First, water droplets of dew have to be wiped off the car's windows. (2) Within moments of being in the car, the inside of the windows steam up. (3) To avoid having a nasty accident, the driver turns the car's hot air blower to maximum and directs it to the windscreen. Explain, in scientific terms, what is happening in these three stages.

● (1) During the night the air has cooled, losing some of its capacity to carry water vapour, resulting in water condensing on the car's windows. (2) A human being is constantly blowing out warm, moist air. When in a car, this air quickly comes into contact with the cold air of the car and the cold glass of the windows. The breathed-out air cools and so its water vapour condenses out onto the glass. (3) Warming the inside of the car enables the air to carry more moisture, and so less water condenses out. Also, blowing the hot air across the glass warms it, keeping the air near to it warm and so able to carry more water vapour, rather than having it condense onto the glass. Water therefore evaporates off the windscreen and becomes invisible water vapour.

Just as excess heat from your body can be removed by evaporating sweat from your skin, heat from the warmed surface of the Earth can be removed by the evaporation of water. The evaporated water vapour is then carried upwards by convection. This means that convection plays a double role in transporting energy: transferring the thermal energy involved in the faster gas particle motion of warm air, and transferring the latent heat of evaporated water. As part of the process of convection, this rising warm air cools, and so eventually the water vapour condenses again. This condensation process is what forms clouds, which are composed of a great many tiny droplets of water. Each of these droplets has condensed around tiny aerosol particles (e.g. dust), known as **condensation nuclei**. When a droplet reaches a certain mass, it begins to fall, ultimately reaching the ground as a raindrop or, at temperatures below freezing, as hail or snow.

The condensation of the rising water vapour occurs once the surrounding air has fallen to the dewpoint temperature (see Section 2.2.4). Because the temperature falls with increasing height in the troposphere, the dewpoint temperature corresponds to a particular height, that of the cloud base. Although not always obvious from the ground, the cloud base is usually quite flat, as shown in Figure 3.13. The structure above the cloud base can be quite complicated as it depends on the air motions involved in convection. Speeded-up movies of clouds show these structures to be bubbling, as the rising air reaches its maximum height and begins to return downward again. All this motion is usually confined to the troposphere, but on occasion it can overshoot and protrude into the usually calm, neatly layered stratosphere.

Figure 3.13 The flat cloud-base, where water vapour evaporated from the surface condenses back to liquid water. Convecting motions give the upper part of the clouds their bubbling appearance.

3.3.4 Horizontal motions in the atmosphere

So far, we have considered only the vertical motion of air, but a moment's thought should be enough to convince you that vertical motion can't take place without some horizontal motion. Consider a rising parcel of air taking part in convection. At a particular moment, it occupies a volume at a particular height. Moments later, it will have moved upwards a little. The volume it previously occupied is certainly not a vacuum because even a small difference in pressure causes air to move towards low pressure very quickly. The obvious answer is that air from below, also involved in convection, has simply moved into this volume now. In fact, as might have already been obvious to you in the first place, air is moving upward in a continuous stream. But what happens at ground level, where there is no air below? The answer, of course, is that air is drawn in from the sides, causing horizontal motion. In this way vertical motion, in this example convection, must include horizontal motion too, as we implied in Figure 3.11.

A difference in pressure between two points causes air to move between them, from high pressure to low pressure. Even a small pressure difference compared to the local atmospheric pressure can cause a fast wind. Pressure differences can be caused by differences in temperature, or differences in heating. We will now consider a few other examples of how heating causes pressure differences.

Draughts

On a still, cold winter's day, most people like to turn up the heating in their house, and seal it from the atmosphere outside by closing all doors and windows. Fires, radiators or any other kind of heater heat the air around them, and then this air moves around the house (by convection), keeping it warm inside. If the house is well sealed against the outside (no draughts) then the amount of air inside the house won't change as the air heats up. Let's consider the perfect gas law: $p = (N/V)kT$. We have just said the amount of air in the house hasn't changed, so N is fixed, but as we heat up the house the temperature, T, increases. This means that the pressure, p, increases. So, when a window or door is opened, there will be a mini-wind of warm air blowing out of the house, some of which is replaced by cold air entering — the dreaded draught! Of course, this is only part of the story of domestic heating (loft insulation, heat conduction through the walls aren't mentioned), but it should at least convince you that heating air encourages it to move, and importantly for heating a building, heated air moves away from the source of the heat.

Sea and land breezes

Imagine an island on a clear day when there is not too much wind due to large-scale weather patterns. Before going further, bear in mind an important fact: land absorbs and releases heat more quickly than the sea. Throughout the day, the temperature of the ground rises faster than that of the sea. Also, the ground loses energy more quickly, through emission of infrared radiation, heating the air above its surface more than the sea can heat the air above it. The air above the land rises and, if the air is unstable, undergoes convection. After cooling at the top of its ascent, the air moves out over the sea and sinks back to the surface. The cooler sea air must take part in this convection by moving onto the island in place of the air that has risen. There is therefore a continuous wind blowing onto the island from the sea. The story doesn't end there, however. When the Sun sets, the land cools faster than the sea, and so at some point during the night, the situation reverses. Consequently, air blows out to the sea. These phenomena are known as the **sea breeze** and **land breeze**, and are a classic demonstrations of how heating differences can lead to air movement: see Figure 3.14, which also shows the cloud formation that accompanies the sea breeze.

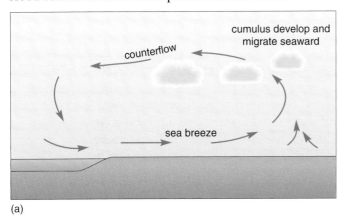

(a)

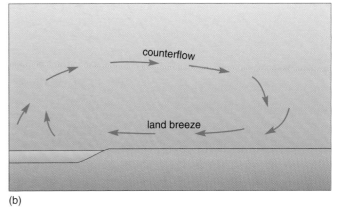

(b)

Sea breezes do not necessarily develop just over islands. For example, weather patterns in the gulf of Mexico during summer are often dictated by a sea breeze situation. Smaller-scale examples can also be found around the south and east coasts of Britain. In this case, sea breezes extend up to heights of 1 km and blow (move inland) at speeds of around 4–$7\,\mathrm{m\,s^{-1}}$ (10–15 miles per hour) penetrating 50 km (30 miles) inland by mid-evening. Similar **lake breezes** are also found over large lakes, such as the Great Lakes in North America. It's worth noting that due to the Earth's rotation (actually the Coriolis effect, described in Section 4) sea breezes can sometimes gradually change direction over a few hours, becoming more parallel with the shore.

Figure 3.14 Sea and land breezes over an island: (a) a sea breeze in the early afternoon; (b) a land breeze at night.

3.3.5 Explaining the temperature structure of the atmosphere

We are now in a position to explain the rather complicated temperature structure of the atmosphere, though we will confine our interest to the troposphere and the stratosphere. First, a quick recap. Temperatures are highest at ground level, and then fall with a typical lapse rate of about $6\,^{\circ}\mathrm{C\,km^{-1}}$. At the top of the troposphere, this lapse rate falls to around zero; this is at the tropopause, which is at a typical height of 10 km (Figure 1.10). Above this is the stratosphere, where the lapse rate is either close to zero or negative. Typically, the lapse rate remains close to zero for about 10 km; this is an isothermal (constant temperature) layer. Above this, the temperature increases with height up to the top of the stratosphere, the stratopause, at a height of typically 50 km. Bear in mind that the heights given in this description vary with both latitude and time of year. For example, the height of the tropopause can be lower than 8 km at the poles in winter, but as high as 20 km at the Equator. The reason that the temperature structure is so much more complicated than that of pressure, and so variable depending on season and latitude, is that it depends on the rates of energy transfer in and around the atmosphere. In summary, the basic form of the temperature structure from ground to stratopause is therefore: falling, constant, then rising again.

Since the atmosphere is mostly transparent to solar radiation at visible wavelengths, most of the energy it brings goes into heating the surface. The infrared radiation emitted by the warmed surface is readily absorbed by the air, i.e. the air is opaque to it, so the ground preferentially heats air close to the surface. *This is why the air temperature in the atmosphere is greatest near the surface.* The heated air then moves upwards, carrying energy as both thermal energy and latent heat, the latter being from water evaporated from the surface. Under unstable conditions — which are commonly found in the troposphere — this air rises, resulting in convective motion, carrying energy from the surface to the higher layers.

The maximum height of convective motion depends on where the temperature stops dropping with height, called a **thermal inversion layer**, because the temperature switches from decreasing to increasing. Thermal inversion layers often occur in the troposphere and effectively act as lids on the air below them, and unfortunately for some large cities, concentrate pollution below them. The tropopause is the ultimate upper limit on convection, because it is at the bottom of a permanent thermal inversion layer. It therefore acts like a lid on the entire tropospheric weather system. The temperature structure of the troposphere is therefore mainly dictated by the convection that takes place within it. The temperature is highest at the base where it is heated, and lowest at the top where cool, convected air pauses before descending again.

In the stratosphere, the temperature either stays constant or else increases with height, in contrast to the falling of temperature with height in the troposphere. The stratosphere is layered, with little vertical movement, whereas the troposphere is continually being mixed by convection. The essential reason for these differences is that the stratosphere is heated directly by solar radiation, whereas the troposphere is heated from below. Although the density of air in the stratosphere is 100–1000 times lower than the density of air at sea-level, the presence of certain gases means that it can strongly absorb certain wavelengths of solar radiation. The lower density also means that it takes much less energy to raise a given volume of stratospheric gas to a particular temperature than it would for the same volume of tropospheric gas.

One reason for the stratosphere's ability to catch energy from the incident solar radiation is its high mixing ratio of ozone (mixing ratios were explained in Box 1.1). The ozone molecule readily absorbs certain wavelengths of ultraviolet solar radiation, and its presence in the stratosphere serves to protect us from this harmful output from our Sun. Although the mixing ratio of ozone molecules peaks in the stratosphere (between 20 and 30 km), the fact that the air at the top of the stratosphere is at a lower density, and the fact that it is first to encounter the incident radiation, means that the temperature is maximum at the stratopause. There the temperature can be close to 0 °C. In lower parts of the stratosphere, the heating effect is diluted because much of the ultraviolet part of solar radiation has already been absorbed and the air is at a higher density. So, the tropopause, isolated from convective warming from below, and with the whole stratosphere above it absorbing uv radiation, sits at a temperature minimum of typically −55 °C (the value depends on latitude and season).

Having given a broad-brush explanation of the typical temperature structure of the atmosphere, it is worth remembering that there are many exceptions and complications. For one thing, we have said nothing about the variations because of the Earth's rotation, the seasons or location on the Earth, which are addressed in the next subsection.

3.4 Latitude and seasons

This is a short summary section, as the topic of latitude and seasons is covered by a DVD activity, which you should do now.

Activity 3.1 Latitude and seasons

This activity illustrates how the Earth's position and orientation in space determines the amount of solar radiation we receive at different places on the Earth, during different seasons. It also considers the effect of the atmosphere and the effect on observable quantities such as temperature and rainfall.

The climate and its change with the seasons depend on many complicated, interlinking factors. However, there is one basic factor that underlies all others: **latitude**. Latitude is the angle that measures a place's location with respect to the poles and the Equator. The Equator has a latitude of 0°, the North Pole has a latitude of 90° N and the South Pole has a latitude of 90° S.

As a general rule, locations on Earth closer to either of the poles receive less solar radiation than those closer to the Equator. The explanation for this is based on the angle between the direction of the solar radiation and the surface, or, in everyday language, the height of the Sun above the horizon. In equatorial regions of the Earth, the Sun is almost overhead around noon. In polar regions, the Sun never climbs high in the sky even in summer, though it can remain above the horizon (i.e. there will be continual daylight) for up to six months of the year. In winter it can spend up to six months below the horizon.

A low-lying Sun means that the incident radiation is spread over a greater area of the surface, as shown in Figure 3.15. This 'dilution' of radiation per unit area is so important that it outweighs the effect of continual day in polar summer. Also, as can be seen in Figure 3.15, solar radiation has to traverse a greater depth of the atmosphere at locations further from the Equator. This means that atmospheric absorption is increased. You can see evidence of this effect in the form of red sunsets and sunrises — the blue light is scattered out of the line of sight, leaving mainly red light travelling to your eye.

The Earth rotates around its **axis**: an imaginary line running through the Earth that joins the poles. If the Earth's axis were perpendicular to its orbit around the Sun, then we would have no seasons at all. Seasons occur because the axis of the Earth is fixed at a tilt of about 23.5° to the perpendicular of the Earth's orbit (this angle does vary, but not on time-scales that need concern us here). As shown in Figure 3.16, this tilt means that the poles spend parts of the year (the time taken for one complete orbit around the Sun) tilted both towards and away from the

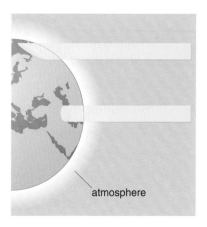

atmosphere

Figure 3.15   This diagram illustrates that the same-sized beam of solar radiation is spread over a larger surface area near the pole, than at the Equator of the Earth. Notice also that the beam travels through more of the atmosphere near the pole.

Sun. When a **hemisphere** (each half of the Earth, either north or south of the Equator) is tilted towards the Sun, it receives more solar radiation; this is summer. When a hemisphere is tilted away from the Sun, it is winter in that hemisphere. For summer at a particular place, the tilt of 23.5° towards the Sun is, in effect, like moving it to a latitude 23.5° closer to the Equator. Similarly winter is like moving the place 23.5° away from the Equator, towards the pole. A latitude decrease of 23.5° is like moving London to Northern Africa, so seasonal variations are clearly very important.

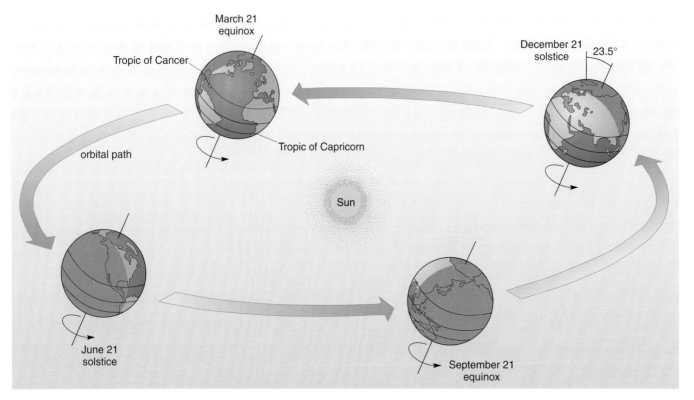

Figure 3.16   The tilt of the Earth explains the seasons as each hemisphere is tilted towards and away from Sun, as the Earth orbits it.

Latitude may be fundamental to understanding how climates vary around the world, but it is by no means the whole story. For example, London and Moscow are at similar latitudes (51.5° N and 55.8° N, respectively) but have quite different annual mean temperatures of 11.0 °C and 4.1 °C, respectively. The UK spans a latitude range of roughly 50° N to 60° N, but has a much milder climate than that of the same latitude band in either inland Europe, or North America. The key difference here is in how energy is transported through the global environment. In this case, the UK benefits from the heat carried to it by the Gulf Stream which, for example, allows palm trees to grow in frost-free areas of western Scotland!

3.5 Summary of Section 3

1   Energy can take on many forms. In the atmosphere the important forms are: kinetic, thermal, gravitational potential energy, radiative and the energy of latent heat.

2   Radiation can be absorbed, which involves energy exchange, and scattered where only the path of the radiation is changed.

3   Solar radiation, being from a body of about 5800 K, has a spectrum that peaks in the visible, at wavelengths where the atmosphere is mostly transparent; it is therefore mostly absorbed by the surface. Terrestrial radiation from the surface and atmosphere, which are at a typical temperature of 290 K, peaks in the infrared, a range of wavelengths at which the atmosphere absorbs strongly.

4   The peak wavelength of radiation emitted by an object at a given temperature is given by Wien's law, and the total energy emitted is given by the Stefan–Boltzmann law.

5   An overall energy balance, or steady state, is achieved where the incoming solar radiation is balanced by the emission of terrestrial radiation into space. The surface temperature of the Earth is higher than the value dictated by this balance because the atmosphere traps energy near the surface by absorbing infrared radiation; this is the greenhouse effect.

6   Heating a column of air under stable conditions leads to an expansion, with air rising on average, resulting in a new steady-state structure where the energy imparted is stored as gravitational energy. Under unstable conditions, where the ELR is greater than the SALR or DALR, convection occurs, in which warm air heated at the surface rises, cools and then sinks down again in a circular fashion.

7   Convection, as well as involving the kinetic energy of bulk air motion, can carry energy in two forms: sensible heat (the thermal energy of the air) and latent heat (energy held by water vapour that has been evaporated).

8   The temperature structure of the atmosphere can be explained in broad terms by realizing that the troposphere is heated from below by the surface, and that the stratosphere is heated by solar radiation being strongly absorbed at certain wavelengths, particularly the ultraviolet.

9   Day and night occur because of the Earth's rotation and the seasons occur because the tilt in the Earth's rotation axis is fixed as the planet orbits the Sun.

Question 3.1

Consider a cube-shaped mass of air at a temperature of 300 K, that is 1 km wide, 1 km in length and 1 km high and is moving at 3 m s^{-1}. Calculate (a) the kinetic energy of the air; (b) its thermal energy (*hint*: use the pressure); and (c) the total energy radiated by it in 10 seconds. (d) Compare these energies and then suggest where the radiated energy must come from. What feature of the equations used accords with your suggestion? Take the mass density of air to be 1.20 kg m^{-3} and the mass of a typical air particle to be 5.00×10^{-26} kg.

Question 3.2

A patio is made of black and grey concrete slabs that are squares of side 50 cm. The black slabs have an albedo of 0.1 and the grey slabs have an albedo of 0.4, and solar radiation delivers 780 W to every square metre of the surface. (a) How much energy does each slab absorb? (b) What are the temperatures of the slabs? (c) If the relative humidity is 100%, the local ELR is $6\,°C\,km^{-1}$ and the SALR is $4\,°C\,km^{-1}$, explain how the emitted energy is carried away.

Question 3.3

Pipes carrying hot water at a typical temperature of 60 °C are often lagged (covered in insulating foam). List the ways in which this situation is similar to the Earth's surface and its atmosphere, and also the ways in which it is different.

Learning outcomes for Section 3

After working through this section you should be able to:

3.1 Apply the concept of energy conservation and use the formulae associated with kinetic energy, gravitational potential energy and thermal energy. (*Question 3.1*)

3.2 Explain the distinction between absorption and scattering. (*Question 3.2*)

3.3 Apply Wien's law and the Stefan–Boltzmann law to calculate respectively the peak wavelength and total emitted energy of radiation. (*Questions 3.1–3.3*)

3.4 Calculate the typical temperature of the Earth by balancing the incoming power of solar radiation with the outgoing energy of terrestrial radiation. (*Question 3.2*)

3.5 Explain the greenhouse effect in relation to why the Earth temperature thus calculated is far less than the surface temperature, and what the terms steady state and positive and negative feedback mean in this context. (*Question 3.3*)

3.6 Define the terms stable and unstable with respect to atmospheric conditions and decide which applies, given the appropriate lapse rates. (*Question 3.2*)

3.7 Explain how, depending on stability conditions, heating of air in the atmosphere can lead to either storage of gravitational potential energy, or convection. (*Question 3.2*)

3.8 Explain the typical temperature structure of the atmosphere in terms of the heating processes involved.

3.9 Explain day and night and seasonal and latitudinal variations, in terms of the Earth's rotation and orbit around the Sun. (*Activity 3.1*)

The global weather machine

4.1 Introduction

Meteorologists identify atmospheric motion on a variety of scales of both space and time, from the extremely extensive global features down to the microscale, such as local gusts. We are concerned here with describing the principal aspects of the atmosphere's global circulation that become apparent when, for example, wind, temperature and humidity are averaged over a month, season or year over the Earth.

It is a fact that the most fundamental basis of our planet's weather is the Sun. Switch it off and the atmosphere and oceans would grind to a halt over weeks and months. It is of course effectively switched off across 50% of the Earth at any one instant at night, but keeps things going during daytime hours by supplying vast amounts of heat to the Earth–atmosphere–ocean system. The Earth does supply heat to the atmosphere, via volcanic activity, but this is an extremely small contribution indeed. The Earth loses heat to space incessantly, so the Sun is clearly critical in keeping our planet hospitably warm, as are the all-important greenhouses gases.

One fundamental drive for such very large-scale patterns is the low to high latitude (low being generally within or nearer to the tropics) difference in the incoming solar radiation that is absorbed by the Earth and atmosphere, and outgoing terrestrial radiation that is lost to space (Figure 4.1).

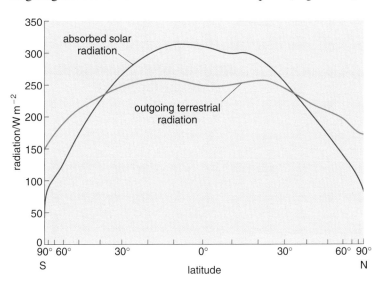

Figure 4.1 The variation with latitude of the solar radiation absorbed by the Earth's surface and atmosphere and outgoing terrestrial radiation. Values are averaged over the year and scaled according to the area of the different latitude bands.

These two quantities vary with latitude and, on an annual average basis, equal each other at just one latitude, around 40°. The annual average picture is of a radiative excess equatorward (towards the Equator), i.e. more solar radiation being absorbed than terrestrial radiation being lost, and a radiative deficit poleward (towards the poles). This means that at higher latitudes the terrestrial radiation lost to space outweighs the absorbed solar radiation over the year.

If the Earth had no atmosphere or ocean then it is likely that its surface in the low-latitude region of excess would become hotter and hotter till the outgoing terrestrial radiation balanced the incoming solar radiation (as discussed in Section 3). Similarly, the surface in the higher-latitude deficit zone would cool down till the outgoing radiation weakened sufficiently to balance the incoming solar radiation. In reality, this radiative equilibrium doesn't occur, because the atmosphere and ocean respond as fluids being driven to flow in such a way that they work to transport the excess heat of the tropics into higher latitudes where the deficit reigns.

The large-scale flow pattern that is related to the Earth's radiation budget means that massive amounts of heat are transported towards the poles, in both hemispheres, tending to offset the radiative imbalance. Temperatures in higher latitudes are elevated by such heat transport and suppressed in tropical latitudes by heat export. There are of course regional variations of temperature, within both tropical and non-tropical regions, because thermal energy is ducted away as heat in particular areas, by frontal depressions and tropical cyclones (discussed later); both act to carry warm, moist air towards higher latitudes.

There are, however, very significant variations indeed between the extreme seasons. The radiative excess across the low-latitude region remains broadly the same all the year round, but such values over the higher-latitude region of radiative deficit vary enormously from the summer to the winter. This latter variation is related to the stark difference between polar day and polar night, with the attendant intense cooling in the winter and much higher temperatures in the summer. The variation through the year of incoming solar radiation is illustrated in Figure 4.2: the changes over and near the poles are clearly significant. The variation through the year of the amount of terrestrial radiation lost to space is much less dramatic.

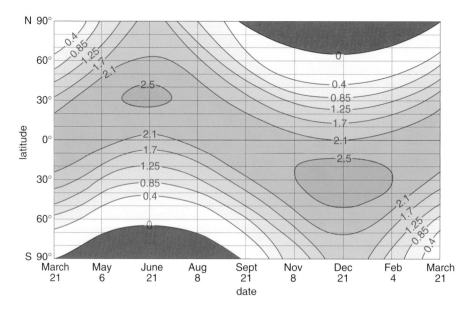

Figure 4.2   Seasonal and latitudinal variation of the daily incoming solar radiation. The numbers on the contours are energy per unit area divided by 10^7 J m^{-2}. The figures account for the absorption of the atmosphere. Note that this is not a map, but a plot of how incoming solar radiation varies with both latitude and time of year.

The upshot of this strong seasonal variation is that the excess/deficit radiation gradient across latitude — and the related temperature gradient at the surface and within the troposphere more generally — are much larger in winter than in summer. In fact, the difference between the hottest and coldest latitudes in winter is about twice that in the summer. This means that the atmosphere and ocean are required to transport more heat away from low latitudes in the winter than the summer. People in the UK are familiar with the fact that winds are stronger in the winter than the summer, which is a local indication of the global influences at work. Frontal depressions are windier in the winter because they are transporting heat more vigorously on their stronger winds, of which more later.

This section describes and explains the major components of the global circulation of the atmosphere and oceans and how they are related. It links these factors to a selection of observed climates around the Earth to give them a more practical definition.

4.2 Global temperature and pressure

4.2.1 Mean surface temperature

Figure 4.3 shows the temperature variation over the Earth's surface. In January (Figure 4.3a) Northern Hemisphere temperatures at high latitudes have fallen to 0 °C or below. The coldest regions are inland regions such as in Canada, Eastern Europe and Siberia. At this time, southern latitude temperatures are at their highest. The highest values are again inland, such as in Australia, and southern parts of Africa and South America. The picture in July (Figure 4.3b) is reversed, as it is winter in the Southern Hemisphere and summer in the Northern Hemisphere. The picture is of course not just a mirror image because there is a lot more 'dark brown' in the Northern Hemisphere.

○ Why is there a lot more dark brown in Figure 4.3b as compared to Figure 4.3a?

● There is more land in the Northern Hemisphere than the Southern. Since land warms faster than sea (Section 3), this means that there are more higher-temperature regions (dark brown) in the northern summer.

○ Referring to Figure 4.3c, where are the biggest temperature changes (either positive or negative) and where are the smallest changes?

● The largest changes correspond to land, especially inland regions. The smallest changes correspond to oceans.

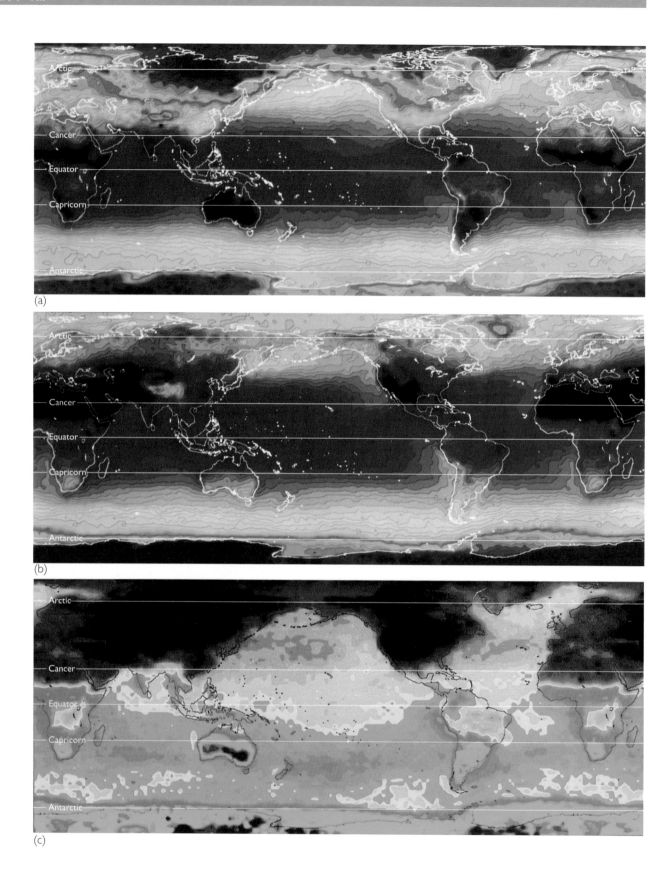

(a)

(b)

(c)

Figure 4.3 Daytime surface temperatures as measured from space. (a) January temperatures; (b) July temperatures; (c) the difference between January and July temperatures. For panels (a) and (b), regions with temperatures below 0 °C are shown green and blue, the highest-temperature regions are shown by red and dark brown and regions of intermediate temperature are shown yellow. In (c), areas of greatest difference (temperature in July being higher) are red and dark brown, areas of smallest difference (temperature in January being higher) are blue and green, while yellow denotes regions of intermediate change. The lines between the different shades of colour are **isotherms**, contours on which temperature has the same value (note the contours are wrapped around the world, so do not always appear to be closed).

4.2.2 Mean sea-level pressure: January mean pattern

Mapping the time-averaged mean sea-level pressure illustrates clearly the scale, location and intensity of the major highs and lows. These are the features that are associated with characteristic weather types. Doing this for the extreme seasons of winter and summer exhibits the changes in these factors that can be related closely to other features, such as the surface wind. The kinds of large-scale features that become apparent are the dramatic seasonal changes over southern Asia and West Africa (i.e. monsoons, discussed later) and the changes in both the location and the intensity of frontal cyclones (depressions) in the mid-latitude ocean areas.

Figure 4.4 illustrates the global pattern of average, mean sea-level pressure and winds during the northern winter/southern summer (January). If such values are averaged, then we see very large-scale features of the atmosphere's circulation. This process of averaging irons out the day-to-day travelling highs and lows to leave us with a picture of the 'climate' of the month. This is not to say of course that these travelling disturbances are not important.

Figure 4.4 The prevailing winds at the Earth's surface in January and the position of the **Inter-Tropical Convergence Zone (ITCZ)** where the Northern and Southern Hemisphere wind systems meet. The regions of high (H, red) and low (L, blue) pressure are also indicated.

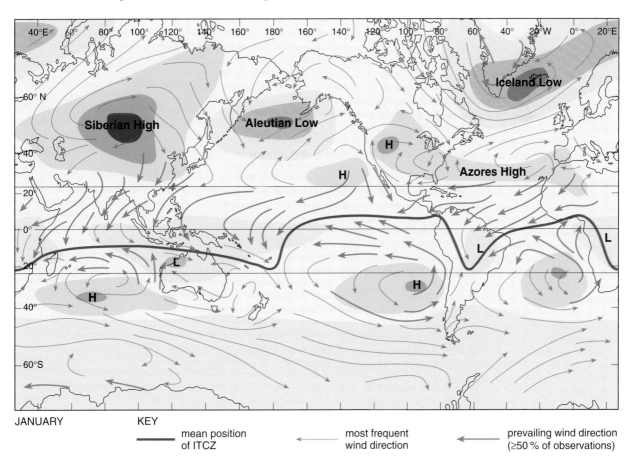

JANIUARY

KEY

——— mean position of ITCZ

←—— most frequent wind direction

←—— prevailing wind direction (≥50 % of observations)

In contrast to what we have learned so far, that air experiences a force in the direction from high to low pressure, we can see in Figure 4.4 that flows seem to form *around* high- and low-pressure regions. This is because of additional forces acting on the airmasses, such as that due to the rotation of the Earth (which we will explain properly in due course). However, note the extensive highs or anticyclones that are located both across the subtropical oceans of both hemispheres and over the mid- to high-latitude wintertime continents, i.e. land regions experiencing wintertime in either hemisphere—in this case, the Northern Hemisphere. These areas of high pressure are actually source regions of air, in that at low levels, i.e. in the lowest kilometre or so of the troposphere, the air spirals slowly out from them (Figure 4.4). This occurs in a clockwise sense in the Northern Hemisphere, and anticlockwise in the Southern Hemisphere.

This notion of highs being source regions is very important; some are sources of cold, dry air—from central Asia for example—while others are sources of warm, moist air, over the subtropical oceans.

At this point it is important to define the terms, as used in meteorology, that identify various parts of the world. These terms roughly correspond to latitude bands, but are somewhat vague because of seasonal variations. **Tropical** refers to the region between the centres of the subtropical anticyclones (high-pressure regions). As we shall see, these regions are influenced by the Trade Winds (see later) and ITCZ (Figure 4.4). The region varies seasonally because the anticyclones migrate, but corresponds roughly to the latitudes between the **Tropic of Capricorn** (23.5° S) and the **Tropic of Cancer** (23.5° N). The **subtropics** refer to the regions either side of the Equator, that extend from the Tropic of Capricorn or Cancer to about 35° S or 35° N, respectively. **Extratropical** refers to all regions that lie outside the tropics. The **mid-latitude** zones are those extratropical regions that lie within approximately 35° and 65° of latitude.

Subtropical anticyclones

A **subtropical anticyclone** is a semi-permanent high, located around 30° S or 30° N of the Equator. In the Northern Hemisphere, these highs are located over the subtropical North Pacific and North Atlantic, where the latter is often known as the Azores High (Figure 4.4). They occur throughout the depth of the troposphere where the air is sinking gently from upper to lower levels. This **subsidence** is necessarily linked to, and indeed supplies, the air spiralling out in the lowest layer. It is associated generally with dry weather at the surface and low relative humidity in great depth through the troposphere. Perhaps not too surprisingly, the atmosphere in the lowest kilometre or so over the lower-latitude oceans is rich in moisture and tends to often have extensive but quite shallow stratiform cloud (Figure 4.5). This means that although anticyclones are linked to dry weather, it can quite often be cloudy.

These deep anticyclones are semi-permanent features of the global circulation; on any weather map they are there, although in the northern winter they are generally shifted somewhat towards lower latitudes than in the summer. The deep sinking motion is associated with adiabatic compression and warming, so this type of anticyclone is called a warm high, simply because at any tropospheric level the air is warmer in the high than in the surrounding region. They are also termed 'dynamic', because they are permanently linked to deep subsiding motion.

Figure 4.5 Extensive stratiform cloud under the South Atlantic subtropical anticyclone.

As the dry air sinks within such highs, it warms adiabatically (because the air is not saturated), which means that its relative humidity decreases simultaneously. The dry adiabatic process means that because the unsaturated air is being compressed as it sinks into continuously higher pressure, it warms at the quite rapid rate of 9.8 °C km^{-1} (the DALR). If we assume that there is no more water vapour evaporated into the air (a very likely situation, given the circumstances) and that none is lost from it via condensation (also very likely), then as the drybulb temperature increases rapidly, it means that the relative humidity must fall dramatically. This is because, although the absolute humidity stays more-or-less the same, the saturation value increases significantly as the drybulb temperature goes up.

A typical vertical profile up through such a high over the subtropical ocean reveals a moist low-level layer over the sea surface within which the temperature falls off with height. This is capped by a thermal inversion layer through which the drybulb temperature increases very rapidly and the dewpoint temperature decreases very rapidly. Such a temperature profile is known as a **dry or subsidence inversion**. (Figure 4.6).

This discussion holds too for the subtropical oceanic highs in the Southern Hemisphere: those over the South Indian, South Atlantic and Pacific Oceans.

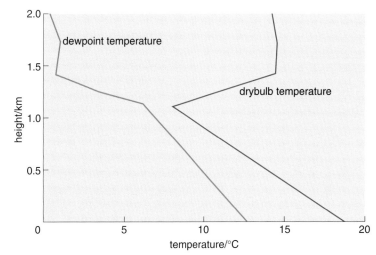

Figure 4.6 A dry inversion, also known as a subsidence inversion.

Continental anticyclones

The higher-latitude wintertime continental interiors are characterized by strong radiative cooling to space during prolonged hours of darkness. The efficacy of this outgoing terrestrial radiation is enhanced by the fact that the air across these extensive tracts of North America and Asia is far from any moisture source and is thus generally very dry indeed.

The intense cooling, which starts in the autumn, leads to frigid air in the lowest layer of the troposphere, to increased air density (the reverse of the situation in Section 3.3.1) and to the genesis of a shallow anticyclone. These are cold highs; they are very important features of the global wintertime circulation in the lowest two kilometres or so of the troposphere but are so cold that the vertical rate of change of pressure up through them is relatively rapid. This means that by a few kilometres up, they are no longer present (Figure 4.7). Even though they are shallow, they are still linked to sinking motion down through a few kilometres, and are sources of intensely cold air at the surface. Unlike the warm highs, however, they are seasonal. Because they are so dry and cold, there is very little if any cloud associated with them. This combination of prolonged clear skies, dry air in depth and relatively short daylight hours means that extremely low surface temperatures can often be experienced within these areas. This is particularly true of the Siberian High, because it is easily the most extensive and most divorced from sources of warmer, moister air (Figure 4.4).

Figure 4.7   A vertical cross-section through a cold anticyclone. (750 hPa corresponds to a height of roughly 2.3 km.)

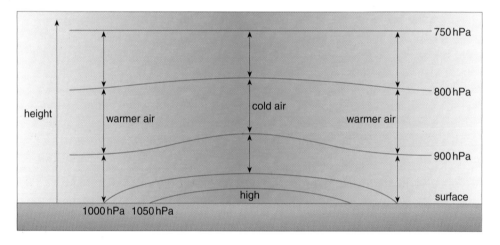

Mid-latitude cyclones

The air that flows clockwise out of the 'source regions' represented by highs has to go somewhere. It ends up flowing anticlockwise (in the Northern Hemisphere) and clockwise (in the Southern Hemisphere) into low-pressure areas (cyclones) which are thus in a sense 'sink' regions across which the air ascends through the troposphere. There are two such low or cyclonic centres in the northern winter.

The Iceland Low is located very close to that island on average (Figure 4.4). This doesn't mean that there is a low-pressure system there every day of the winter, but clearly low-pressure systems must be seen very frequently in the region. What it does indicate is that, on average, the travelling lows of the North Atlantic reach their minimum sea-level pressure (or are deepest) at that location.

In addition, the elongated pattern of the lower pressure that stretches northeastwards from the Icelandic minimum towards the Barents Sea, northeast of northern Norway, indicates that there is a very important depression track there.

Such travelling low-pressure disturbances, often in the form of **frontal depressions** (Box 4.1), generally track from southwest to northeast, transporting a lot of warm, moist air right up into the Arctic Basin. Such systems do of course follow other tracks in the northern wintertime — witness the bad storms we see in the British Isles occasionally in the autumn and winter. The process of averaging many Januaries simply smooths out such features, unless we look at a particularly stormy individual January for Britain and other parts of western Europe. In such a case, we would see the 'signature' of low tracks in our region. Figure 4.4 is a summary of many Januaries.

Box 4.1 What are frontal depressions?

Firstly, we should be clear about some terminology. As you know, the word 'depression' is synonymous with 'low' and with 'cyclone'. Frontal depressions are almost always travelling weather systems that move across mid-latitude ocean basins and elsewhere. They typically last from 3–5 days and track quite a few thousand kilometres during this period, from the North American coast right across to the British Isles, for example. There are 'non-frontal' depressions, some of which will be discussed in detail later in this section, e.g. tropical cyclones. The term 'frontal' refers to the fact that such systems are characterized by boundaries that separate air of different weather properties. For example, a **warm front** is the leading edge of tropical maritime air that has streamed into the mid-latitude North Atlantic from the Azores High (Figure 4.8a), most often as a southwesterly current that moves north-eastwards. **Tropical maritime** air is warm, moist air that originates in the subtropical anticyclones over the oceans. The principal source regions are the North Atlantic and the North Pacific in the Northern Hemisphere, and the South Atlantic, South Pacific and south Indian Ocean in the Southern Hemisphere. A **cold front** is the leading edge of polar maritime air (Figure 4.8b) that has swept across the North Atlantic, for example, ultimately from a cold anticyclone across North America. Not only are there temperature contrasts across such fronts, but they are the leading edges of moist and dry air respectively. **Polar maritime** air is a cool, moist airmass, which in the Northern Hemisphere, often originates as polar continental air over the wintertime North American or higher Asian landmasses. As this cold dry air flows over the adjacent Atlantic and Pacific Oceans, it is warmed and moistened from below and destabilized over the North American or Asian landmasses and is cooled as it moves out over the Atlantic and Pacific Oceans. In the Southern Hemisphere, there is a continuous zone of polar maritime air between approximately 40° and 60° S.

The two airstreams, the tropical and the polar, are often found to be juxtaposed across the mid-latitude North Atlantic and Pacific, and the corresponding latitudes of the southern ocean basins. Figure 4.9 illustrates a typical situation whereby these two streams flow in opposite directions on either side of a line. We term the separating line a 'front': in fact, in this situation over the mid-latitude oceans it is called the 'polar' front, because along part of its length it demarcates the forward edge of the cold, polar air that flows towards lower latitudes. The warm and moist air that runs alongside the colder and drier air is less dense than the latter, so rides up and over it, across a very shallow sloping 'frontal' surface. This 'surface' is in fact three-dimensional: it has a depth of about 1 km, so it is really like a thin sheet.

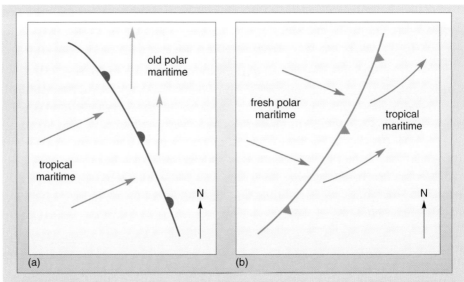

Figure 4.8 (a) A warm front, (b) a cold front and the associated surface airmasses.

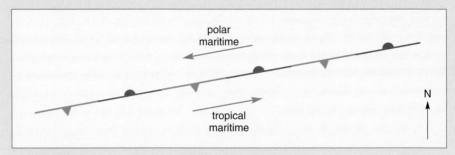

Figure 4.9 Opposite airflows on either side of a front.

The colder, drier air on the polar side of the front therefore undercuts the overriding tropical maritime air; the separating front slopes typically at between 1 in 60 and 1 in 120. It really is a shallowly sloping zone.

The temperature and humidity contrast exists through the depth of the troposphere, not just at the surface. The existence of a persistent thermal and moisture contrast in such depth is associated with a wind that increases with height until it climaxes in the form of the 'polar front jet stream' (Box 4.3) in the upper troposphere. This feature is part and parcel of a frontal system; the two phenomena are inextricably linked.

The first sign of a frontal depression's development is the appearance of an open wave (Figure 4.10a and b) with distinct warm and cold fronts, and a strong cyclonic circulation (anticlockwise in the Northern Hemisphere).

As time progresses, the frontal depression, low or cyclone travels most often northeastwards (southeastwards in the Southern Hemisphere) as a 'wave' that runs along the front — rather like the wave you can make run along a line of thin rope by a deft flick of the wrist. The cold front moves faster than the warm one and therefore gradually catches it up (Figure 4.10c). This occurs first of course where the two fronts join. It is here that the warm, tropical maritime air is first lifted off the surface, so that only cool or cold air remains. This process of lifting the warm, moist air up and away from the surface is that of occlusion

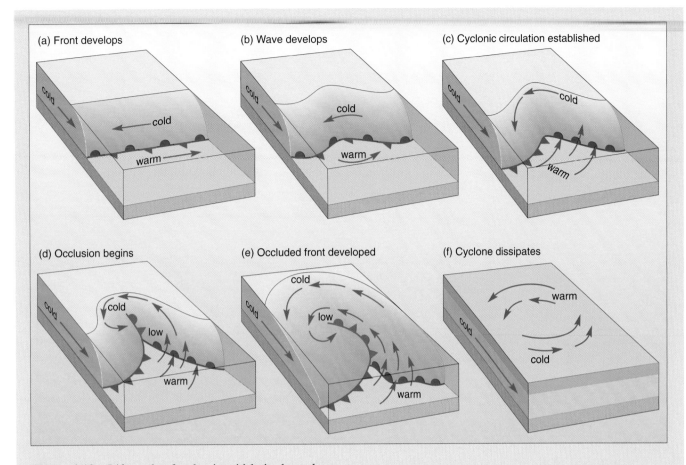

Figure 4.10 Life cycle of a classic mid-latitude cyclone.

(from the Latin *occludere*: to shut up) (Figure 4.10d). It continues through the lifetime of the frontal system so that the warm sector, i.e. the extensive region between the warm and cold fronts, gradually shrinks until the system is fully occluded (Figure 4.10e). Figure 4.10f illustrates the situation just after the cyclone has dissipated.

Because the majority of Atlantic frontal cyclones form off the eastern coast of North America, they are well occluded by the time they reach the shores of Europe, simply because it takes them a few days to reach our region. This isn't however always the case. Some systems develop not far from northwest Europe and run across the British Isles, for example, as young open waves (those that are not occluded at all).

The North Atlantic **storm track** (track of depression centres) for an individual month (Figure 4.11) illustrates why the far northern reaches of the northeast Atlantic are ice-free all year-round. Not only does the warm, moist air penetrate far north there, but the ocean circulation is such that large amounts of heat-transporting warm water get into the same region through the wide expanse of the ocean between Norway and Iceland.

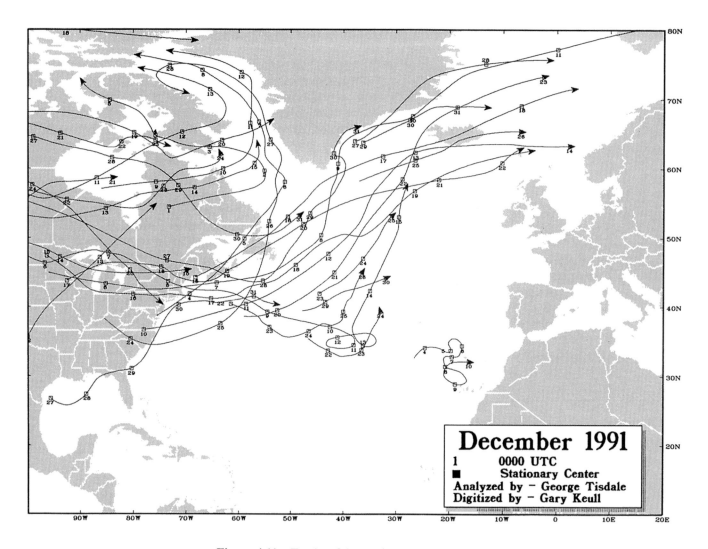

Figure 4.11 Tracks of depression centres at sea-level over the North Atlantic in December 1991. The small box on each track indicates a centre's location at 00.00 UTC on the day of the month given by the number next to it. The arrowhead at the end of the track emphasizes the direction of the low's movement.

There is a second wintertime pressure minimum in the North Pacific (Figure 4.4), known as the Aleutian Low after the long chain of Alaskan Islands it is near. There is also a storm track, but it is less extensive than its North Atlantic counterpart because there is less 'opportunity' for the travelling depressions to move into the Arctic Basin via the very narrow Bering Strait between Alaska and Russia. In addition, in contrast to northwest Europe, the Rockies are a marked barrier to the northeastward penetration of lows into the continental interior.

The southwest to northeast orientation of both these storm tracks is related in part to the orientation of the large temperature contrasts between the east coasts of North America/North-West Atlantic and of Russia/North-West Pacific. The mean isotherms (lines of constant drybulb temperature) stretch out parallel to both coasts which are themselves aligned southwest to northeast (Figure 4.3). The travelling lows develop along this temperature contrast and tend to run along it northeastwards.

One very important difference for the atmospheric and oceanic circulation in the Southern Hemisphere — in middle and higher latitudes — is that there is no equivalently orientated thermal contrast between chilled continent and warmer ocean. There is, instead, the contrast between the circumpolar continent of Antarctica and the surrounding Southern Ocean. The temperature contours are thus mainly parallel to lines of latitude. This means that the travelling lows track in general rapidly from west to east around the flank of the Antarctic, producing a very elongated west-to-east pressure minimum with no sign of any distinct minima — in contrast to the Northern Hemisphere. This also means that these cyclones, although vigorous (in the Roaring Forties — see later), do not transport much heat into the highest of latitudes. The same is true for the circumpolar ocean currents.

The Equatorial Trough

Moving towards the Equator from the subtropical highs leads to an elongated pressure minimum that in January lies mainly in the Southern (summer) Hemisphere (Figure 4.4). This broad and rather shallow (not very deep or low values) region of low pressure is termed the **Equatorial Trough** — not that it lies on the Equator in many places at all, but stretches around the tropics in generally low latitudes. It dips furthest south into the summer hemisphere across the heated continents of South America and Southern Africa.

The Equatorial Trough is not easily noted as a distinct minimum on Figure 4.4, but is traced out by the ITCZ. It is shown as the continuous line around low latitudes, and is where the Trade Winds flow smoothly together, into a pressure minimum. The **Trade Winds** are northeasterly in the Northern Hemisphere and southeasterly in the Southern Hemisphere, flowing from subtropical anticyclones (latitude ranges 30–40° N and 30–40° S, respectively) to the ITCZ.

4.2.3 Mean sea-level pressure: July mean pattern

Subtropical anticyclones

The subtropical anticyclones are still present across the oceans and are more intense in the northern summer than its winter (Figure 4.12). This feature is associated with more vigorous circulation within these warm highs, with stronger sinking motion and higher surface pressure. Note that these anticyclones shift towards the poles in the summer and towards the Equator in the winter. In the Southern (winter) Hemisphere, they have strengthened somewhat, with higher central pressures, and have shifted nearer to the Equator.

Continental anticyclones

In contrast to the continents of the Northern Hemisphere, those in the Southern Hemisphere narrow towards higher latitudes. This means that although the higher latitude tracts of South America, Southern Africa and Australia do cool in the winter (July for example), they do not possess their own distinct cold highs. Rather, there are mean ridges over them that connect with the oceanic anticyclones into a continuous belt of high pressure round the world (Figure 4.12).

Although low temperatures occur over these continents in their winter, they are nowhere near as low as over the much more extensive northern continents in

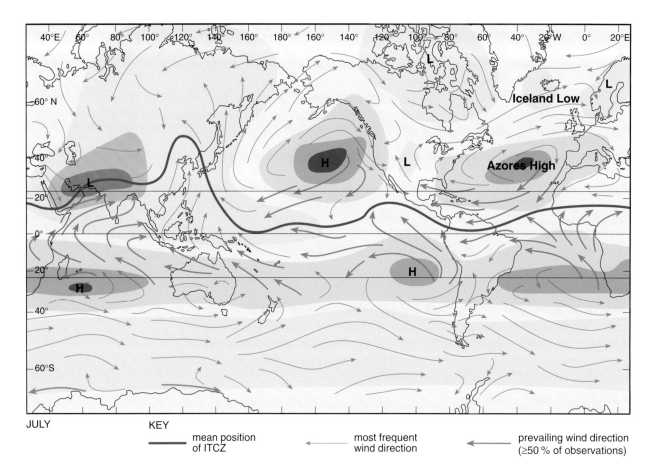

JULY   KEY

	mean position of ITCZ		most frequent wind direction		prevailing wind direction (≥50 % of observations)

Figure 4.12   The prevailing winds at the Earth's surface in July and the position of the ITCZ where the Northern and Southern Hemisphere wind systems meet. The regions of high (H, red) and low (L, blue) pressure are also indicated.

their wintertime, with the exception of the very high-latitude Antarctic. The Southern Hemisphere is so much more oceanic than the Northern that it is does not experience the large annual range of temperature that the latter experiences.

The annual range of temperature (indicated by Figure 4.3c) is simply the mean temperature of the warmest month minus that of the coldest month. The oceans have a small range because of the large thermal capacity of water — they absorb and release thermal energy slowly compared to the less heat-conserving continental surfaces. This effect is crucial in understanding the workings of the climate.

Mid-latitude cyclones

Over the summertime North Atlantic there is still a sign of the Iceland Low (Figure 4.12) although it is clearly shallower than in the winter and its associated storm track is less obvious. There are nevertheless cyclonic disturbances that run across this region but with less vigour than in the winter. There is little sign of the Aleutian Low, while the weather pattern of parts of the summertime continents has 'flipped' from the extensive cold anticyclones to the presence of broad low-pressure features across Scandinavia and the Canadian Archipelago, for example.

In contrast, the Southern (winter) Hemisphere circumpolar ocean is characterized by an extensive belt of low pressure. Like the summertime pattern, this reflects the persistent west-to-east tracking of depressions around the flank of the Antarctic continent. In the winter, frigidly cold air flows off the extensive continental glacier of Antarctica, enhancing the thermal contrasts around its flank.

○ What is the reason for the dramatic change over, for example, the USA from low to high pressure between summer and winter? What phenomenon (discussed in Section 3) is this like, but on a much larger scale?

● Like other mid-latitude continents, the land surface heats up rapidly during the spring and into the summer. This heating leads to a fall in pressure and the presence of an extensive region of low pressure. In contrast, the autumn and winter are times when the land surface chills quickly, leading to the development of high pressure. The situation is like a kind of 'mega' sea or land breeze.

The Equatorial Trough

In the northern summer this feature lies entirely within the Northern Hemisphere (Figure 4.12) in response to the strongly heated, extensive continents and warmer tropical sea surface. Its northernmost excursion is over southern Asia where the most significant feature is the low that stretches from Saudi Arabia to northeast India. Notice how this marked depression is in complete contrast to the pattern there in the northern winter (Figure 4.4).

4.3 Surface wind

4.3.1 What influences the wind direction?

We have seen that the basic driving force of the wind is the horizontal pressure gradient. The steeper this gradient, the greater the force on the flow down the gradient from high towards low pressure.

If we lived on an Earth that didn't spin, then the wind would blow directly from high towards low pressure at right angles to the isobars. The simplest kind of large-scale circulation pattern would be one of air rising along a strongly heated Equator, flowing towards the poles in the upper troposphere, sinking to the surface in high latitudes, and returning to low latitudes across the surface (Figure 4.13). The astronomer Edmond Halley published a truly pioneering map of the Trade Winds over the tropical oceans in 1686, suggesting that there could well be ascent where the Trades converge in low latitudes. This idea was developed by George Hadley in 1735, who proposed that there should be an upper returning current flowing towards the poles in either hemisphere. This flow would then sink to the Earth's surface at some latitude and flow back towards the Equator as the Trades. The complete cycle of motion in the north-to-south vertical plane is now known as the Hadley circulation, formed of two **Hadley cells**.

The same link would be seen on such a weather chart for any day too. However, the west-to-east spin of the Earth introduces an important phenomenon that deflects the air flow. This is termed the **Coriolis effect** (see Box 4.2). This effect, together with the influence of friction at and near the Earth's surface, means that the winds blow not exactly parallel to, but more-or-less parallel to, the isobars. In fact, they cross the isobars at a relatively small angle that depends on the frictional force applied to the wind, which depends on the roughness of the surface. This is a small angle over the sea (around 10–15°) and a larger one over land (around 20–30°). Although the sea is rough in the sense that it gets choppy in strong

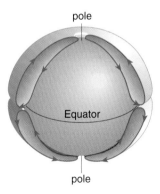

Figure 4.13 Hypothetical wind system on a non-rotating Earth.

winds, it is in fact much smoother than the land surface with, for example, its hills and valleys, its changes in vegetation and its extensive built-up areas. The cross-isobar flow is such that the air flows into lows and out of highs.

Box 4.2 A diversionary deflection—the Coriolis effect

An important fact, which must be taken into account when we wish to understand why the wind blows from a certain direction, is that we are moving round on a rotating Earth. The latitude/longitude grid is of course 'fixed' to it too and we must remember that it also moves around with the Earth. This fact means that when we view air moving across the Earth's surface (or at any level), we must take into account the fact that we are on a spinning planet.

So what does this mean exactly?

Well, imagine that you are a scientist working at the North Pole and that you are returning samples in a rocket intended to land at a target site in the sea to the south of London, UK. This means of course that your job should be easy: you just aim due south along the appropriate line of longitude, say the Greenwich Meridian. Let's suppose that you launch the rocket and that you have calculated it will take two hours to reach its target. After the launch, you look at the automatic tracking radar's output and are surprised to see that it is a little to the west of the Greenwich Meridian.

After an hour your concern is heightened by the fact it is no longer over the northern Norwegian Sea, half way to its destination, but quite a bit to the west. You think about this a little then suddenly realize that during that hour the Earth has spun about its axis through 15° of longitude (that is, one twenty-fourth, or one hour's worth, of its full circle). You then draw a diagram to explain to yourself what has happened (Figure 4.14).

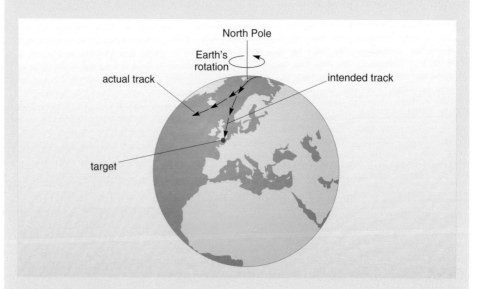

Figure 4.14 The Coriolis effect is demonstrated by a hypothetical rocket fired south from the North Pole. It does not simply follow a line of longitude, due to the effect of the Earth's rotation.

This influence on the track of projectiles was formalised by the French mathematician Gaspard-Gustave de Coriolis (1792–1843). It not only influences missiles of course, but any particle that moves freely with respect to the Earth. This is where the meteorologist and winds come in.

If it were a wind blowing from the North Pole as a northerly (i.e. due southwards), to a weather observer under its influence, but some distance from the Pole, it would be seen as a northeasterly. It would have been deflected to the right of the track that it would have taken on a non-rotating Earth, just like the rocket in Figure 4.14.

We can imagine that there is a Coriolis force which, it turns out, increases with latitude and is proportional to the speed of the object experiencing it. So wind, i.e. moving air, apparently experiences several forces. One is due to the pressure gradient. Another is the Coriolis force, which is always directed to the right of the wind (at right angles) in the Northern Hemisphere and to the left in the Southern, no matter what the wind direction. The end result is that, when all the forces are balanced, the wind tends to blow parallel to the isobars rather than across them.

4.3.2 Surface winds: January mean pattern

Figure 4.4 illustrates the relationship between the monthly patterns of mean sea-level pressure and those of the mean surface wind. It is immediately apparent that highs or anticyclones really are 'source' regions of different **airmasses** at the surface and in the lower troposphere. The air spirals out to influence the weather in more distant regions than just those directly underneath them.

Subtropical anticyclones

The North Atlantic Azores High supplies air in all directions, but most significantly, in terms of linking up to low pressure/cyclonic regions, towards the north and the south.

On its poleward flank, the warm, moist air blows mainly northeastwards towards northwest Europe as southwesterly winds. These are by far the most common mild winds we see in wintertime in the British Isles, when we may see temperature maxima up to typically 12–13 °C in many places. The air from this source is termed tropical maritime (introduced in Box 4.1) as an expression of its lower-latitude oceanic origin; it is often cloud-laden with extensive stratus and stratocumulus (Figure 4.15).

On the equatorward side of the high, the air flows vigorously towards lower latitudes as the North-East Trade Winds. These warm and very humid currents blow towards and 'feed' the ITCZ with huge amounts of very warm and very humid air. This warming and moistening come from sensible (due to contact) and latent (due to evaporation) heating as the air flows across increasingly warm tropical ocean towards the lowest of latitudes.

Figure 4.15   Tropical maritime airstream with stratus moving northeastwards over the Atlantic west of the Iberian Peninsula (Spain and Portugal). The extensive layer cloud is in the depression's warm sector.

The same situation, broadly, occurs on either flank of the North Pacific anticyclone. The tropical maritime air flows north and northeastward to feed the depressions of the mid-latitude North Pacific while the North-East Trades blow towards and into the Pacific's ITCZ.

Continental anticyclones

These sources of very cold and dry air are obviously extremely influential in determining the winter weather and climate of regions within their sphere of influence. The Siberian High supplies such air very widely indeed—to the south, winds blow out across the tropics and are reduced in strength as they go. The northeasterlies that characterize the winter monsoon over central southern

Asia have their root in this anticyclone. They flow over and around the Himalayan Massif; to the east they are responsible for the cooler conditions that spread south across tropical China and southeast Asia as occasional, notable outbreaks. A **monsoon** is a wind that persists in a particular direction during a season, but changes direction, often reversing, at a change of season. The explanation of a monsoon lies in the differential heating of land and oceans. The colloquial meaning arises from the rain that occurs in Asia during the reversal of the wind from a northeasterly to a southwesterly.

The Siberian High is also linked to frigid, dry weather in western Russia and eastern Europe on occasion. It is not true that really cold conditions with easterlies in a British winter are due to air from Siberia. That region is to the east of the Urals, which are located more or less along 60° E, far too far to the east for the air to come whipping across many thousands of kilometres. Less dramatically perhaps, the really cold air in easterlies comes from highs over Scandinavia or possibly northwest Russia. Such days can nevertheless be very cold in Britain, with heavy snow showers along the east coast (Figure 4.16).

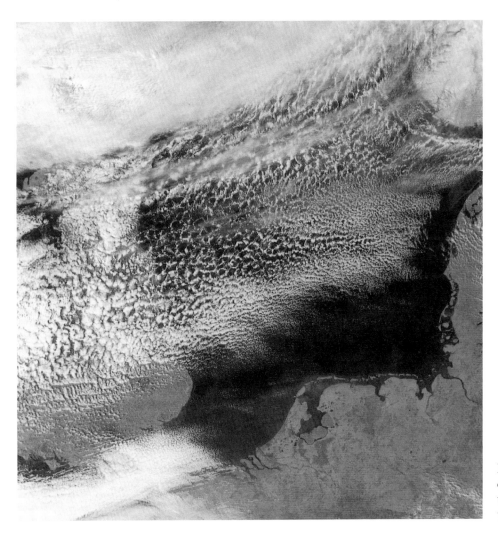

Figure 4.16 Lines of cumulus clouds and snow showers across the North Sea within the wintertime easterly flow.

Air also circulates out of the Siberian High across the Arctic Ocean to the north and it spills across the relatively warm northwest Pacific Ocean where it is intensely warmed and moistened. In this region, northwesterly winds blow commonly across the Asian coast from Kamchatka in the north to Japan in the south. These cold, dry winter winds cross the coast and suddenly encounter a water surface that is much warmer, on their way to flowing into the Aleutian Low (in the North Pacific).

The North Atlantic has a similar region that lies off the east coast of North America, roughly from the St. Lawrence River in the north to the mid-Atlantic states in the south.

Indeed, the air that flows off this region as commonplace northwesterlies in the winter sometimes reaches the British Isles as cool and moist weather with many showers. Just like the Pacific off northeast Asia, the Atlantic 'downstream' of these outbreaks of cold air from the wintertime continent of North America pumps vast amounts of heat into the chilly air. This modification of what starts out as a very cold, very dry polar continental airmass is enormously significant for European weather. Figure 4.17 illustrates the way in which air from this North American (or occasionally Greenland) source flows on average towards northwest Europe; it arrives at the UK as polar maritime air that is cool, relatively moist and has a tendency to be populated by convective clouds (cumuliform) with showery weather.

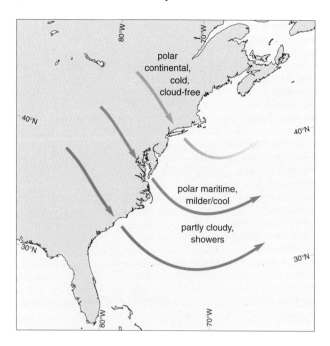

Figure 4.17 Modification of polar continental air into polar maritime air across the western North Atlantic.

It is possible to see this kind of airmass on most wintertime satellite images of the North Atlantic (Figure 4.18) — the air is rich in scattered convective cloud that streams most often with components towards the south and towards Europe.

Figure 4.18 The polar maritime air is present over the Atlantic, east of the eastern seaboard of the USA. Its presence is made visible by the lines of cumulus clouds that have formed within it as the intensely cold polar continental air flowing off the USA is heated strongly from below by the warm waters of the western North Atlantic.

Figure 4.4 shows that Britain lies in a region where there is a mixture of polar and tropical maritime air; indeed a good deal of western Europe experiences this classically variable weather in the winter. There are fluctuations of cloud type, temperature, humidity and changes from prolonged rain to showers as part of the natural variability of conditions in this part of the world. The same is true for those who live in Alaska, British Columbia and, to some extent, the Pacific northwest of the USA. Their winter weather can be a real mixture, and for the same reasons.

These kinds of weather contrasts that are related to the frequent change in airmass type over some mid-latitude regions must also occur round the circumpolar Southern Ocean. This is certainly true in the wide belt that is known traditionally as the Roaring Forties, where it is so windy that the strength of the flow often reaches or exceeds gale force.

Mid-latitude cyclones

These appear of course as minima in the monthly average, mean sea-level pressure maps and in the wind patterns where the low-level flow circulates anticlockwise into the low centres. This is apparent in both the Iceland and Aleutian Lows (Figure 4.4), but less so for the Southern Ocean simply because the systems there are much more mobile, and so are less distinct on the mean pattern.

Inter-Tropical Convergence Zone (ITCZ)

This 'zone' stretches around low latitudes as a more-or-less continuous line of deep cumulus cloud. Its average January location (Figure 4.4) highlights the fact that it migrates furthest south across the heated southern continents; it is associated with the simultaneous migration of the Equatorial Trough.

The 'convergence' of the ITCZ expresses the fact that the North-East and South-East Trade Winds flow strongly towards each other along a line or zone (ITCZ). The observations show that, as they do, the speed within the currents falls markedly along the line of the flow. Over the equatorial Atlantic for example, the strength of both Trade currents can decline from say 15 knots down to 5 knots as the air streams towards the line of the ITCZ. This is significant, meaning that the air must ascend in association with this horizontal deceleration (Figure 4.19).

Figure 4.19 The deceleration of the low-level winds that flow into the ITCZ. The slowing down of the air is associated with it 'piling up' and with its ascent.

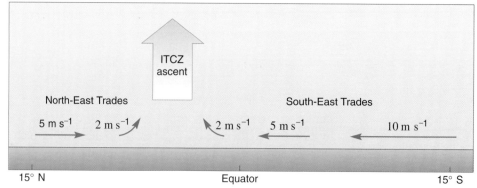

This zone is a very significant component of the global atmospheric circulation, since the low-level convergent airstreams are heavily moisture-laden, having evaporated vast quantities of water vapour from the warm tropical ocean across which they flow.

The deep ascent (updrafts that penetrate to great heights) that characterizes the ITCZ is thus strongly linked to very heavy precipitation along most of its length. The 'most' signifies that in some places across the low-latitude oceans, the sea surface temperature in the vicinity of the ITCZ is too cool to initiate deep convective cloud. The warmth of the sea or land surface is critical in determining whether deep convection will occur. Across the eastern equatorial Pacific, for example, the surface is anomalously cool for its latitude, so it rarely sees convective cloud.

The location of the ITCZ in January is related to the fact that regions affected by it experience one very significant major precipitation season in the year, in and around January. This is true of countries across parts of southern Africa such as Zimbabwe and Zambia. Such rain is life-giving for those areas.

4.3.3 Surface winds: July mean pattern

Subtropical anticyclones

The air that blows out towards the poles from the subtropical anticyclones in the summertime North Atlantic and North Pacific spreads its influence further north than in the winter (Figure 4.12). The highs themselves migrate polewards in the summer and lead to generally more 'settled' weather. The impact of this shift on the seasonal variation of weather is very obvious in the Mediterranean region, for example. The long, sunny and mainly dry summers are associated with the shift of the subtropical high over this area. In addition, good summers in Britain and adjacent parts of northwest Europe are normally associated with the slightly anomalous northward spread of this anticyclone, with its attendant sunny skies and warm winds.

Note that the Trade Winds, which blow equatorward in both hemispheres, flow strongly towards the ITCZ (as they do in the northern winter).

Continental anticyclones

Continental anticyclones do not exist in the wintertime Southern Hemisphere because of, in essence, the very much smaller extent of mid- to high-latitude continents. The winds over part of Australia indicate the presence of a small region of anticyclonic circulation (Figure 4.12), but in the main, the continents do not play a significant role in shaping the cold season's circulation.

Mid-latitude cyclones

In the Northern Hemisphere, mid-latitude cyclones are clearly very much less influential in the summertime wind pattern than in the winter one. There is an indication of cyclonic flow in the vicinity of a much shallower Iceland Low but no semblance of such circulation across southwest Alaska. The Iceland region is characterized essentially by easterlies to the north and westerlies to the south, which is an expression of the cyclonic circulation (anticlockwise) around the low.

As in its summer, the Southern Ocean experiences marked westerly flow right around the belt that is some 20–30° of latitude wide and to the north of the Antarctic. The frontal depressions track generally eastwards to produce the elongated zone of Roaring Forties; their 'mobility' all the way round this belt means that we see no distinct pressure minimum — unlike the Northern Hemisphere pattern.

Box 4.3   What are jet streams?

The existence of rapidly-moving air high in the atmosphere had been appreciated a long time before the era of flight. This was based on observations of how rapidly high cirriform (cirrus-type, Section 2.2.5) cloud could move across the sky. The first time that such strong winds were encountered directly was when *Dauntless Dotty* led 110 B29 bombers on the first high-level raid on Japan on 24 November 1944. The aircraft flew westwards across the North Pacific into the teeth of a strong westerly jet stream that proved to be a massively strong headwind, making navigation extremely difficult and the bombadiers' work virtually impossible.

Jet streams are elongated, narrow bands of rapidly flowing air, which occur under particular circumstances in the troposphere and lower stratosphere. There is one, known as the Polar Front Jet, that is part and parcel of a frontal cyclone in mid-latitudes.

We know that fronts are shallowly sloping zones that occur through the depth of the troposphere, separating airmasses with different thermal and humidity characteristics. We are also aware that barometric pressure decreases upwards and that the rate at which this happens is larger in cold air than in warm air (see Section 3.3.1).

Let's now consider a real situation whereby we have a south-to-north vertical cross-section from the subtropics to higher latitudes in which the sea-level pressure varies gently from 1021 hPa at 30° N to 1014 hPa at 75° N (Figure 4.20). This means that there is a relatively light wind from the west at sea-level, because there is a weak horizontal pressure gradient (a difference of 7 hPa across the 45° of latitude).

On a non-rotating Earth, the steepening pressure gradient would be associated with a strengthening wind from subtropics to higher latitudes. The spin of the planet does, however, mean that the flow is deflected to run parallel to the isobars, i.e. into the cross-section as a due west wind. The rule, expressed by the Dutch scientist Buys Ballot, states that if you stand with your back to the wind (in the Northern Hemisphere at a properly exposed site) then low pressure will lie to your left and high pressure to your right. Conversely, the lower pressure to the north here means that the wind must blow into the section, from the west. For clarity, we continue our jet stream discussion assuming a non-rotating Earth.

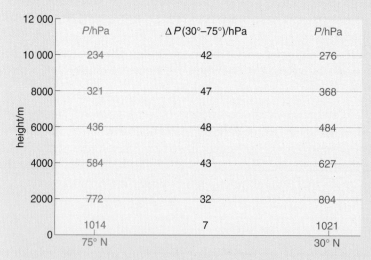

Figure 4.20   The decrease of pressure towards the pole and the associated steepening of horizontal pressure gradient with height.

We know that the troposphere is warmer in the tropics than in higher latitudes, so that the rate at which pressure decreases up through the warm air is slower than in the cold air. Let's assume therefore that two weather balloons are released simultaneously from the subtropical and the higher-latitude location and that we are sent data about what the pressure is at 2000 m above mean sea-level.

The real data tell us that above the lower-latitude site the pressure is 804 hPa, while above the higher-latitude one it has fallen to 772 hPa. This means of course that we now have a horizontal pressure gradient of 32 hPa over the 45° of latitude, which has developed simply because of the horizontal thermal contrasts in the troposphere. The cross-section highlights the gradual increase of the horizontal pressure gradient into the upper troposphere, where it is some seven times steeper than at mean sea-level (Figure 4.20).

So, we can see how a temperature gradient, with colder air towards the pole, can force a wind to blow. Fronts therefore are zones across which there is a large temperature (and humidity) contrast in depth. This means that, using the argument above, there must be an increase of wind speed with height in a frontal zone, and that if the front is aligned west-to-east (the temperature gradient south-to-north), then the wind will increase in strength from the west. Its speed doesn't go on increasing; it reaches a peak in the upper troposphere, above which the different thermal gradients in the lower stratosphere lead to a decrease in strength.

This peak in the wind speed is the jet core, i.e. the maximum within the polar front jet stream. Speeds in the jet core are higher in winter than in summer, basically because the frontal thermal contrasts are so much more marked in the winter.

The polar front jet stream meanders a lot (Figure 4.21), in unison with the varied positions of its associated frontal depressions. In contrast, the subtropical jet stream is geographically more 'fixed'; it is associated with thermal contrasts in the upper troposphere near the poleward limit of the Hadley cells, which we will describe shortly.

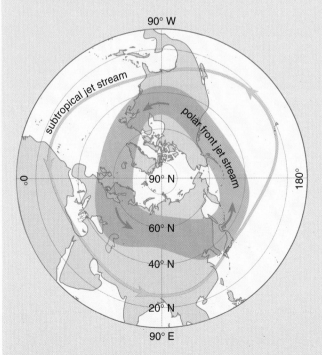

Figure 4.21 The polar front and subtropical jet streams. The figure shows the belt of winter activity of the northern polar jet stream (blue tone), and the mean winter position of the axis of the northern subtropical jet stream (yellow tone).

The Inter-Tropical Convergence Zone

The month of July is within the period during which the ITCZ has its northernmost excursion; during the height of the northern summer, it is found entirely within that hemisphere (Figure 4.12).

As in the southern summer, the ITCZ brings critically important rainfall to most regions that it lies across. There are some regions, such as the eastern Pacific, where the sea surface temperatures are often below the crucial value to set off convective cloud and showers. Elsewhere, however, its presence is life-giving.

Perhaps the most obvious feature is the marked low across the Arabian Peninsula and northern India. This is related to a cyclonic wind pattern with easterlies to the north and very moisture-laden southwesterlies to the south. These flow strongly across the Arabian Sea and Bay of Bengal to supply the summer monsoon. Note the contrast in the surface wind direction between this southwesterly and flow from precisely the opposite direction during the northern cool season (Figure 4.4), during which cooler, dry, cloud-free air circulates across from the northeast as the prime component of the winter monsoon. Recall that the term 'monsoon' simply means a reversal of the surface wind from summer to winter. There are other 'monsoons', such as that over parts of West Africa, where southwesterlies in July change to northeasterlies in January (Figure 4.4).

4.3.4 Global vertical circulations

So far we have looked at the wind (and pressure) patterns across the horizontal surface of mean sea-level. We have noted that in anticyclones the air descends — as part of the continuous flow of mass within such a feature — and that it ascends within cyclones. So, is this up and down motion organized into large-scale patterns like the mean sea-level winds are?

Meteorologists don't in fact measure the up and down part of the air flow. Instead, they have to deduce it by studying how the horizontal patterns of flow change strength and direction with height in the troposphere. They do this to analyse whether, for example, the low-level convergence that occurs in the ITCZ is capped in the upper troposphere by divergence (Figure 4.22). (**Convergence** is where air flows into a particular region and **divergence** is where air flows out of a particular region.) This must be so — there is a layer some few kilometres deep in the upper reaches of the ITCZ where the air streams away towards the poles, linked intimately to the huge amount of air that is transported upwards in the ITCZ thunderclouds.

Meteorologists look at the monthly or seasonal means of these vertical currents, which complete the pattern at the surface. They look from pole to pole in an extensive, global vertical cross-section, which shows essentially three 'cells' that exist in this north-to-south vertical plane.

January three-cell pattern

Meteorologists call this a pattern of the 'meridional wind' because it is that part of the wind direction that is parallel to lines of longitude or 'meridians'.

Figure 4.23 illustrates the presence of two troposphere-deep overturning Hadley cells within tropical latitudes. One of these cells depicts deep sinking motion above and down to the surface across a belt from 30–15° N, while the other

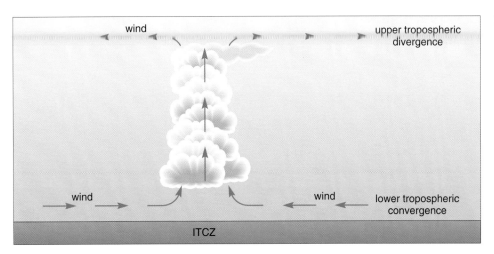

Figure 4.22 A vertical cross-section through the ITCZ.

influences a region from 15–30° S. This deeply subsiding air ties up with the presence of the subtropical oceanic anticyclones.

At low levels, some of the air from these two regions flows towards the Equator to converge (as the ITCZ) near 10° S. Above this convergence zone is deep ascent that is capped by flow towards the pole. In addition to these two cells, which are by far the most powerful in the amount of mass transported, there are two weaker ones that occur in both hemispheres.

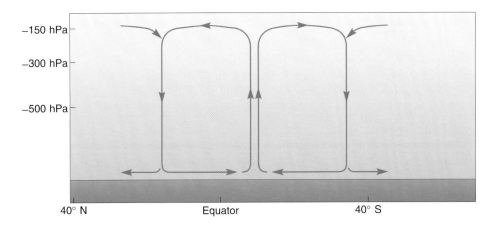

Figure 4.23 A vertical cross-section through the Hadley cells.

In the Northern Hemisphere, some of the air that sinks near 30° N flows northwards at low levels and ascends in a zone that lies around 60° N. This surface flow over the oceans is the tropical maritime air that is an important component of the circulation in mid-latitude cyclones. The ascent is not as deep as in the ITCZ, because the troposphere is shallower in mid-latitudes. It is the mark of frontal ascent in fact. Aloft, there is evidence of some flow south towards the Hadley cell's sinking branch, and north towards a region of subsidence over the highest latitudes. The mid-latitude cell is known as the Ferrel cell (Figure 4.24) after the American scientist William Ferrel, who did fundamental work on the deflection of air that moves across the rotating Earth.

The third, and weakest, cell is the polar cell (Figure 4.24), which is characterized by weak sinking over the polar region and weak easterly outflow back towards mid-latitudes. An example of such outflow is the polar maritime air across the wintertime mid- and high-latitude oceans.

Figure 4.24 A vertical cross-section of the three-cell model (January).

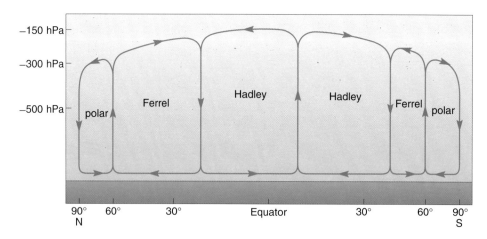

These features are mirrored in the Southern Hemisphere's summer. The Ferrel cell's ascent is centred across a belt from 50–65° S (somewhat further poleward than the matching area in the winter hemisphere).

July three-cell pattern

The total of six cells globally is preserved into the northern summer, as it is in fact year-round. There are however important differences compared to the January pattern.

The two Hadley cells are present (Figure 4.25), but the deep ascent within the ITCZ has shifted into the Northern Hemisphere to lie between 5 and 15° N. In the summer hemisphere the sinking motion has migrated towards the pole to between 25 and 40° N (to include the Mediterranean region). The zone of frontal ascent on the flank of the Ferrel cell (Figure 4.25) is weaker and shifted slightly north compared to January. The polar cell (Figure 4.25) is also weaker than in the winter, with sinking over 70–80° N; there is evidence of very weak ascent over the polar area.

Figure 4.25 A vertical cross-section of the three-cell model (July).

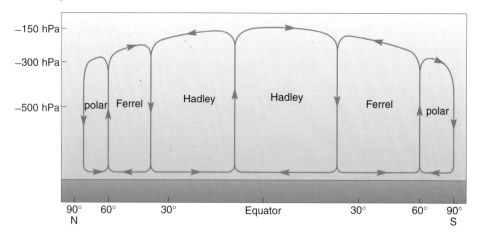

In the winter Southern Hemisphere, the descent in the Hadley cell spreads markedly across the 10–35° S belt, while the frontal ascent zone is stronger than in January but still essentially around the Southern Ocean between 50 and 65° S.

The southern polar cell exhibits marked descent over the wintertime Antarctic, with attendant strong and frigidly cold flow off the continent towards the cyclones of the circumpolar ocean.

4.3.5 High- and low-index flow

As we have seen, jet flow develops on the boundary between cold polar and warm tropical air. With the general decrease in temperature towards the pole, there is a marked westerly component to the flow through the troposphere outside the tropics (Figure 4.21).

Sometimes this westerly flow is more or less parallel to a line of latitude, when the circulation is said to be in a **high-index** regime. Here 'high' means that the flow is very **zonal** or parallel to latitude lines, with frontal depressions moving more or less due west-to-east.

In contrast, the mid-latitude circulation can gradually evolve into what is termed a **low-index** regime, when there are large-scale excursions of cold polar air equatorwards and warm tropical air polewards (Figure 4.26). This is termed a meridional circulation, as branches of the flow run more-or-less parallel to lines of longitude. Frontal lows tend to form on the downstream limbs of the troughs, and thus move northeastwards. A further development of such meridional circulation is that 'cut-offs' can form; that is, pools of cold or warm air become isolated and tend to be slow-moving some distance away from their source region (Figure 4.27).

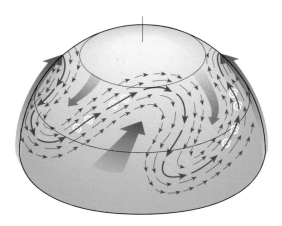

Figure 4.26   A low-index regime where meridional circulation is developing.

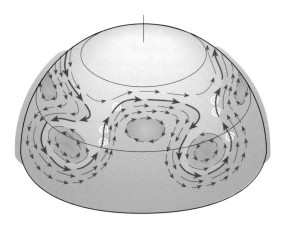

Figure 4.27   As in Figure 4.26, but now large cells of polar and tropical air become cut off. Cyclones (i.e. depressions, shown here as white swirls of cloud) form along parts of the front.

4.4 Precipitation and evaporation patterns

The pressure and wind patterns are strongly related to each other, and in turn can both be linked to the contemporaneous precipitation and evaporation distribution across the Earth's surface. This is because precipitation is produced by the ascent of moist air, within the ITCZ, for example, or across the zone of frontal convergence in mid-latitudes. Evaporation is large in regions where dry air blows strongly across relatively warm ocean surfaces.

Figure 4.28 illustrates the annual zonal mean precipitation and evaporation: the term 'zonal mean' refers to the procedure often undertaken to summarize just the latitudinal variation of weather measures. This is done by averaging a whole range of annual precipitation (or evaporation) totals within latitude strips to derive one average annual value for the strip, or 'zone', in question. These zones cover the Earth from pole to pole and may typically be 10–15° wide.

Figure 4.28 Annual zonal mean evaporation and precipitation.

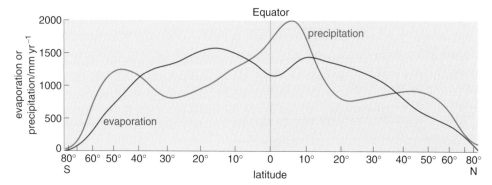

Although such summaries are useful, an important drawback is that they mask the important west-to-east variation in whichever variable is being depicted.

4.4.1 Precipitation

The notable features of global precipitation are listed below.

1 The precipitation maximum of about 2 metres around 5° N. This is the signature of the convergence of water vapour into the ITCZ, deep ascent, and common torrential rainfall. It is not on the Equator because over the whole year the ITCZ migrates further into the Northern than into the Southern Hemisphere. This is related to the much more extensive, strongly heated tropical continents to the north of the Equator.

2 The two mid-latitude precipitation maxima of about 1 metre. These are related to water vapour convergence and ascent along frontal zones. The value is higher in the Southern Hemisphere because there is more evaporation within the Roaring Forties of the Southern Ocean.

3 Two subtropical minima of precipitation between 20° and 30° S and between 20° and 30° N. These are around 700–800 mm and mask very significant west-to-east variation — from the extremely wet summertime monsoon in India to the year-round parched Sahara.

4 Two precipitation minima at the highest latitudes of both hemispheres. These very small values are expressions of both the extremely low water vapour content of very cold air (so very little potential to produce significant

precipitation) and the fact that there are rarely the atmospheric mechanisms to lift and cool the air to produce cloud and then precipitation.

4.4.2 Evaporation

There are three major features of the annual zonal mean evaporation pattern:

1   Two maxima in the tropics between about 10° and 20° S and between 10° and 20° N. These regions are characterized by the strong, persistent flow of the North-East and South-East Trades. They evaporate massive amounts of water vapour from the warm seas across which they flow.

2   The near-equatorial minimum is linked to the presence of deep convective cloud in the ITCZ. Although the total evaporation is high (near 1200 mm per year), it is a 'local' minimum because of the reduced insolation under the ITCZ cloud band and possibly due to the lighter winds close to that zone.

3   The two minima across each hemisphere's highest latitudes. This is related to the very cold air there and its severely reduced capacity both to stimulate evaporation and to contain any significant amount of water vapour.

An interesting picture emerges if the annual zonal mean precipitation total (P) is subtracted from its 'matching' evaporation total (E). The resulting graph (Figure 4.29) reminds us that there are regions of the Earth's surface across which, taking the annual view, $E > P$ or $P > E$. This means, for example, that the surface between about 10° and 40° S and 10° and 35° N is where more evaporation occurs than precipitation falls. It is therefore a 'source' region of water for the atmosphere. In contrast, the regions polewards of 40° S and 35° N and between 10° N and 10° S are where there is more precipitation than evaporation. These are termed 'sink' regions of water for the atmosphere because, over an average year, round the latitude strip (zone), more water is supplied to the surface than is lost from it by evaporation.

The interesting thing about this picture is that the source and sink areas must somehow be connected to each other. It's not sustainable to have a region that year after year experiences more evaporation than precipitation.

Figure 4.29 illustrates the zonal mean sea surface salinity between 70° S and 50° N. **Salinity** is the mass mixing ratio (usually expressed as parts per thousand, ppt) of salt in water. There is a clear relationship between the saltiness of the ocean surface water and the source and sink regions of water (Figure 4.29). Note that where $E > P$, the salinity is some 35.6 parts per thousand, while it lies between 33.3 and 34.3 parts per thousand in areas of $P > E$.

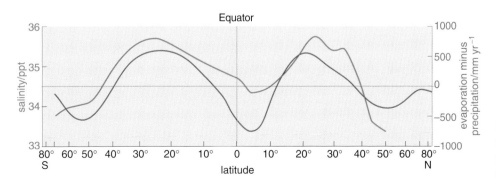

Figure 4.29   Annual zonal mean evaporation minus precipitation, and sea surface salinity.

4.5 The ocean circulation

We have discussed aspects of the global fluid that is the atmosphere. The world's oceans are obviously another type of global fluid that is in fact, like the atmosphere, transporting huge amounts of energy around as it circulates on a large-scale.

The surface currents of the world's oceans (Figure 4.30) are mainly driven by the winds that blow across the water surface, known as **gyres**. We therefore see very large gyres underneath the year-round subtropical anticyclones in which warm tropical water is driven northeastward towards higher latitudes (as the Gulf Stream/North Atlantic Drift and the Kuroshio, for example). In contrast, cooler water is drawn equatorward within the same gyre, as the Canaries, Benguela, Peru and California Currents.

The heat transported on the Gulf Stream/North Atlantic Drift flow is very significant to the weather and climate across much of Europe. The waters are anomalously warm for their latitude (refer back to the average temperature map in Figure 4.3) in this region, as are the related air temperatures. The seas are ice-free all year round, right up to the north of Norway. In contrast, although there is large heat transport in the North Pacific Ocean too, the Bering Strait is a very significant block to the progress of such warm water into the Arctic Ocean. In the Southern Ocean, the surface currents are essentially a mirror of the persistent Roaring Forties; that is, the oceans drift towards the east around the whole belt.

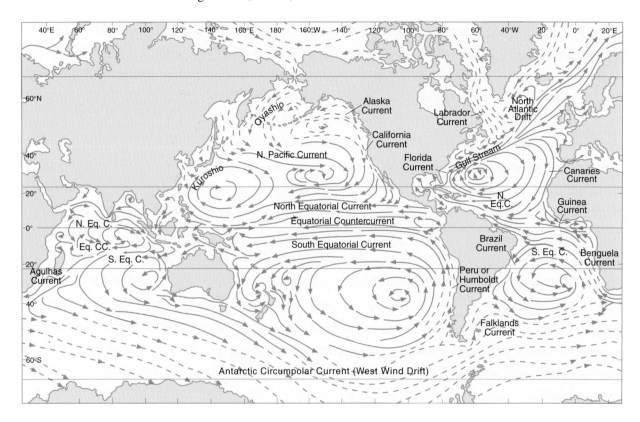

Figure 4.30   The global surface current system in the northern winter; this is the long-term average pattern — at any one time, the pattern differs in detail. There are local differences in the northern summer, particularly in regions affected by monsoonal reversals. Cold currents are dashed. Note that even strong currents, such as the Gulf Stream, rarely exceed speeds of a few metres per second.

Figure 4.31 shows the rates of energy transport by the atmosphere and ocean required to maintain the observed radiative balance. The maximum transport by the atmosphere occurs near 40° S and 40° N, while for the ocean it is nearer to 20–25° S and 20–25° N. It is interesting to note that at their peaks, the atmosphere and ocean heat energy transport rates are about the same (3–4 × 10^{15} W annual mean).

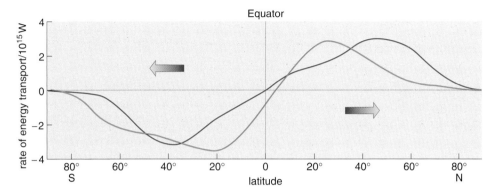

Figure 4.31 Annual mean rates of energy transport by the atmosphere (red curve) and the oceans (blue curve) that are required to maintain the observed low to high latitude temperature difference. Positive values denote energy transport northwards, negative values denote transport southwards.

4.6 The El Niño Southern Oscillation (ENSO)

At the end of each year, around Christmastime, the temperature of the coastal waters off northern Peru and Ecuador regularly increases, as a warm current invades the region for a few weeks, before the cold Peru Current takes over again for the vast majority of the year. This is the period when fish migrate to follow the nutrient-rich cooler waters elsewhere, and when the local fishermen settle down to undertake maintenance of their vessels and nets. This routine event is termed El Niño, which is Spanish for 'the little boy'. It relates specifically to the birth of Jesus at Christmastime, when this annual but short-lived warming of the sea occurs.

However, a much stronger warming occurs in this region every two to seven years. It lasts for several months and often leads to very bad flooding along the normally arid coastal regions of Ecuador and Peru. It was during the 1960s that meteorologists and oceanographers linked such dramatic events to a similar warming across the central and eastern equatorial Pacific Ocean. In addition, they realized that there was also a link between the warming and a large-scale, pan-tropical-Pacific variation in the mean sea-level pressure, known as the **Southern Oscillation (SO)**. The intertwining of the oceanic and atmospheric phenomena (El Niño and Southern Oscillation, respectively) is the basis of the term **ENSO**.

The long-term pattern of mean sea-level pressure across the tropical Pacific Ocean has two principal pressure centres: the low across northern Australia and part of Indonesia and the high over much of the tropical South Pacific. The SO is related to changes in the pressure across these two features: meteorologists take two stations as the 'keys' for their pressure variation over time. These are Darwin in northern Australia, and Tahiti which lies roughly midway between Australia and Peru. The difference between the monthly average mean sea-level pressure at these two places is known as the SO Index (SOI) (Figure 4.32). When the SOI is high, i.e. when the Tahiti minus Darwin pressure difference is large, then the Trade Winds blow strongly across the tropical Pacific from east to west (Figure 4.33). This situation is related to the long-term average distribution of tropical sea surface temperatures right across the Pacific, where the warmest waters occur in the far west with very much cooler waters on the eastern side. The sea is warm enough (around 27–28°C or more) to spark off cumulonimbus clouds and torrential rain which characterizes the western sector. The ocean is so cool in the east that it doesn't experience much rain.

Figure 4.32 Smoothed values of the Southern Oscillation Index. Negative values indicate warm ENSO phases.

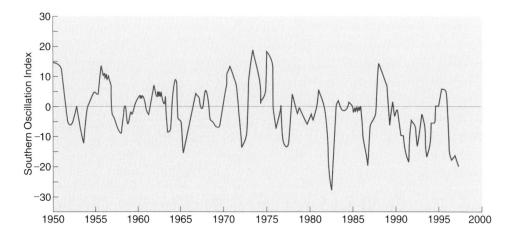

Figure 4.33 Conditions during a high-index phase of the Southern Oscillation. Red to blue denote high to low sea surface temperatures; the buff colour indicates land; white areas represent regions where temperature is not specified.

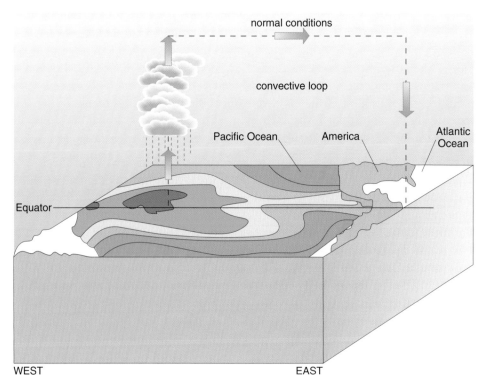

There are times, however, when there is a gradual change over these two pressure centres, such that the high weakens (its pressure falls) over the same period as the low fills (its pressure increases). This is the low-index phase, when the horizontal pressure gradient between Tahiti and Darwin is at a minimum. This change has a great consequence for the Trades: they become very much weaker, or can even reverse into westerlies (Figure 4.34). When the gradual change from a high- to a low-index regime occurs, the relaxation of the surface winds stimulates the slow eastward migration of the warm water across the low-latitude Pacific Ocean. The extensive patch of anomalously warm water is 'shadowed' by cumulonimbus clouds, which are formed by the surface heating and bring torrential downpours to Pacific island groups that normally don't experience such extremes.

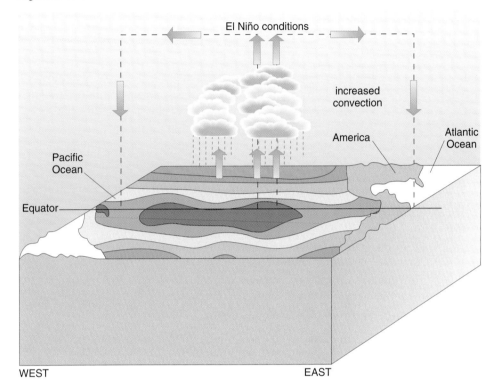

Figure 4.34 Conditions during a low-index phase of the Southern Oscillation (El Niño). Colour coding as in Figure 4.33.

The very tall thunderstorm clouds transport a lot of heat up into the upper troposphere. This process enhances the horizontal temperature gradients at these levels, which in turn both strengthens the jet streams and extends them towards the east. Both these effects mean that an El Niño often leads to unseasonably dramatic weather outside its arena across the tropical Pacific.

As the warm water migrates eastwards, the ocean around Indonesia and northern Australasia becomes cooler. The absence of water that is critically warm enough to develop heavy showers means that this sector suffers anomalously dry conditions, with the increased risk of drought and forest or bush fires. The El Niño of 1982–1983 was estimated to have cost between £7000 and £8000 million in losses and up to 2000 lives globally, all in association ultimately with anomalous weather.

The El Niño of 1997–1998 was a very dramatic example of the phenomenon. It started up quite suddenly in the tropical Pacific during May of 1997 and proved to be the most extensive and the warmest for at least a century. The nature and extent of the weather anomalies that the 1982–1983 event led to, formed the basis of quite confident predictions of how the 1997–1998 El Niño would evolve, and thus which nations should be made aware of the high risk of flooding or drought, for example. Interestingly, it turned out to be not all bad news for the United States. Although the winter of 1997–1998 produced unseasonably damaging storms, in Florida and California for instance, many parts of the northern USA benefited from the marked mildness, because of the reduced energy demand for heating and other purposes.

The tropical Pacific Ocean is monitored virtually constantly from the **geosynchronous** satellites (satellites that orbit the Earth so that they stay above a fixed point on its surface) and by a special array of moored and drifting buoys that sense not only atmospheric parameters but the upper layer of the ocean too.

Today we know that El Niños are expressions of large-scale atmosphere–ocean interactions that can have much wider and significant impacts on the atmosphere's circulation than just within the tropical Pacific region. We know also that, although the problem of seasonal prediction is a very tough nut to crack, when an El Niño is developing there is an important degree of reliability in the location and extent of the weather anomalies that might occur some few months later. In addition, this phase is related to stronger circulation over Indonesia and northern Australia, with much wetter conditions than average.

4.7 A selection of climates

The atmosphere's global circulation displays marked patterns of highs and lows, which change location from summer to winter, and change intensity too. The associated movement of regions characterized by ascending or descending air in the troposphere ties in broadly with climates that are either wet or dry, perhaps seasonally or all year round.

We will now examine the climates of several locations moving from the North Pole to the South Pole, through the UK, to illustrate the way in which the seasonal changes of the global circulation explain the observed progression of monthly mean precipitation and temperature through an average year.

Isfjord Radio (Svalbard), 78° 04′ N, 13° 38′ E, 7 metres

Svalbard lies in the Greenland Sea, well within the Arctic Circle, and roughly midway between northeastern Greenland and northwest Russia (Figure 4.35). It is positioned therefore in the region through which a large number of frontal cyclones track as they finish their journey of some few thousand kilometres northeastwards over the North Atlantic. This means that, for its latitude, it is relatively mild — though the period of time when the mean temperature is above freezing runs on average from about early June to late September (Figure 4.36a). The coldest air comes in the wintertime flow which ranges in direction from westerly through northerly, round to southeasterly, direct from the frozen Arctic Ocean or across sea ice between Svalbard and Greenland or northwest Russia. The mildest conditions of course occur in the summer on winds from a southerly quarter. Although the station is on one of a group of islands, the annual range of temperature (the difference between the mean temperature of the warmest month and that of the coldest month) is quite large at 17.1 °C. This is essentially because for a good proportion of the year the islands are surrounded by extensive sea ice and so are continental then. The term continental refers to how locations some distance from the sea display a larger annual temperature range than those that have more 'maritime' climates on or very near the coast. For example, sites some distance from the sea are more susceptible to frost in the winter, and stronger heating in the summer (recall Box 2.3).

Isfjord's precipitation is reasonably well spread through the year, although there is a distinct maximum from August to around November (Figure 4.36a). This is associated with the shift of frontal cyclones towards the pole in the summer and evaporation in the open stretches of the Norwegian Sea to the southwest. The distinct minimum in the spring is linked, for one thing, to the extensive sea ice at that time of year, and to the absence of any relatively local evaporation. For much of the year, the precipitation falls as snow or sleet.

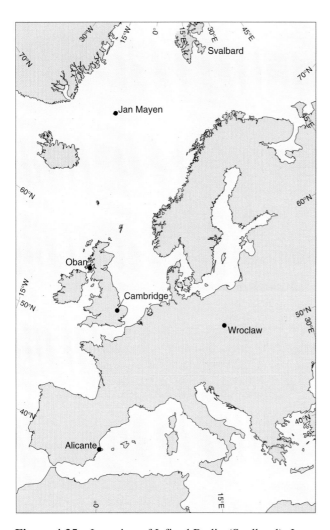

Figure 4.35   Location of Isfjord Radio (Svalbard), Jan Mayen, Oban, Cambridge, Wroclaw and Alicante.

Figure 4.36   Mean monthly temperature and monthly rainfall totals at ten weather stations in different parts of the world: (a) Isfjord Radio (Svalbard); (b) Jan Mayen; (c) Oban, UK; (d) Cambridge, UK; (e) Wroclaw, Poland; (f) Alicante, Spain; (g) Tessalit, Mali; (h) Accra, Ghana; (i) Harare, Zimbabwe; (j) Cape Town, South Africa.

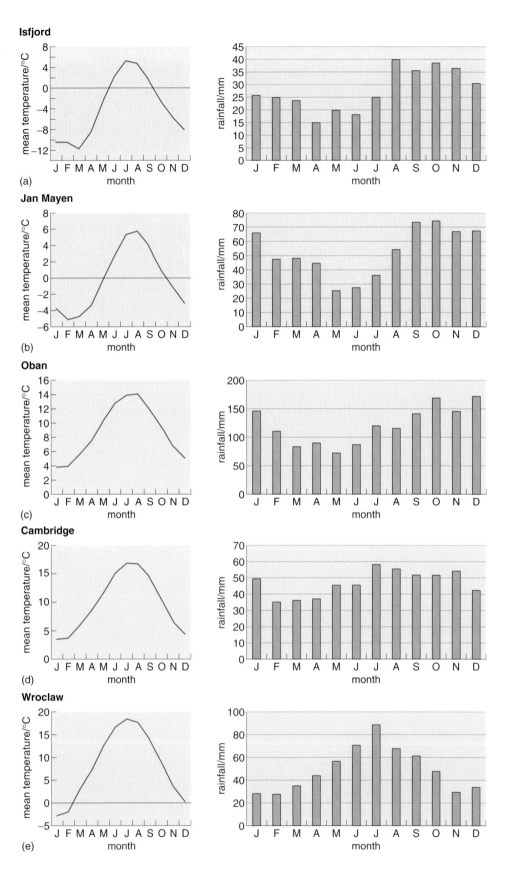

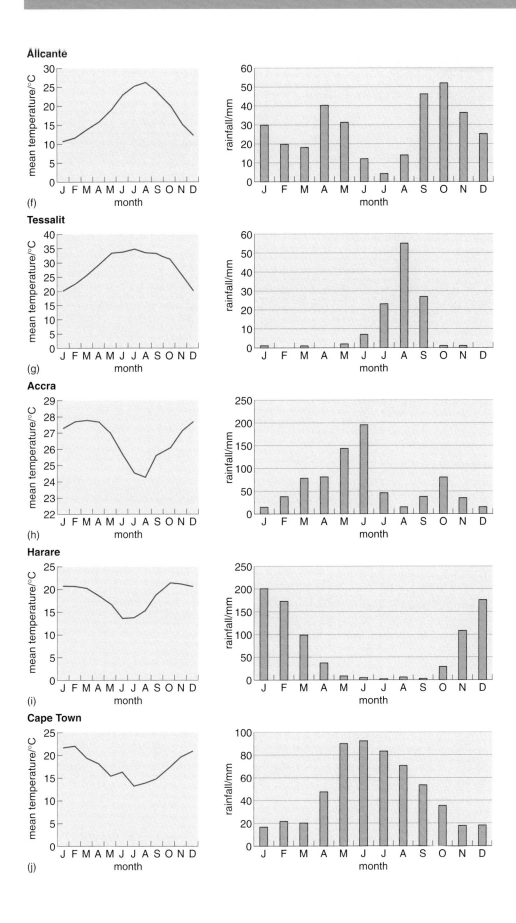

Allcante

(f)

Tessalit

(g)

Accra

(h)

Harare

(i)

Cape Town

(j)

Jan Mayen, 70° 59′ N, 8° 20′ W, 23 metres

This island lies about 1000 km southwest of Svalbard within the Arctic Circle, to the northeast of Iceland (Figure 4.35). Like all the region over and to the east of the Norwegian Sea, it is relatively mild for its latitude. The winter months are some 5–6° C warmer (Figure 4.36b) than at Isfjord, partly because of the smaller influence of sea ice, while the late summer and autumn months are some 2–4° C milder, partly because of the closer proximity of mild waters. Its smaller annual temperature range of 10.8° C bears witness to the comparatively smaller role played by sea ice in the late winter and spring, when it reaches its most extensive.

Like further north, the precipitation is well spread through the year, but with a springtime minimum and a maximum from September to January. The wettest/snowiest conditions are related mainly to the passage of frontal cyclones, although during the cold season a lot of non-frontal lows form over the waters of the Norwegian Sea, affecting these islands and Norway in particular. Jan Mayen's mean annual total of precipitation (628 mm) is roughly twice that at Isfjord (337 mm).

Oban, UK, 56° 25′ N, 5° 30′ W, 69 metres

This western Scottish site is some 1500 km south of Jan Mayen (Figure 4.35), near sea-level. It has an annual range of temperature of 10.1° C, which is virtually the same as Jan Mayen's, although the overall temperature levels are of course higher (Figure 4.36c). The mean wind direction here is southwesterly, with generally very mild conditions all the year round. The coldest air in the wintertime come streaming south across the Norwegian Sea, while the highest summer temperatures are associated with a gentle southeasterly flow that has been heated across a long land track. Oban does not see the high summertime maxima experienced in parts of southern England, however, because it is too far removed from the prime source of strongly heated air over France or southern England itself. Like many coastal sites, the warmest month tends to be August rather than July, because of the influence of the sea's annual temperature cycle on the warmth of the air — the sea reaches its peak temperature around September in the Northern Hemisphere. Jan Mayen shows the same subtle maximum (Figure 4.36b).

The mean annual total of precipitation is 1451 mm, about twice that in the much cooler Jan Mayen. A good deal of Oban's rain comes from frequently passing frontal cyclones — either widespread within warm sectors, or more scattered but heavy in the showers that follow cold fronts. The wettest months are September to January. Springtime minimum rainfall is, like further north, in part linked to smaller evaporation totals from the North Atlantic when this ocean is around its coolest, and in part to the relatively higher frequency of anticyclones during that period.

Cambridge, UK, 52° 12′ N, 0° 08′ E, 12 metres

Cambridge's mean annual temperature range, 13.5 °C, is larger than Oban's. This is because it is some distance inland and, even on the scale of the British Isles, is more continental. For example, it sees more frosts in the winter and higher maxima in the summer months than Oban, or indeed coastal locations not far to its east, such as Great Yarmouth, Cambridge's hottest month on average is July, although only marginally (Figure 4.36d). This subtly higher value than August's is more typical of an inland station compared to one on the coast like Oban.

The annual mean precipitation total of 558 mm is much less than half that of Oban. The eastern side of the British Isles is dry. Many areas have yearly values that are 700 mm or less, from low-lying eastern Scotland down to extensive areas of southeast England. A very significant proportion of the rain in Britain comes from frontal depressions and is deposited on the higher ground on the western side of the island. This is where the systems first make contact with the land and where orographic enhancement (Box 2.7) is an important factor in the substantial totals. The lower regions to the east are affected by the 'scavenging' of much of the frontal lows' water by the higher ground, and are consequently drier overall. This is not to say that heavy rain does not occur in Cambridge, though.

The months with the largest falls are July and August, bearing witness to the relatively stronger summertime heating inland, and the larger risk of thundery rain there than, say, in Oban. The fairly even spread of rain from month to month points to the fact that depressions do provide rain for the area, but in generally smaller quantities than in western Britain.

Wroclaw, Poland, 51° 08′ N, 16° 59′ E, 116 metres

Wroclaw, in southwest Poland, is at about the same latitude as Cambridge but is very much more continental (Figure 4.35). Its mean annual temperature range is 21.1 °C, with significantly colder winter months: both January and February have mean values below zero (Figure 4.36e). Poland's winter is more dominated than the Atlantic shores of Europe by the presence of anticyclones, with their associated radiative cooling in clear skies and the occasional transport of cold air from Scandinavia or Russia, for example. July is the hottest month.

Wroclaw's mean temperatures are higher than those of Cambridge from around mid-April until mid-September, during which time it is some 1–2 °C warmer.

The mean annual precipitation total of 584 mm is just about the same as Cambridge's, with a similar month-to-month progression. Snow is more frequent here of course, and the summertime maximum precipitation consists of greater totals than at Cambridge, but is also mainly generated by thunderstorms. The winter and spring falls are produced principally by frontal systems that moved across Europe from the Atlantic Ocean. A depression that produces rain or snow in Cambridge in, say, December may well deposit more snow a day or two later in western Poland.

Alicante, Spain, 38° 22′ N, 0° 30′ W, 81 metres

Although beside the Mediterranean (Figure 4.35), Alicante has a fairly large annual temperature range of 15.2 °C, and its warmest month is August (typical of a coastal site). The larger range is related to it being susceptible to very hot air from the Spanish interior in the summer, and quite cold air from the same source in the wintertime. The hot summers are linked to the poleward migration of the Hadley cell in the summer season, with its descending branch spreading generally cloud-free skies across the entire Mediterranean. The wintertime temperatures are relatively high too at this quite low-latitude site.

Alicante's annual rainfall total of 328 mm is fairly typical of coastal Spain, with a very marked dry period from June to August inclusive (Figure 4.36f). This, like the high temperatures, is produced by the high pressure and deep subsidence of the Hadley cell as it shifts into Mediterranean latitudes. Larger totals of rain occur from September through to May; this rain tends to be showery and short-lived, except in the winter when the occasional depression either runs into the region from the Atlantic, or forms not far to the east over the western Mediterranean.

Tessalit, Mali, 20° 12′ N, 0° 59′ E, 494 metres

Tessalit is a quite high station, deep in the heart of the Sahara (Figure 4.37). Its annual range of temperature is 13.4° C, with intensely hot conditions (mean values above 30 °C) from May until October (Figure 4.36g); the hottest month is July. Even the coolest month, January, has a mean temperature of 20.1 °C, bearing witness to the year-round prolonged sunshine which is more intense during the summer months. The region is dominated for a large proportion of the year by anticyclonic conditions, except during the period when cloud and rain produced by the ITCZ affect northern Mali.

The very small annual total of rain is 118 mm, most of which (105 mm) falls in July to September. This desert region's rainfall is notoriously unreliable compared to, say, that at Cambridge or Alicante. It depends almost entirely on the annual migration of the ITCZ to its northernmost limit in the summer — and on how moisture-rich its associated showery weather is. As the Hadley cell moves north into the Mediterranean, so the ITCZ shifts north into the southern and middle part of the Sahara. Tessalit, like all similarly located sites, depends critically on this one short-lived rainy season.

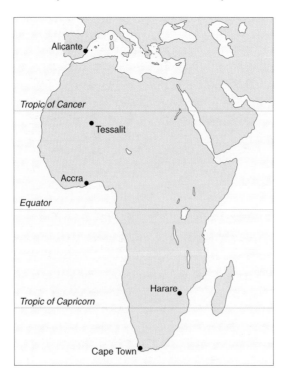

Figure 4.37   Locations of Alicante, Tessalit, Accra, Harare and Cape Town.

Accra, Ghana, 5° 36′ N, 0° 10′ W, 65 metres

Moving south towards the Equator leads to the coastal city of Accra (Figure 4.37). It has a very small annual range of temperature, 3.4 °C, with March the warmest and August the coolest months (Figure 4.36h), even though it is (just) in the Northern Hemisphere. This is clearly different from all the other climates described so far.

The small annual range of temperature is typical of low-lying low-latitude coastal sites, where the insolation does not vary strongly through the year and the sea plays an important role in influencing air temperature. The coolest months of July and August occur because as the ITCZ moves north into the Sahara (to affect Tessalit, for example), air that streams in behind it spills across the Equator from the cooler waters of the tropical South Atlantic during its winter. The warmest months of December to April are those when, generally, the air comes from the southern Sahara.

The rainfall exhibits two maxima, which is typical of places that lie between the extreme northward and southward seasonal locations of the ITCZ. Tessalit sees August as the wettest month. As the ITCZ moves north to influence that area, it crosses Accra, producing the wet season in May and June. On its way south, the ITCZ crosses the area again, to provide more thundery downpours mainly during October. The annual total of 787 mm comprises principally rainfall from this migratory feature.

Harare, Zimbabwe, 17° 50′ S, 31° 01′ E, 1492 metres

This elevated city lies in the region where the ITCZ attains its southernmost location during the year. Its altitude means that although it lies well within the tropics (Figure 4.37), its mean temperature is a pleasant 20.0 °C or just over, during the warmest period from October to March (Figure 4.36i). The coolest months are June and July, during the southern winter.

The rainfall progression through the year is determined strongly by the southward passage of the ITCZ, so that Harare's wet season occurs from December to February. The region is very dry from May to September, when high pressure tends to dominate the circulation. Harare's rainfall total of 848 mm is much larger than Tessalit's — at approximately the same latitude in the opposite hemisphere — mainly because the ITCZ is supplied with a lot more water from the oceans, in particular to the east of Zimbabwe. There is no matching source for the ITCZ in Mali.

Cape Town, South Africa, 33° 56′ S, 18° 29′ E, 12 metres

This site samples the climate of the extreme southern region of Africa, in what could be termed Mediterranean latitudes (Figure 4.37). It has a relatively small annual range of temperature, 8.4 °C, with the warmest months during the southern summer from December to February (Figure 4.36j). Like its Northern Hemisphere coastal counterparts, its warmest month is February, not January as would occur in the South African interior. The winter temperatures are influenced by cooler air brought to the region by travelling frontal cyclones that track from west to east across the South Atlantic to the South Indian Ocean.

Rainfall reaches a peak in association with these depressions, with a marked wet season that stretches from April to September. Once the warmer conditions begin to dominate, from air that comes from the continental interior, the rainfall totals decline significantly.

4.8 Summary of Section 4

1   The latitudinal distribution of annual average incoming solar and outgoing terrestrial radiation is fundamental in driving the atmosphere and ocean system. The pattern, averaged over the year, is one of radiative 'excess' broadly within the tropics and 'deficit' outside the tropics.

2   The regions of radiative excess and deficit force energy transport towards the poles by both atmospheric and oceanic motions, away from the hot tropics into the much colder regions outside the tropics. The annual picture masks very significant seasonal variations, such that the latitudinal winter gradients of radiative excess/deficit are very much stronger than those in the summer. This means that the surface temperature difference between tropical and polar latitudes is approximately twice as large in winter as in summer. Consequently, wind speeds in mid-latitudes are about twice as large in winter as in summer.

3   Anticyclones (highs) serve as 'source' regions for airmasses, particularly when they are slow-moving or stationary over broadly uniform surfaces such as tropical oceans or wintertime high- and mid-latitude continents. They occur across such areas over whole seasons or longer, so that the air that spirals out at low levels possesses the properties of the underlying surface. Tropical maritime air flows out as a warm, moist current, while polar continental air is very cold and very dry.

4   Cyclones (lows or depressions) are very much more mobile than anticyclones in general. They are extensive areas into which different airmasses converge and ultimately mix. Frontal systems are examples of an atmospheric mechanism that carries out some of the heat transport associated with the radiative imbalance outlined in 1 and 2 above. They transport warm, moist air towards the poles and upwards, and cold, drier air towards lower latitudes and downwards. The former processes act to cool air while the latter tend to warm it.

5   The Trade Winds, the Equatorial Trough and the ITCZ are all important features of the tropical atmosphere. The changes in their intensity and geographical location with season are very important determinants of tropical climates. The ITCZ in particular supplies a good deal of the rainfall experienced across the tropics, through the very deep convective clouds with which it is associated.

6   Monsoons are sub-continental-scale changes in the atmospheric circulation, related specifically to a seasonal reversal in the wind direction. The two most marked examples are the Asian and the West African monsoons; the gradual changes in the pressure and wind patterns are linked to the onset of the life-giving rains in those areas.

7   Air motion in the vertical plane from pole to pole (the north-to-south and vertical components of the observed winds) form well defined cellular circulations that extend through the depth of the troposphere. There are three such cells in each hemisphere, which are more or less symmetric across the Equator. The most influential is the Hadley cell, which occurs between 30° N and 30° S; its deeply sinking air is associated strongly with the location of the world's hot deserts. The convergence between the two Hadley cells in the lowest latitudes is the ITCZ. The mid-latitude Ferrel cell and the polar cell complete the three-cell model in each hemisphere.

8   The large-scale motion of the atmosphere exhibits gross features in such a way that sometimes the flow pattern is more or less westerly (or easterly) around much of mid-latitudes. This condition is known as high-index flow and is one when there may be a period of very disturbed weather across the British Isles, with one depression after another running in from the Atlantic. In contrast, other periods of time may see the large-scale pattern to be one of enormous wavy north–south excursions of the flow in mid-latitudes. This is a low-index regime, in which warm air penetrates quite far towards higher latitudes and colder air flows well into lower latitudes.

9   The large-scale pressure/wind patterns are intimately linked to global precipitation and evaporation fields. The zonally averaged annual graphs of precipitation and evaporation exhibit latitude bands within which precipitation exceeds evaporation (source regions with respect to the surface) and vice versa (sink regions with respect to the surface). These regions are maintained by the horizontal transport of water by the atmosphere, ensuring that the water evaporated from the tropical oceans by the Trades, for example, supplies the massive rainfalls produced by the ITCZ.

10 The El Niño Southern Oscillation (ENSO) occurs once or twice a decade across the tropical Pacific Ocean. It grows simultaneously with a slow change in the mean sea-level pressure difference between the Equatorial Trough in the vicinity of northern Australia and the subtropical anticyclone in the South Pacific. When the difference is large, the Trades blow strongly in their more typical fashion across the low-latitude Pacific. When the difference is much weaker, the Trades weaken significantly or even change to a westerly wind. The latter event starts a train of events linked to the slow eastward migration of very warm water across the equatorial Pacific. This warm anomaly is shadowed by very deep convective cloud which is sparked off by the surface warmth, producing torrential rainfalls in normally dry areas. The anomaly reaches right across to the western coast of South America over a few seasons, producing dramatic flooding.

11 The unusually located, deep thunderclouds and heavy rain influence flow at high altitudes in such a way that jet streams can be strengthened, leading to weather anomalies very far from the Pacific. The enhanced flow produces unusually wet summer conditions in northern Argentina and Uruguay, for example, as well as many seasonal anomalies, especially within the tropics.

12 The atmosphere's large-scale circulation (and the ocean's to some degree) influence the nature and distribution of the Earth's climates. The seasonal shift and change in intensity of the mobile frontal depressions, of the subtropical and continental anticyclones, and of the ITCZ, for example, all play critical roles in the climates of particular locations across the Earth.

Question 4.1

Explain how the atmosphere and ocean circulations in the mid- and high-latitude North Atlantic result in Murmansk, just south of 70° N in northwest Russia (Figure 4.38), being ice-free all the year round. (Murmansk is Russia's only ice-free major port on its northern shore.)

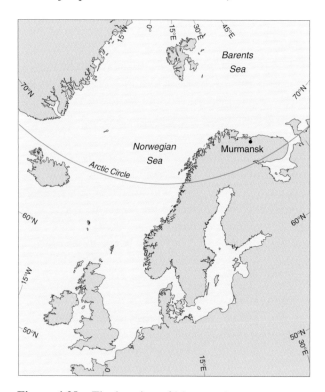

Figure 4.38 The location of Murmansk.

Question 4.2

The winds measured at the surface across the British Isles during the winter are, generally, twice as strong as in the summer. Why is the British winter so much windier than the summer?

Question 4.3

Why are tropical climates classified on the basis of the progression of their monthly mean precipitation rather than temperature?

Learning outcomes for Section 4

After working through this section, you should be able to:

4.1 Describe how average annual and seasonal incoming solar and outgoing terrestrial radiation vary with latitude. (*Question 4.1*)

4.2 Explain the link between the regions of radiative excess and deficit and the associated poleward heat transport carried out by the atmosphere and oceans. (*Question 4.2*)

4.3 Explain the difference between cold and warm anticyclones, how both act as source regions for airmasses and describe their seasonal variation. (*Question 4.2*)

4.4 Understand the way in which frontal cyclones form, and describe the basic nature of warm, cold and occluded fronts.

4.5 Describe the evolution of a 'classic' frontal cyclone.

4.6 Describe and explain the basic types of weather associated with selected airmasses. (*Question 4.2*)

4.7 Understand the basic link between barometric pressure and wind patterns, including the importance of horizontal thermal contrast. (*Question 4.2*)

4.8 Describe the nature of monsoons and other tropical circulation features such as the ITCZ and Trade Winds.

4.9 Understand and explain the physical/dynamical link between the three-cell model and surface climate features.

4.10 Explain the relationship between the pattern of zonally averaged annual mean evaporation and precipitation and the associated horizontal transport of water by the atmosphere.

4.11 Describe the nature of the Southern Oscillation and its link to El Niño.

4.12 Understand the evolution of an ENSO event and its link to seasonal anomalies on a wider geographical scale.

4.13 Describe the nature of, and explain the reasons for, the climate of selected areas of the world as represented by mean monthly precipitation and temperature. (*Question 4.3*)

Comments on activities

Activity 1.1

The values that are highlighted in bold in Table 1.3 are those from the data in Figure 1.13. The other values are from Table 1.2. Even if the pressure values you calculated differ slightly from the ones shown in the table, you should notice that for every 5500 m ascended, the pressure drops by a factor of about 2. This would also be true for the density if the temperature were constant, because then the pressure would be proportional to the density. As it is, the temperature varies with height, and so the density structure is more complicated than that of pressure. Comparison of your calculated values with the data from Table 1.2 (non-bold) should show a rough correspondence.

Table 1.3 Calculated values (**bold**) and typical measured values (non-bold) of atmospheric properties at different heights.

Site	Height/m	Pressure/hPa	Temperature/K	Density/kg m^{-3}
mean sea-level	0	1013	—	1.23
	0	**1010**	**288**	**1.27**
Mont Blanc	4810	550	—	0.75
	5500	**507**	**252**	**0.73**
Mount McKinley	6190	460	—	0.65
Mount Everest	8850	315	—	0.48
cruising Boeing 747	11000	225	—	0.35
	11000	**253**	**216**	**0.43**
	16500	**127**	**216**	**0.21**
cruising Concorde	18 000	70	—	0.13
	22 000	**63**	**218**	**0.11**
	27 500	**32**	**223**	**0.05**
	3 3000	**16**	**228**	**0.03**

Answers to questions

Question 1.1

The perfect gas law tells us that *for a constant temperature, T*, the number density, *N/V*, and pressure, *p*, are directly proportional to each other, i.e. $p = C(N/V)$. However, we know that the temperature does change with height, but the fact that this temperature variation has little impact on the density profile tells us that it cannot have much of an effect. We can see why by considering the ranges involved.

The range of pressure from sea-level to the top of the stratosphere is from about 1000 hPa to roughly 1 hPa. The ratio of maximum to minimum pressures is therefore about 1000. The temperature range is from 15 °C (typical sea-level) to about −60 °C (the minimum is at the tropopause, *not* the stratopause: see Figure 1.10). However, the temperature in the perfect gas law is expressed in kelvin; the maximum and minimum temperatures are therefore 288 K and 213 K. This gives a ratio of only 1.35, very much smaller than the ratio of pressures. So although the temperature difference between −60 °C and 15 °C may seem like a lot (the difference between life and death in fact!), it is very small compared to the difference in pressures when calculating density in the perfect gas law.

Question 1.2

(a) The force on the platform is the product of the mass and acceleration (Newton's second law, Equation 1.4): $F = ma = 10\,\text{kg} \times 9.81\,\text{m s}^{-2} = 98.1\,\text{N}$. (b) The pressure is obtained by dividing the force by the area (Equation 1.5): $p = F/A = 98.1\,\text{N}/(20\,\text{m}^2) = 4.91\,\text{N m}^{-2} = 4.91\,\text{Pa}$. (c) If the area is halved, the force remains unchanged, but the pressure is doubled, to 9.81 Pa.

Question 1.3

From Table 1.2, the mass densities of air at sea-level and at 18 000 m are $1.23\,\text{kg m}^{-3}$ and $0.13\,\text{kg m}^{-3}$ respectively. We can convert these to number densities knowing that the average air particle is $5.0 \times 10^{-26}\,\text{kg}$ in mass; we simply divide each mass density by this particle mass to get $2.46 \times 10^{25}\,\text{m}^{-3}$ and

$2.60 \times 10^{24}\,\text{m}^{-3}$, at sea-level and at 18 000 m respectively. These are the numbers of *all* the particles in $1\,\text{m}^3$ of air; so to calculate the number of ozone molecules in $1\,\text{m}^3$, we multiply by the appropriate mixing ratios:

at sea-level, number of ozone molecules
$= 5 \times 10^{-9} \times 2.46 \times 10^{25} = 1.23 \times 10^{17}$;

at 18 000 m, number of ozone molecules
$= 6 \times 10^{-6} \times 2.60 \times 10^{24} = 1.56 \times 10^{19}$.

So, despite the number density of air being almost 10 times smaller at 18 000 m, the increase in the mixing ratio by a factor of over 1000 times means that there is more than 100 times more ozone in an air sample from 18 000 m as compared to sea-level.

Question 1.4

At the end of Section 1.3.2, it was stated that the pressure falls by a factor of 2 for every 5500 m ascended in height. 16 500 m is exactly three times 5500 m, so the pressure is halved three times, i.e. the pressure has dropped by a factor of $2 \times 2 \times 2 = 8$.

Question 1.5

(a) The force at the base of the column is simply the sea-level pressure there multiplied by $1\,\text{m}^2$:

$$1013\,\text{hPa} \times 1\,\text{m}^2 = 101\,300\,\text{Pa} \times 1\,\text{m}^2 = 101\,300\,\text{N}$$

(b) 101 300 N is the force exerted by the mass (i.e. the weight) of all the air above the $1\,\text{m}^2$ area. We know that force exerted by a mass due to gravity is mass times the gravitational acceleration, i.e. $F = mg$. So

$$m = F/g = 101\,300\,\text{N}/(9.81\,\text{m s}^{-2}) = 10\,330\,\text{kg}$$

(c) The pressure at the top of the column, at a height of 11 km (according to Figure 1.13) is about 255 hPa. So the mass above the column is

$$25\,500\,\text{Pa} \times 1\,\text{m}^2 /(9.81\,\text{m s}^{-2}) = 2600\,\text{kg}$$

(d) The total mass of the air column is therefore $10\,330\,\text{kg} - 2600\,\text{kg} = 7730\,\text{kg}$, which is about 75% of the mass of the atmosphere above the base of the air column.

Question 2.1

There are a number of important factors that determine the daytime maximum temperature. One is the *time of year*. This is important because it obviously defines the length of daylight and therefore the potential duration and intensity of solar radiation at a specific site available for warming the air.

Another is the *minimum temperature from the previous night*, as this is the temperature from which the warming must start. This factor is partly related to the airmass that affects the place during the day. If there is a wind, the *direction* it is bringing air from is significant: from the north at any time of year means cool or cold; from the west at most times means 'about average' temperatures, because the most common direction is west to southwesterly; from the south indicates generally above-average temperatures, while easterlies generally bring colder than average conditions in the winter and warmer than average in the summer.

How *cloudy* the weather is of course influences the duration and intensity of sunshine. A rather more subtle effect is related to the clarity of the atmosphere: more solar radiation gets through to the surface in *cleaner air*. The *wind strength* also plays a role in the sense that in high winds, warm or cold air for instance moves more quickly into a region than when the winds are light.

Question 2.2

Frost occurs when the drybulb temperature falls to 0.0 °C at the surface or in the air. The air cools on frosty nights when it is generally calm and clear because the ground cools quickly by emitting infrared radiation. So it is air near the Earth's surface that is most likely to fall to 0 °C or below.

The infrared cooling process does take some time, however, and takes more time for air above the surface. So, the required chilling to bring down the screen temperature to 0 °C (at a height of 1.25 m) does not always occur. This is the reason that ground frosts are more common than air frosts: it takes some time for the chilling to be transmitted up through the lowest few metres of the atmosphere.

Question 2.3

The mass density (ρ) of water is $1000 \, kg \, m^{-3}$. A fall of 20 mm of rainwater onto one square metre is a volume (V) of $20 \, mm \times 1 \, m^2 = 20 \times 10^{-3} \, m^3$. So the mass of rainwater is $m = \rho V$ or $10^3 \, kg \, m^{-3} \times 20 \times 10^{-3} \, m^3 = 20 \, kg$.

Question 2.4

(a) The daily mean temperature is calculated from daily values of maximum and minimum temperature simply by taking the arithmetic average (mean) of the two, i.e. we add them together and divide by two.

(b) To calculate a weekly or monthly mean temperature, it is necessary only to average a week's or a month's individual daily means.

(c) The daily *range* of temperature is simply the maximum minus the minimum value.

[The range tends to be small when a day is cloudy and windy, and large when it is sunny, calm and clear at night.

Note that it is also instructive, if you keep a reasonably long record of some few months for example, to look at the frequencies of maxima and minima within bounds that you can define. You can then see how common minima of between, say, 0.0 °C and −1.5 °C are, then −1.5 °C and −3.0 °C, etc. This way, if the record covers a number of years (ideally), you will be able to estimate the likelihood of a particular minimum or maximum value occurring in a given month, say, at your site. You'll be building up a useful 'climate' of your site.]

(d) The matter of daily rainfall total is somewhat different. It doesn't rain every day, so for a sequence of consecutive daily falls of, say, 1.2, 0.0, 0.0, 7.4, 0.1, 0.0 and 0.3 mm during a week, the daily average 'fall' of 1.29 mm masks the fact that it didn't rain on three of the days.

[What is done normally is that the rain is simply totalled, to 9.00 mm in this case, and summarized by stating that there were two 'raindays' (days during which 1 mm or more rain fell).

Just like the temperature data, you can also make up a data set of the frequency of daily rainfall values in order to look at the likelihood of totals exceeding certain critical values. If you don't measure rainfall, you can use broadsheet newspaper weather data or visit the Met Office's website.]

Question 2.5

There are two main problems:

(1) Fixed physical features such as hills and buildings can show up on radar scans, but are a nuisance because they reflect radar signals and 'mask' any precipitation falling behind them. Radar beams are elevated at very shallow angles to avoid 'echoes' from such features.

(2) This orographically enhanced rainfall is related to the way in which a precipitating layer cloud at say 2 km up (the seeder) washes out extra rainfall from a lower cloud (feeder) that is located over a hill. Radars have no problem monitoring the upper precipitation but cannot detect the lower and very important *extra* amount of rain that's being washed out. This is overcome by having recording raingauges in such hilly areas to monitor the actual fall at the surface. If this did not happen in hilly areas, the radar would seriously underestimate what can be flood-producing heavy rain in hills.

Question 2.6

(a) The south-facing aspect means that the garden will be exposed to marked solar heating through the day, particularly during the summer. During the daytime the heating of the air just above the garden depends on the albedo of the grass and tarmac. The latter is very much darker than the grass and will therefore absorb significantly more solar radiation. This in turn means that the air over the tarmac heats up more than that over the grass, leading to a higher maximum temperature in the screen there.

As the Sun goes down, and the air stays calm, the surface cools quickly everywhere. As this cooling progresses during the hours of darkness, the cooled air flows slowly downslope towards and into the depression at the bottom of the garden. Its bowl shape means that the cold air can become pooled in it, acting as a reservoir of chilly air. Its temperature minimum is lower than over the tarmac back up the slope.

(b) The large temperature differences in (a) are crucially dependent on the right, quiescent, weather conditions. They vanish if the day in question is cloudy throughout — which suppresses the daytime warming and night-time cooling — and if it's windy too. These conditions lead to the effective mixing of the air through a deep layer, to minimize any surface temperature changes during the 24 hours.

Question 3.1

(a) The volume of the air is

$$V = (1000\,\text{m} \times 1000\,\text{m} \times 1000\,\text{m}) = 10^9\,\text{m}^3$$

Its mass m is calculated by multiplying its mass density ρ by V:

$$m = \rho V = 1.2\,\text{kg m}^{-3} \times 10^9\,\text{m}^3 = 1.2 \times 10^9\,\text{kg}$$

The kinetic energy of this airmass is therefore

$$\begin{aligned}E_K &= \tfrac{1}{2}\,mv^2 \\ &= 0.5 \times 1.20 \times 10^9\,\text{kg} \times (3.00\,\text{m s}^{-1})^2 \\ &= 5.40 \times 10^9\,\text{J}\end{aligned}$$

(b) Box 3.1 tells us that multiplying pressure (which can be thought of as a kind of energy density) by volume gives us a rough estimate of thermal energy. To calculate pressure we must first convert the mass density to a number density, N/V (see Section 1). From Equation 1.1:

$$\begin{aligned}N/V &= \rho/(5.00 \times 10^{-26}\,\text{kg}) = 1.20\,\text{kg m}^{-3}/(5.00 \times 10^{-26}\,\text{kg}) \\ &= 2.40 \times 10^{25}\,\text{m}^{-3}\end{aligned}$$

We can now use the perfect gas law (Equation 1.3) to give us the pressure:

$$\begin{aligned}p &= (N/V)kT \\ &= 2.40 \times 10^{25}\,\text{m}^{-3} \times 1.38 \times 10^{-23}\,\text{J K}^{-1} \times 300\,\text{K} \\ &= 99\,400\,\text{J m}^{-3}\end{aligned}$$

Note that $1\,\text{J m}^{-3} = 1\,\text{Pa}$, emphasizing that pressure is a kind of energy density (we can also see this from the fact that $pV = NkT$, which is the total number of particles in the volume, N, times the average energy of a particle at temperature T, i.e. kT).

If we now multiply the pressure by the volume we can obtain the thermal energy estimate, E_T:

$$E_T = pV = 99\,400\,\text{J m}^{-3} \times 10^9\,\text{m}^3 = 9.94 \times 10^{13}\,\text{J}$$

(c) The Stefan–Boltzmann law (Equation 3.4) tell us that the rate of energy emitted R by an object of surface area A, at temperature T is

$$R = A\sigma T^4$$

where $\sigma = 5.67 \times 10^{-8}\,\text{W m}^{-2}\,\text{K}^{-4}$. Here, we have a block of air that has six sides, each of area $1000\,\text{m} \times 1000\,\text{m} = 10^6\,\text{m}^2$. Therefore $A = 6.00 \times 10^6\,\text{m}^2$. The power, i.e. energy emitted per second, is therefore:

$$R = 6.00 \times 10^6 \, \text{m}^2 \times 5.67 \times 10^{-8} \, \text{W m}^{-2} \, \text{K}^{-4} \times (300)^4$$
$$= 2.76 \times 10^9 \, \text{W}$$

The energy emitted in ten seconds is therefore

$$E = R \times 10 \, \text{s} = 2.76 \times 10^{10} \, \text{J}$$

(d) Conservation of energy tell us that energy cannot simply appear and disappear. The amount radiated must come from energy possessed by the gas. In 10 s the gas radiates more energy than it has kinetic energy, but only a fraction of its thermal energy. We must conclude that the radiated energy comes from the thermal energy. The fact that the temperature appears in both the perfect gas law and the Stefan–Boltzmann law also accords with this conclusion.

Question 3.2

(a) Each slab has an area A of $(0.5 \, \text{m})^2 = 0.25 \, \text{m}^2$. The rate at which solar radiation energy falls on this area, R_S, is

$$R_S = 780 \, \text{W m}^{-2} \times 0.25 \, \text{m}^2 = 195 \, \text{W}$$

The black slab has an albedo of 0.1. This means it scatters a fraction $0.1 R_S$ back into the atmosphere, absorbing the remaining $0.9 R_S$:

$$R_b = 0.9 \times 195 \, \text{W} = 176 \, \text{W}$$

Similarly, the grey slab absorbs $0.6 R_S$:

$$R_g = 0.6 \times 195 \, \text{W} = 117 \, \text{W}$$

(b) Rearranging the Stefan–Boltzmann law (Equation 3.4) gives us

$$T^4 = R/(A\sigma) = R/(1.42 \times 10^{-8} \, \text{W K}^{-4})$$

Replacing R with the value R_b gives the temperature of the black slab (after hitting square-root twice on the calculator):

$$T_b = 334 \, \text{K} = 61 \, °\text{C}$$

Similarly, the temperature of the grey slab is

$$T_g = 301 \, \text{K} = 28 \, °\text{C}$$

Clearly, having black slabs in a patio on a sunny day could be quite uncomfortable on your bare feet!

(c) Since the ELR is greater than the SALR, the conditions are unstable and convection can take place. The infrared radiation emitted by the patio is absorbed in the air above it, this air then rises, carrying the energy away as thermal energy.

Question 3.3

Similarities:

- At 60 °C, the pipe emits radiation in the infrared, like the Earth's surface. (You can check this by applying Wien's law.)
- The lagging absorbs this radiation, just as the atmosphere absorbs infrared radiation.
- The lagging re-emits this as infrared radiation, giving some back to the pipe, hence trapping energy as the atmosphere does.

Differences:

- The pipe is supplied with hot water and not heated by an external source of radiation as is the case with the Earth.
- The lagging only radiates energy; it is not a gas or a liquid, so cannot undergo convection, as can happen in the Earth's atmosphere.

Question 4.1

The mid- and high-latitude North Atlantic are linked to the Arctic Ocean by a broad west-to-east 'gap' between Scandinavia and Iceland/Greenland. This 'opening' contrasts very markedly with the extremely narrow connection between the North Pacific Ocean and the Arctic Basin that is the Bering Straits.

The Atlantic Ocean circulation is such that the warm current that flows far north as the North Atlantic Drift transports a huge amount of heat into the Arctic Basin. This water is anomalously very warm for its latitude; it washes the shores of Scandinavia, and right around the North Cape (the far northern tip of Norway) to the Kola Peninsula (extreme northwest Russia).

The atmosphere's circulation across the mid- and high-latitude North Atlantic is very frequently characterized by the southwest-to-northeast tracking of frontal depressions. These weather disturbances often form off the eastern shores of North America and run thousands of kilometres towards western and northwestern Europe. The mean southwest wind that is associated with these systems is a symbol of their role of transporting heat from low to higher latitudes — as warm or mild air that often penetrates into the flanks of the Arctic basin via the Norwegian Sea.

This means that air temperatures are anomalously high for their latitude in the same region as are sea temperatures; indeed, the warm water is driven into this region in large part by the atmospheric circulation.

Question 4.2

The explanation is linked to the fact that the horizontal pressure gradient is twice as steep or strong in the winter as in the summer. This in turn is largely an expression of the fact that the travelling frontal depressions of the North Atlantic are deeper in the winter, so that the Iceland Low has a lower minimum then than in the warm season.

This relatively local (compared to the global scale) change in intensity of the weather systems that often affect the British Isles is related, in the main, to the very much larger-scale seasonal variation of, in the main, the incoming solar radiation. On an annual average basis, there is a region of the planet, between about 30° N and 30° S, where the incoming solar radiation exceeds the outgoing terrestrial radiation. This zone of excess in each hemisphere is flanked by a region of deficit, where the outgoing terrestrial radiation exceeds the incoming solar.

All this means that the fluid motions of the atmosphere and oceans are such that they partly transport energy away from the excess region and into the deficit one. One key means of doing this is the frontal depression. (The magnitude of the flow of heat that they achieve can be represented basically by the wind strength times the air temperature.)

The annual mean picture of radiative excess and deficit masks very significant seasonal variation. While it is true that the outgoing terrestrial radiation varies somewhat from summer to winter (not much at all in the tropics, however, because temperature doesn't change much there around the year), the incoming solar radiation varies very significantly. Its seasonal change is dramatic in higher latitudes in association with the shift from polar day to polar night.

The upshot is that in the winter, the low-to-high latitude difference in the size of the excess and deficit is very big indeed, while in the summer it is much smaller. This means that the atmosphere has to work harder to transport more heat towards the pole in the wintertime than in the summer. The result is then that such large-scale 'forcing' leads to deeper frontal systems in winter, and therefore we observe much windier conditions on average.

Question 4.3

Except where climates are particularly arid within the tropics, the changes of month-to-month average temperature round the year are not very marked. Most sites do not have an obvious cold and warm season like the British Isles, for example; in the main this is because the variations in the elevation of the Sun, and in the total duration of sunshine through the year, are not very great. There are differences in monthly mean temperature, but they are most often not significant compared to the seasonal/monthly variation in precipitation totals.

One critically important feature of the tropical atmosphere is the ITCZ. It is characterized by the presence of very deep convective clouds which produce a significant proportion of the rainfall experienced at tropical locations. It is a migratory feature that shifts towards the pole in the summer season and towards the Equator in the winter. Such north-to-south or south-to-north movement is most extensive over the tropical continents where the seasonal variation in heating is more marked than over the tropical ocean. The heating pattern is linked to the location of the Equatorial Trough/ITCZ.

The ITCZ reaches its northernmost migration (over Africa, for example) during August, at which time it produces the one rainfall maximum experienced during the year across the southern flank of the Sahara. It reaches its southernmost location across that continent during February, when it provides the wet season in Zimbabwe for instance.

Between these latitudinal extremes, it is broadly the case that stations see two rainfall seasons during the year. One occurs as the ITCZ moves northwards and the other when it is shifting in the other direction. Places near the Equator, such as Accra (Figures 4.36h, 4.37), experience two wet seasons during two consecutive months between February and August (northbound ITCZ) and August to February (southbound).

It is the case, therefore, that tropical climates, from the semi-arid flanks of the subtropical deserts to the jungles of equatorial regions, are best defined by their rainfall regimes rather than by any temperature variations they exhibit.

Acknowledgements for Part 1 *Air*

Cover illustration: Simone Pitman/Open University; *Figure 1.1:* Reproduced with permission of Colonel Joe Kittinger; *Figures 1.2, 3.4*: NASA; *Figure 1.7:* © 1998 Everest Live Executive Committee; *Figure 1.8:* Kevin Church/Open University; *Figure 1.9:* Courtesy of The National Meteorological Library; *Figure 1.11*: © Carl R. Nave; *Figure 2.1*: © Crown Copyright. Reproduced with the permission of the Controller of Her Majesty's Stationery Office; *Figures 2.3, 2.14*: © Department of Meteorology, University of Reading; *Figures 2.5, 2.11*: Simone Pitman/Open University; *Figures 2.13, 2.17, 3.13, 4.32, 4.33, 4.34, 4.36 (a)-(j)* : Ross Reynolds (University of Reading); *Figure 2.21:* © European Centre for Medium-Range Weather Forecasts ECMWF http://www.ecmwf.int/; *Figure 2.22:* © Royal Meterological Society; *Figures 2.25, 2.26, 4.18*: Copyright © 2001 EUMETSAT; *Figures 2.29, 2.30*: © European Centre for Medium-Range Weather Forecasts ECMWF http://www.ecmwf.int/; *Figure 4.3*: InterNetwork Inc., NASA/JPL and GSFC; *Figures 4.4, 4.12*: A.H.Perry and J.M.Walker (1977) *The Ocean-Atmosphere System*, Addison-Wesley; *Figures 4.5, 4.15, 4.16*: © NERC Satellite Station, University of Dundee; *Figure 4.11*: NOAA; *Figures 4.21, 4.30*: A. Strahler (1973) *Earth Sciences*, Harper & Row.

Every effort has been made to trace all copyright owners, but if any have been inadvertently overlooked, the publishers will be pleased to make the necessary arrangements at the first opportunity.

PART 2
EARTH

PART 2
EARTH

Nancy Dise, Bill Dubbin and Michael Gagan

Introduction

The Earth's **crust**, the outer solid 'skin' of the Earth, is a fundamental component in the study of environmental science, since it makes two essential contributions to every environmental locality. Firstly, it provides the physical 'framework' within which environmental systems can operate. The underlying rocks and the superficial deposits above them are the result of geological processes taking place both at the surface of the Earth and far beneath it. The *shape* of any environment is its **geomorphology**. Secondly, the rocks of the crust provide the major raw material for the surface deposits that interact with the atmosphere, hydrosphere and biosphere — the soil. The processes that lead to a soil capable of sustaining plant life are complex, but the starting point is the weathering of solid rock into mineral fragments and chemical materials in solution.

It is generally believed that the Earth formed about four thousand six hundred million years ago (4.6×10^9 years or 4600 **Ma**), by the accumulation and fusing together of particles from a hot gas cloud which surrounded the Sun. Shortly afterwards, or even while this process of accretion was still taking place, an intensely energetic process occurred, which separated the different components of the new planet and resulted in a gross structure for the Earth of **core**, **mantle** and **crust**, not dissimilar from the one that exists today (Figure 1.1).

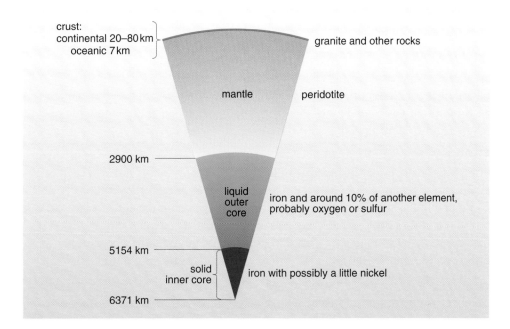

Figure 1.1 The layered structure of the Earth.

The primitive planet was rich in the elements iron, oxygen, silicon, magnesium — and to a lesser extent nickel, sulfur, calcium and aluminium — but the formation of the core removed much of the iron, nickel and other heavy elements, leaving a mantle composition dominated by lighter elements like oxygen, silicon and magnesium (Table 1.1).

Table 1.1 Distribution of major elements in the whole Earth, the mantle and the crust (values are % by mass; only values above 0.5% are included).

Element	Symbol	% Whole Earth	% Mantle	% Crust
oxygen	O	29.5	44.0	45.2
silicon	Si	15.2	22.4	27.4
aluminium	Al	1.1	1.6	8.0
iron	Fe	34.6	9.9	5.8
calcium	Ca	1.1	1.7	5.1
sodium	Na	–	0.8	2.3
potassium	K	–	–	1.7
magnesium	Mg	12.7	18.6	2.8
sulfur	S	1.9	–	–
nickel	Ni	2.4	–	–
others		1.5	1.0	1.7

Immediately after the separation of the core, the outer layers of the Earth would have been molten. However, as the temperature dropped to about 1000 °C, a thin skin of crystalline material, the primitive crust, would have developed on the surface. At this temperature the complex **silicates**, the major rock-forming compounds in which oxygen and silicon combine with other metals, began to crystallize. The rocks of this first-formed crust probably had a composition close to that of the molten rock (**magma**) of the mantle from which they crystallized. So they would have been more like the rocks of today's mantle, with a higher magnesium content than the crustal rocks that formed later.

This early crust is unlikely to have been very stable, and would have evolved as different silicate minerals repeatedly formed, were re-melted, and absorbed back again into the mantle. Some of these crustal minerals were of lower density and so floated to the surface of the magma, where they accumulated to form rocks more like those of the present-day continental crust. The overall composition of the crustal material that crystallized preferentially, quite quickly became different from that of the magma — richer in elements like aluminium and calcium, and denuded in iron and magnesium.

The hot, primitive Earth must also have acquired an atmosphere of volatile materials, including water, hydrocarbons and other gases, but little of this remains. As the surface solidified, a new atmosphere formed as gases, principally water vapour, nitrogen and carbon dioxide, were forced out of the mantle by volcanic activity. At this stage the atmosphere did not include free oxygen, since this was tightly bound in the silicates. Levels of oxygen built up slowly, first as ultraviolet radiation from the Sun cleaved water into its elements (photodissociation), and later through the photosynthesis of carbon dioxide by green plants. As the surface cooled, the water vapour condensed to form the Earth's oceans. Interaction between the rocks, the water and the atmosphere began immediately.

The oldest rocks at the surface of the Earth today are to be found in Greenland, and they have been dated at almost 4000 Ma. They include not only **crystalline rocks**, such as would form directly from magma and volcanic activity, but also **fragmentary rocks**. These rocks can only be formed after the original crystalline rocks have been broken down through processes of weathering. The fragments are transported, mostly by water, and accumulate in deposits, like seashore sand or estuarine mud. Later they are formed into rock by the heat and pressure of deep burial, the process of **lithification**. The great age of these rocks shows that the formation of fragmentary rocks through the weathering of the primitive crust must have begun soon after the crust first solidified, and has continued ever since, as you will learn later in the course.

Question 1.1

'Almost all the Earth's crust is composed of silicate or aluminosilicate minerals.' What does this statement tell us about the principal elements to be found in the Earth's crust?

2 Rocks and minerals

We have used the terms rocks and minerals in the introduction, but it is important to distinguish between these terms. A **mineral** is a naturally occurring material with a limited range of chemical composition and well-defined properties like crystal structure and shape, hardness, and melting temperature range. Some minerals may be familiar to you, like the white veins of quartz, or the golden cubes of iron pyrite. Precious and semi-precious stones are usually minerals too. Most rocks are aggregates of a number of different minerals. Very few rocks are composed of a single mineral; however, **marble** is essentially pure calcite (calcium carbonate), and **quartzite** is almost pure **quartz** (silicon dioxide, or silica), for example. Of the materials illustrated in Figure 2.1, garnet and mica are minerals; slate and granite are types of rock. The mineral mica is one of the constituents of the rock, granite.

(a)

(b)

(c)

(d)

Figure 2.1 Specimens of (a) garnet, (b) slate, (c) granite and (d) mica.

The distinction will become clearer as you examine specimens from the *Rock Kit* and *Mineral Kit* later in the Block.

The first group of rocks we shall consider is the **igneous** (Latin: *ignis*, fire) category of rocks. This type of rock is formed by the cooling and crystallization of magma (Greek: *magma*, a thick ointment). Cooling may take place rapidly at the Earth's surface (Figure 2.2), as with the formation of the first rocks of the primitive Earth; or may occur deep beneath the surface, where crystallization might not be complete for extended periods of time — hundreds or thousands of years. Igneous rocks consist of a mixture of several minerals, and it is their mineral composition that enables us to classify them further.

Figure 2.2 Basalt magma cooling rapidly after emerging from a volcanic vent, Kalapana, Hawaii.

The most common igneous rock at the surface of the Earth is **basalt**. It occurs mainly as the base crustal rock of the ocean floor, the **oceanic crust**, and it forms in vast quantities at volcanic vents far beneath the surface of the deep oceans. This production of new crust is of major importance to the dynamic activity of the Earth, the system of **plate tectonics**, which is discussed later in the course. Basalt is also the major component rock of volcanic islands, but it is found in huge sheets in continental locations as well. **Andesite** is an important igneous rock of the **continental crust**, those rocks mostly of lower density than basalt that form the Earth's landmasses. It derives its name from the volcanic chain of the Andes mountain range down the western side of South America. Another familiar and widely distributed igneous rock is **granite**, which is much used as a building stone.

The magma that finds its way into the crust is formed by the **partial melting** of rocks in the mantle. Complete melting would ensure that the composition of the solid rock and the magma were the same. If only partial melting takes place, the first to melt are those silicate minerals with lower melting temperatures. This means that the magma becomes depleted in those elements present in larger proportions in the higher-melting point silicate minerals. This is a major reason for the differences in the percentage composition of the crust and the mantle, which is clearly to be seen by comparing the data in columns 4 and 5 of Table 1.1.

○ Compare the two sets of data for mantle and crustal composition (Table 1.1) and identify the elements showing a depletion in the crust with respect to the mantle. What does this suggest about the composition of minerals in the mantle rocks?

● Both magnesium and iron show a notable decrease in their percentage contribution between the mantle and the crust, indicating that minerals found in mantle rocks are richer in iron and magnesium than are those of the crust.

Magma also undergoes a change of composition as it moves within the Earth. Some of the silicate minerals with higher melting temperatures may crystallize preferentially as magma rises into cooler parts of the crust, a process known as **fractional crystallization**. As a result the magma may lose some of its iron, magnesium, aluminium and calcium. Conversely, as hot liquid magma passes into or through rocks that are already in place, it can dissolve out elements like potassium, which is too large to be accommodated in the regular crystal structures of mantle minerals, increasing the levels of these elements in the magma.

○ Where are these processes taking place today?

● In volcanoes, as the magma rises towards the surface.

One of the effects of this interaction is that some less-common elements may be concentrated in crustal rocks. These elements dissolve in the magma, and often crystallize out irregularly in the crystal lattices of more common minerals. This is why they occur in the Earth's crust at a much higher proportion than would be expected from their overall occurrence in the Earth's chemical composition. Alternatively, they may be left in solution as the silicate minerals crystallize, becoming more concentrated and forming valuable metallic **ores** when they do solidify.

The occurrence of igneous rocks at the Earth's surface is not solely the result of volcanism. Geological processes have exposed rocks originally formed at depth (Figure 2.3), and some of these exposures result in dramatic features of the landscape. The high moors of Devon and Cornwall (e.g. the north west area of the Teign catchment) rest on an extensive mass of granite formed by magma cooling at depth that is now exposed at the surface. Some of the weather-resistant cores of ancient volcanoes are now exposed as volcanic plugs, as are the remnants of subterranean horizontal and vertical volcanic 'plumbing' systems (Figures 2.4 and 2.5).

Igneous rocks are a major type of crustal material, but because of their mode of formation they are not now as abundant at the Earth's surface as another group of rocks. Most of the rocks that you will encounter exposed at the surface, not only in the British Isles but also throughout the world, come from the group that includes **sandstones**, **shales**

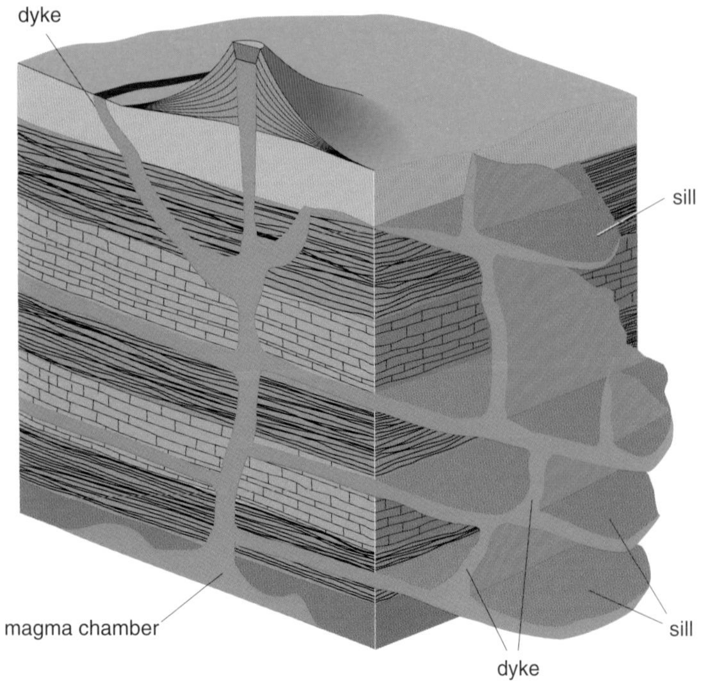

Figure 2.3 Schematic diagram showing the subterranean system of a volcano.

Figure 2.4 Salisbury Crags, a horizontal sheet of magma (sill) that solidified below the Earth's surface, now forms a dramatic geological feature above Edinburgh.

Figure 2.5 Vertical sheets of solidified magma (dykes), left when the softer overlying rock has weathered away.

(or **mudstones**), **limestone** and **the Chalk** (a particular fine-grained form of limestone, which is widespread in the south east of England). These are fragmentary rocks, formed from the weathered fragments of more ancient rocks, or rocks resulting from chemical and biological processes, that have been compressed and cemented together by lithification. Since they share the common feature that they were originally laid down as beds of **sediment** that accumulated in rivers, lakes or seas, they form the class of **sedimentary rocks**.

There is one further major rock type, and it too may be included in the broad category of crystalline rocks — the **metamorphic rocks**, so called from the Greek *meta,* after, and *morphe,* form, because they have undergone a process of *change*. The characteristic appearance of metamorphic rocks arises from their mode of formation. Originally metamorphic rocks were of igneous or sedimentary origin, subsequently subjected to intense heat or pressure, often both, through burial or other geological processes. The forces acting on the original rocks have changed their crystalline components into different minerals that are more stable under the new conditions. These new minerals often form in differentiated layers, giving metamorphic rocks a characteristic layered or banded appearance. Pressure, acting from a certain direction, encourages crystal growth at right angles to that direction, resulting in the 'squeezed out' appearance of the crystals.

2.1 Examining rocks from the DVD *Air/Earth*

Box 2.1   *Using the Rock Kit from the DVD Air/Earth*

The *Rock Kit* opening screen shows a menu of numbered rock specimens. For reference, a full list is provided in the appendix. You can always return to this screen by clicking on the button Rock Kit in the sidebar.

Click on any picture and a separate screen opens with a picture of that rock specimen, which can be rotated by placing the pointer (a Hand icon) anywhere in the screen, holding down the mouse button, and moving the mouse from side to side, or up and down.

There is a row of icons underneath the picture — a Microscope, one or more Projector slides, and a Video camera. When the Video camera is red rather than black, you know that the specimen on view can be rotated and manipulated, as described above.

Clicking on a Projector slide icon gives another view of the rock sample. Note that some of the projector slide icons are empty: only those with 'rock specimens' in them will give a new picture. These specimens cannot be rotated using the hand icon, but a new icon, Label, is available.

Selecting Label gives information about the specimen on view, and points out particular features. At the same time a quiz window, Name that rock appears on the right hand side of the screen.

A click on Loudspeaker opposite each rock name will tell you how to pronounce the name of the rock (if you have the computer sound switched on!), and if you select a rock name with the mouse, a panel below the rock picture will tell you whether your identification is correct, and give you further information about the rock being studied. If you choose incorrectly, you are given further hints towards selecting the correct answer. However, in this course you will usually be told which rock or mineral is being studied at the time.

The Microscope icon generates a view of a **thin section** of the rock. A very thin slice of rock is transparent to light, and so it can be viewed under the microscope. What is seen is a cross-section of the mineral grains. These have a characteristic appearance, so it is easy for a trained geologist to identify the constituent minerals of the rock specimen. The thin sections of the *Rock Kit* again have labels pointing out the important minerals. A click anywhere in this screen takes you back to the initial view of the rock sample.

These computer-based teaching materials were first prepared for use with the OU Level 2 Course S260 *Geology*. We shall not be using all the rocks and minerals in the two *Digital Kits*, but if you have time, the items we do not investigate are there to be explored for your own interest.

Throughout this section we shall be using two parts of the DVD *Air/Earth*, first the *Rock Kit* and later the *Mineral Kit*. Later still we shall use the part on *Earth Materials*. It will therefore be useful to carry out your study with your computer available for use. If you are unable to view the DVD material at the same time as studying the text, do ensure that you have worked through the DVD exercises before moving onto the *Soils* section of this book.

Activity 2.1   Igneous rocks from the *Rock Kit*

Igneous rocks are well represented in the *Rock Kit*, but we shall look first at those we have mentioned already, together with **peridotite**, which is the principal rock of the mantle, rich in magnesium and iron.

Load the DVD *Air/Earth* and open the *Rock Kit* section.

Look in turn at the four rocks; peridotite [23], basalt [3], granite [13] and andesite [1], in the *Digital Kit*, both in the hand specimen, and as viewed as a thin section under the microscope.

How would you *describe* each of these rocks? Concentrate on the features that they have in common. Look for example at the *texture* of the rocks, their *colour* and their *constituents*. A geological analysis is *not* wanted here. You may need to make a few notes before you move on to the next specimen.

Basalt and andesite are igneous rocks that formed at or near the surface, whereas granite and peridotite formed at depth. What feature of the rocks' appearance can be linked to this statement?

Allow about 30 minutes for this activity.

Comments on activities begin on p.306

Activity 2.2   Sedimentary rocks from the *Digital Kit*

The fragmentary nature of the sedimentary rocks can be clearly seen in the specimens from the *Rock Kit*. Selected for examination are specimens of two major types of sedimentary rock, sandstone and limestone. These two fragmentary rocks are formed in a quite different way from the crystalline rocks. The fragments are brought together by the geological process of **sedimentation** and converted into rocks, by a lithification process in which the spaces between the fragments are filled with a mineral **cement**.

From the *Rock Kit* on the DVD *Air/Earth* look at the rocks sandstone [10] and limestone [22].

What differences do you observe between these two specimens, and between these two and the crystalline rocks you examined earlier?

Allow about 20 minutes for this activity.

Activity 2.3   Metamorphic rocks from the *Digital Kit*

Again two specimens, this time of metamorphic rock, are selected for examination, one each from the two major subdivisions, **gneiss** and **schist**.

As a last excursion into the *Rock Kit* in this section, have a look at the rocks gneiss [4] and phyllite [11].

Although these two rocks show some of the characteristics of the igneous rocks you examined earlier, what would you identify as the significantly different feature of these metamorphic rocks?

Allow about 25 minutes for this activity (including Activity 2.4).

Activity 2.4 (optional) Additional metamorphic rocks from the *Digital Kit*

Now look at the two other rocks mentioned earlier, marble [17] and quartzite [5].

How do these rocks differ from those examined above? Can you suggest a reason why no obvious alignment of crystals is apparent in these two rocks?

2.2 Rocks of the Teign catchment

The geology of the Teign catchment is laid out on the geological map (Figure 2.6).

In detail the geology is complex, but on a more basic level it can be broken down into four divisions. Dartmoor Granite dominates the northwest of the catchment, surrounded by a broad swathe of sandstones and mudstones, as we pass from the high ground towards the coast. Bordering the coast are the younger and coarser breccias and sandstones, which form prominent coastal cliffs and escarpments between Teignmouth and Torquay. In the southern part of the valley are found slates, shales, sandstones and limestones.

The highly variable topography of this last region results from the contrasting resistance to weathering of the massive limestones and the soft slates. The slates have been worn down to what is now a low plain, broken by many isolated hillocks, marking more resistant igneous intrusions.

○ Where might you expect to find metamorphic rocks in this valley area?

● As metamorphic processes require heat or pressure, or both, metamorphosed sediments are most likely to occur close to the edge of the granite intrusion.

This region is indeed called the **metamorphic aureole**, and here we find that some mudstones have been baked.

In the Bovey Basin, which extends south-eastwards from the edge of the granite to the south of Newton Abbot, the underlying rocks are covered by more recent clays and sands. This is the source of the economically valuable **ball clay**, a fine-grained clay, rich in kaolinite, valued for its high plasticity and dry strength, and used particularly in the manufacture of sanitary ware! Almost a quarter of the world's output of ball clay is obtained from the Bovey Basin.

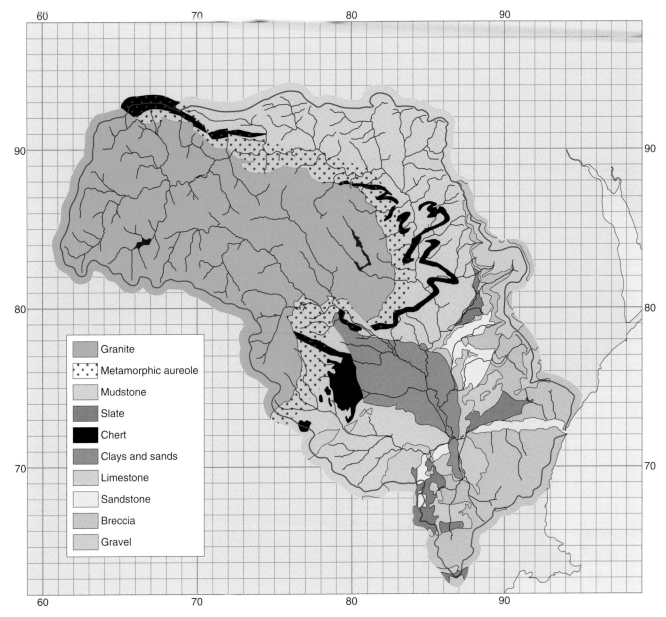

Figure 2.6 Geological map of the Teign catchment.

Legend:
- Granite
- Metamorphic aureole
- Mudstone
- Slate
- Chert
- Clays and sands
- Limestone
- Sandstone
- Breccia
- Gravel

Extensive mining of metallic ores has taken place in the Teign catchment in the past, but no mines are currently active; there has been no commercial mineral extraction since 1975. Sites of earlier workings and the currently active quarries are shown on Figure 2.7.

As a comparison of the two maps clearly shows, most of the **mineralization**, that is the formation of ores, is associated with the granite intrusion. Almost all former mine workings are found within 3–4 km of its perimeter, and both oxide and sulfide ores are present. Tin and iron oxides are found within the granite body. The principal ores to be found in the chert formations to the north and the east of the intrusion are of the metal manganese. Most of the mining has, however, concentrated on sulfides of copper, lead (which includes some silver) and zinc (which may include some cadmium), which are found, accompanied by barium sulfate and calcium fluoride, in the shales and slates on the southeastern flank of Dartmoor.

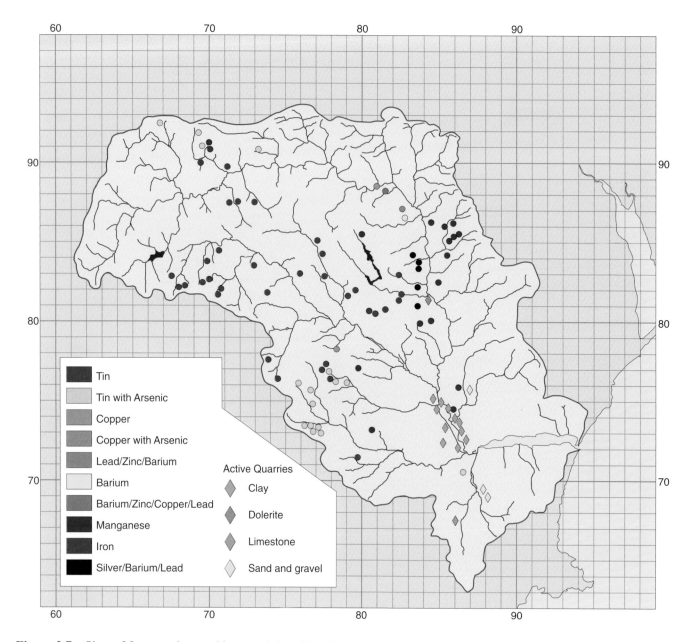

60   70   80   90

90   90

80   80

Tin

Tin with Arsenic

Copper

Copper with Arsenic

Lead/Zinc/Barium

Barium   Active Quarries

70   **Barium/Zinc/Copper/Lead**   ◆ Clay   70

Manganese

Iron   ◆ Dolerite

Silver/Barium/Lead   ◆ Limestone

◆ Sand and gravel

60   70   80   90

Figure 2.7 Sites of former mine workings, and the still-active quarries and ball clay extraction sites.

Question 2.1

Before reading the summary below, outline the characteristics of the three classes of rock discussed in this section.

Question 2.2

From the descriptions of the hand specimens of the three rocks given below, classify each one as igneous, metamorphic or sedimentary (fragmentary).

(a) A cream-coloured rock consisting of rounded glassy grains packed together. The surface is rough and pitted, and some of the grains are detached by scratching with a fingernail.

(b) Larvikite. A rock widely used as a decorative stone because of the iridescence of its polished surface. It consists of large interlocking crystals, grey or bluish-grey, with smaller amounts of an irregularly dispersed dark mineral.

(c) A smooth rock, having a silvery sheen on its undulating surface. Some large dark red crystals stand out from the surface, and the elongated crystals in the fine layers seem to flow around them.

2.3 Summary of Sections 1 and 2

1 Crustal rocks determine the geomorphology of the landscape and provide the major raw material for the soil.

2 The rocks of the Earth's crust are composed of minerals, principally formed from the elements oxygen and silicon, with decreasing but substantial proportions of aluminium, iron, calcium, magnesium, sodium and potassium. The major rock-forming minerals are silicates.

3 Igneous rocks, typically ocean floor basalts, and andesite and granite of the continental crust, are formed by the cooling and crystallization of magma.

4 Weathering and erosion of pre-existing rocks, resulting from interaction with water and the atmosphere, lead to the formation of sediments. Subsequent compression and cementation (the lithification process) give rise to the fragmentary (or sedimentary) rocks, which include sandstones and shales (or mudstones). Limestones are usually formed from fragments of ancient life forms.

5 Metamorphic rocks, gneiss, schist and slate, are former igneous or sedimentary rocks, transformed by heat and pressure. They frequently contain minerals different from those in their source rock, often distributed in layers.

Igneous rocks

3.1 Classification and properties of igneous rocks

Although all igneous rocks originate from magma, which is derived ultimately from the rocks of the Earth's mantle, our examination of the igneous rocks of the *Rock Kit* shows that they are very different in appearance. One obvious feature is the difference in size of the component crystals, which was ascribed to the manner and rate of cooling. This difference allows us to form a simple classification of igneous rocks into (a) **fine-grained**, with crystal sizes typically less than 1 mm in dimension, due to rapid cooling at or near to the Earth's surface; and (b) **coarse-grained**, with average crystal dimensions greater than 1 mm. Alternatively the rock types can be grouped as **extrusive** — rocks that crystallized at the Earth's surface, like volcanic lavas; or **intrusive** — rocks that both cooled and crystallized slowly deep within the Earth. Fine-grained basalt and andesite were clearly cooled rapidly and come into the extrusive group, while coarse-grained peridotite and granite must have cooled much more slowly, and meet the criterion for intrusive igneous rocks.

Although the classification we have just explored has proved a useful starting point for field geologists, using a criterion like grain size might seem rather superficial. A more rigorous classification is surely needed, perhaps based on mineral or chemical composition rather than appearance. Environmental scientists might need to know how the underlying rock could affect the character and fertility of a soil derived from it, for example. Chemical composition can be determined by grinding up a specimen and analysing it for silicon, magnesium, iron and other elements. Such analysis has, of course, been carried out, and some typical examples are given (Table 3.1) for peridotite and the three crustal igneous rocks we have examined already.

Table 3.1 Percentage composition by mass of some elements (as oxides) in igneous rocks.

Element	Oxide	Basalt	Granite	Andesite	Peridotite*
silicon	SiO_2	48.5	72.2	59.0	45.1
aluminium	Al_2O_3	19.4	14.5	17.4	3.3
iron	FeO/Fe_2O_3	9.66	2.76	6.63	8.0
magnesium	MgO	5.12	0.72	3.54	39.8
calcium	CaO	12.0	1.86	7.08	2.6
sodium	Na_2O	2.53	3.72	3.50	0.34
potassium	K_2O	0.25	4.11	1.44	0.02

* Estimate for total upper mantle composition, since peridotite is the principal rock found in the upper mantle.

○ How do these data justify the claims already made that crustal rocks contain more potassium and less magnesium than mantle rocks?

● All three crustal rocks contain considerably higher percentages of potassium oxide than the very low 0.02% found in peridotite. Similarly the 39.8% of magnesium oxide in peridotite far outstrips the 5.12%, 3.54% and 0.72% found in the other three igneous rocks in Table 3.1.

Since the elemental composition of each of these four igneous rocks is quite distinctive, this enables us to identify each one. However, it is not possible to carry out chemical analyses or make use of thin sections in the field. Instead geologists generally prefer to identify the suite of minerals contained in a hand specimen and to determine in what proportions each mineral occurs, then use this as a basis for classifying rocks. It is not surprising, in view of the differences in the analyses, to find that the nature and proportions of the minerals present in a typical specimen also differ from one rock type to another.

Activity 3.1 Further rocks from the *Rock Kit*

From the *Rock Kit*, examine the rocks **gabbro** [19], **diorite** [9] and **rhyolite** [6]. All these are igneous rocks, like the peridotite, basalt, andesite and granite you examined earlier.

What is the overall appearance of these three rocks? On the basis of their crystal size, do you think they cooled quickly or slowly?

Is the rhyolite likely to be a rock that cooled at the surface, or deep in the Earth's interior; extrusive or intrusive? Can you suggest an explanation for the two sizes of crystals co-existing in the same rock?

Allow about 15 minutes for this activity.

Activity 3.2 The mineral composition of igneous rocks

Return to the *Rock Kit* and use the 'label' function on each rock screen to note which minerals are included in each of the igneous rocks we have previously examined.

You will find there are five groups of important minerals identified in these six rocks: feldspars (described as plagioclase, orthoclase and potassium feldspar); micas (described as biotite and muscovite); pyroxene and amphibole; olivine; and quartz. Do not worry about the *specific* names given to the minerals or what these minerals are like, at the moment. There is an opportunity to examine them in the *Mineral Kit* later.

Fill in Table 3.2 overleaf, putting a tick in the appropriate mineral group column when one of the varieties of that mineral is present. For micas and feldspars, try to name the specific mineral.

Table 3.2   Mineral composition of igneous rocks (to be completed).

Rock type	Rock Kit no.	Extrusive or intrusive?	Feldspar	Mica	Pyroxene/ amphibole	Olivine	Quartz
basalt	3	extrusive					
gabbro	19	intrusive					
andesite	1	extrusive					
diorite	9	intrusive					
rhyolite	6	extrusive					
granite	13	intrusive					

Now compare your table with the completed table given on p. 307.

Allow about 35 minutes for this activity.

○ Which mineral is found in all the rocks of the crust?

● Feldspars are found in more of the common rock types than any other mineral. Indeed, feldspar accounts for almost 70% of the Earth's crust!

○ From your completed Table 3.2 (or Table 3.8 on p. 307) compare the mineral compositions of basalt and gabbro; what do you notice, and what inference can you draw?

● The three minerals plagioclase feldspar, pyroxene and olivine are the principal components of both rocks. A hypothesis that both gabbro and basalt form from magma of the same composition would be a reasonable inference, gabbro being the result of intrusive rock formation, and basalt the extrusive equivalent.

To support this hypothesis we must compare the chemical composition of basalt with that of gabbro, which is now given in Table 3.3. You would not expect perfect correspondence between two rocks taken from different localities, but overall the figures for basalt and gabbro show a pretty good match.

Table 3.3   Percentage composition by mass of some elements (as oxides) in igneous rocks.

Element	Oxide	Basalt	Granite	Andesite	Diorite	Gabbro	Rhyolite
silicon	SiO_2	48.5	72.2	59.0	57.6	47.3	73.9
aluminium	Al_2O_3	19.4	14.5	17.4	16.9	20.1	13.5
iron	FeO/Fe_2O_3	9.66	2.76	6.63	7.9	8.59	2.21
magnesium	MgO	5.12	0.72	3.54	4.2	5.49	0.50
calcium	CaO	12.0	1.86	7.08	6.8	13.2	2.00
sodium	Na_2O	2.53	3.72	3.50	3.4	1.66	3.74
potassium	K_2O	0.25	4.11	1.44	2.2	0.29	4.24

○ Now that we have the chemical composition of the three additional igneous rocks from Table 3.3, can you observe any other similarities in chemical composition?

● The compositions of andesite and diorite also seem to correspond very closely, and so do the compositions of rhyolite and granite.

○ Is there a similar correspondence in the mineral compositions (Table 3.8)?

● It does not seem to be as close as between basalt and gabbro, but nevertheless it proves to be sufficiently similar to enable these rock types to be identified in the field.

From these observations we reach the conclusion that the pairs of rock types basalt/gabbro, andesite/diorite, and rhyolite/granite each have essentially the same composition. The first member of each pair is the extrusive, and the second member the intrusive form of the same rock type. It is therefore possible to fit both extrusive and intrusive rocks onto a single chart (Figure 3.1) that shows how mineral composition changes as we move from type to type. The chart is arranged so that the proportions of silicon, sodium and potassium, more prominent in rhyolite and granite, increase towards the left, and calcium, iron and magnesium, which occur significantly more in basalt and gabbro, increase towards the right.

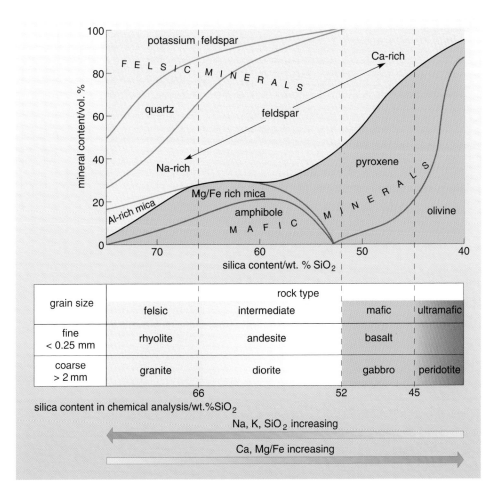

Figure 3.1 Chart to show the mineral composition of the common igneous rocks.

Rocks containing a high proportion of iron and magnesium are called **mafic** rocks, from the initial letters of magnesium and ferrum (iron). For peridotite, with even higher magnesium and iron content, the term **ultramafic** is used. Figure 3.1 shows that these three rocks all contain much pyroxene and olivine. The corresponding term for rhyolite and granite is **felsic**, referring to their high proportion of feldspars and quartz (silica). Andesite and diorite, with compositions lying between the other pairs, are described as **intermediate**. The dominant minerals of this group of rocks are feldspar and amphibole.

3.2 Appearance, chemical composition and properties of minerals

Now that we know about the mineral components of the common igneous rocks, we might reasonably expect that most of these minerals will also be found in the fragments that constitute the sedimentary rocks. We shall next look more closely at the composition and properties of the minerals themselves. When rocks are exposed in different environments, their behaviour can largely be understood in terms of the behaviour of their mineral components.

In terms of abundance in the crust, there are two major families of rock-forming minerals, the silicates and the carbonates, and the greater of these is the silicate family. Silica (silicon dioxide) is the constituent mineral of quartz grains that form the sandstones, and in the rest of the silicate family we find almost all the mineral components of the igneous and metamorphic rocks. These include the feldspars (pale-coloured minerals abundant in felsic rocks), which together with quartz form the felsic group of minerals. Pyroxene, olivine and amphibole are also silicate minerals that are rich in magnesium and iron, and are described as mafic (like the rocks), or **ferromagnesian** minerals. They have characteristically dark colours, which is why mafic rocks are also usually dark in colour. Micas also come from the silicate family.

The carbonate family shows much less variety, being limited mainly to the calcium carbonate (calcite) of the limestones and the Chalk, and the magnesium carbonate of dolomite rocks.

There are other classes of mineral, but they are generally less widely distributed. Some of their names may be familiar, as they include many of the 'useful' minerals — like bauxite (hydrated aluminium oxides; the principal source of aluminium) and iron pyrite (iron sulfide); halides like rock salt (sodium chloride), and sulfates like gypsum (calcium sulfate).

We now need to examine the variety in silicate minerals in more detail, but first we shall look at them using the DVD.

Activity 3.3 Minerals from the DVD *Air/Earth*

The *Mineral Kit* on the DVD *Air/Earth* gives you an excellent idea of the appearance of many of the common minerals, together with other information, and we shall spend a little time exploring it. For reference a full list is given in the appendix.

You will find the procedures for using the *Mineral Kit* from the DVD are the same as those used with the *Rock Kit*, so no further instructions should be needed.

○ Begin by recalling the major mineral components of granite and basalt.

● Granite is mainly composed of quartz and feldspars, with some mica; basalt is largely pyroxene and olivine.

Quartz, feldspar, mica and olivine are investigated through the DVD. As you work through the minerals, fill in the columns of Table 3.4. A calcium/sodium feldspar (named as 'plagioclase' on the DVD) and pyroxene, which are not in the *Mineral Kit*, have been entered already.

Table 3.4   Characteristics of selected minerals (to be completed).

Mineral	Metal ions and silica	Hardness	Colour
olivine			
pyroxene	Mg, Fe, Ca, SiO_2	5–6	almost black
amphibole			
feldspar (plagioclase)	Ca, Na, Al, SiO_2	6	white
feldspar (orthoclase)			
mica (muscovite)			
mica (biotite)			
quartz			

Load the DVD that contains the *Digital Kit* and open the *Mineral Kit* section.

Quartz

Click on mineral III, and note its appearance.

This is a specimen of quartz, indeed it looks rather like a magnified grain of sand.

Click on Label, and then on Quartz in the sidebar.

As with the rock specimens, this will give you more information about the mineral, in particular its hardness, 'H', its chemical formula, and a property called *cleavage*. Cleavage is only mentioned here because it is significant in the processes that lead to the breakdown of crystals, and consequently rocks, when they are exposed in the environment. The planes of cleavage represent weaknesses in the crystal, which can be exploited by the agents of physical and chemical weathering ... but back to quartz!

From the information on the screen, fill in the appropriate columns of Table 3.4 for quartz.

There is a further screen on the DVD for quartz, mineral VII. Access this screen as well, and after exploring it thoroughly, fill in any further information on Table 3.4.

Feldspar

The next minerals to investigate are the feldspars. They are represented on the DVD by a potassium-rich feldspar, called orthoclase, minerals VIII and IX. Explore these two screens and fill in the relevant sections of Table 3.4.

Mica

There are two micas included in the mineral selection on the DVD. Mineral II is muscovite mica and mineral XVI is biotite mica. Examine these two screens and fill in the relevant sections of Table 3.4.

Olivine and amphibole

The mafic minerals are represented by olivine (XVII) (shown both as nodules in a hand specimen of rock and as gem-quality crystals) and amphibole (XIII).

Investigate these screens and use the data to complete Table 3.4.

Finally, check your entries in Table 3.4 with the completed table (Table 3.9) on p. 308.

Allow about 40 minutes for this activity.

3.2.1 Formation of silicate minerals

Silicate minerals may be derived in several ways. Those we have encountered so far have been formed by crystallization from magma, and this is the primary route. There are, however, other ways to form silicates. One is metamorphism, where the application of heat and pressure forces changes in structure at a molecular level, resulting in new minerals that are more stable under such extreme conditions. Another is **alteration**, in which hot **magmatic fluid**, left over when magma has almost completely crystallized, penetrates rocks already in place, dissolving ions and altering the mineral composition. Seawater circulating through rocks, or simply exposure at the Earth's surface, can also bring about alteration. The clay minerals are usually formed by alteration.

Magma is not formed from a single mineral, but from a mixture. Figure 3.2 shows a chart of melting temperatures for the silicates found in igneous rocks. Note that minerals do not have a sharp melting temperature, but melt over a considerable temperature range.

Melting temperature is also critical in determining which minerals will be formed as magma cools. As we might expect, the first minerals to melt are the silicates with lower melting temperatures, so that as partial melting takes place, minerals in the lower part of the chart, like quartz and mica, are the ones that melt, while minerals in the upper part are left as solids. Conversely, as magma cools, minerals with higher melting temperatures, like olivine and pyroxene, begin to crystallize out first.

○ If *partial* melting of a diorite takes place, and the resulting magma is extruded at the Earth's surface, use Table 3.1 and Figures 3.1 and 3.2 to decide what would be the nature of the rock formed.

● Diorite consists largely of feldspar and amphibole, with some mica and quartz. On partial melting, quartz, mica and feldspar will melt before the amphibole. A magma with this composition would generate a rock like rhyolite at the Earth's surface.

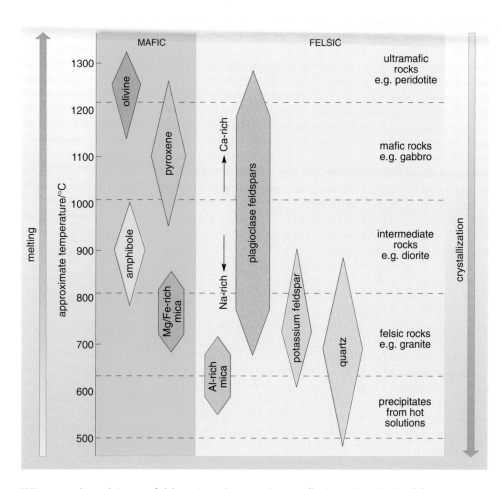

Figure 3.2 Ranges of melting temperature for silicate minerals.

When a mineral (say a feldspar) melts, we do not find 'molecules' of feldspar floating around in the melt. Instead, the **ions** that compose the mineral separate from the crystals and take up an independent existence.

Box 3.1   What is an ion?

Atoms of elements exist in a free state as a central nucleus surrounded by electrons. However, metallic elements in minerals are not found as atoms, but as **ions** from which one or more of the electrons are missing. As electrons are deemed to have a *negative* charge, an atom that loses an electron acquires a *positive* charge in becoming an ion. An ion is represented by a superscript plus or minus sign, together with the number of charges it carries, attached to the atomic symbol; e.g. the calcium ion is shown as Ca^{2+}. Ions with positive charges are called **cations**.

Table 3.5 gives a list of the usual charges on the ions of the common elements to be found in minerals. Carbon and silicon are ascribed charges of 4+, even though no such ions are found on Earth! These are described as *formal charges*. Notice too that iron (Fe) appears twice in the table. This is because iron can form two distinct ions: Fe(II), with two positive charges, and Fe(III), with three positive charges.

Table 3.5 Formal ionic charges of elements present in rocks and minerals.

Element	Symbol	Charge
aluminium	Al	3+
carbon	C	4+
calcium	Ca	2+
iron	Fe(II)	2+
magnesium	Mg	2+
iron	Fe(III)	3+
hydrogen	H	1+
potassium	K	1+
sodium	Na	1+
oxygen	O	2–
sulfur	S	2–
fluorine	F	1–
silicon	Si	4+

Some elements, like fluorine, acquire additional electrons and so have a negative charge. Other species with negative charges are formed by combining several elements to form *complex ions*, like carbonate (CO_3^{2-}) and silicate (SiO_4^{4-}). Ions that have negative charges are called **anions**. It is the attraction between the positively and negatively charged ions that holds the crystals of a mineral together, so-called **ionic bonding**. Materials that have this type of structure are termed **ionic materials** (or less correctly *ionic compounds*).

Within any formula, the sum of the ionic charges must be zero; with no stray + or – left over. A possible formula for the mineral olivine is therefore Mg_2SiO_4, where four plus charges on two magnesium ions, ($2Mg^{2+}$) balance four minus charges on the silicate ion, (SiO_4^{4-}). This rule also applies to ionic materials, whether as solids, when molten, or in solution.

○ What are the ions present in the minerals magnesium carbonate, with formula $MgCO_3$; and calcium fluoride, CaF_2?

● Magnesium carbonate, $MgCO_3$ will give rise to one magnesium ion, Mg^{2+} and one carbonate ion, CO_3^{2-}; calcium fluoride, CaF_2, to one Ca^{2+} ion and two F^- ions.

Many ionic materials are soluble in water, and in the process of solution, ions break free from their crystal and adopt a separate existence, surrounded by molecules of water. When ionic materials melt, their ions similarly become separated from the crystal structures. Magma is therefore composed of a mixture of ions, not separate 'molecules' of minerals.

Table 3.6 Major elements in silicate minerals.

Silicate mineral	Elements
olivine	Mg Fe
pyroxene	Mg Fe Ca
amphibole	Mg Fe Ca Al OH F
feldspar	Ca Na K Al
mica	K Mg Fe Al OH F
quartz	Only silica SiO_2

Magma is therefore a complex mixture of metal cations — principally sodium and potassium (the *alkali metals*), calcium, magnesium, aluminium, iron — and silicate ions. Table 3.6 reminds you which ions are present in the common silicate minerals. Note that amphibole and mica also include the anions fluoride (F^-) and hydroxide (OH^-).

Crystallization from a cooling magma will begin when its temperature drops to the melting temperature of the mineral with the *highest* melting temperature. As a mineral begins to crystallize, it begins to take some of the ions out of the melt. If the melt is rich in magnesium and iron, olivine and pyroxene will crystallize first, taking out these ions, together with many silicate ions. As the magma becomes cooler, calcium, potassium and sodium will compete for the remaining silicate ions to form feldspars. The higher melting, calcium-containing plagioclase will crystallize before the alkali feldspars. If this uses up most of the ions, the rock that is formed, consisting mainly of olivine, pyroxene and plagioclase, will be a basalt.

○ A magma contains very few iron, magnesium and calcium ions, but is rich in sodium, potassium and silica. If it crystallizes deep in the Earth, what minerals will probably be formed, resulting in what rock type?

● This magma would probably yield mainly sodium and potassium feldspars (the so-called 'alkali' feldspars, since sodium and potassium are the alkali metals) and quartz. At depth this would give a granite.

Surprisingly, water is quite important in the formation of silicates. You might think that a rock is completely dry, and a magma at 800 °C would not contain any water either. However, under the intense pressure at depth, water can be present dissolved in magma, with the result that **hydrous minerals** like amphibole and mica, which include OH^- as well as metal ions and silica, can form as magma solidifies. As magma is extruded and pressure is released, water vapour becomes one of the main gases escaping from a volcanic eruption. When minerals begin to crystallize, this water begins to separate out from the rising and cooling magma. Rich in dissolved metallic ions, this aqueous fluid, known as magmatic fluid, becomes an important source of ore minerals and an agent for the alteration of existing rocks.

3.2.2 Structure and physical properties of silicate minerals

The structures of silicate minerals are quite complex, but they are well displayed on the DVD *Virtual Crystals*, and it is not necessary for you to have a deep understanding of the details.

The element silicon is to be found in the same group of the Periodic Table as carbon, and so might be expected to share some of that element's characteristics. In its compounds, carbon forms four single **covalent bonds** (see Box 3.2) to other atoms, directed towards the corners of a tetrahedron (from the Greek -*tetra*, four; *hedra*, base or seat), and in the silicate minerals silicon does the same, bonding to four atoms of oxygen, to form the 'silicate tetrahedron' (Figure 3.3).

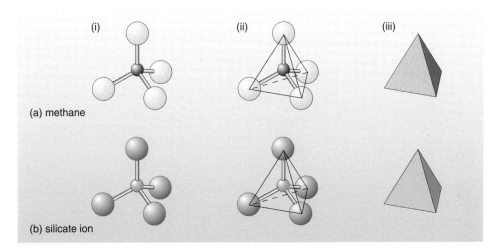

Figure 3.3   Molecular models of (a) methane (CH_4), and (b) silicate (SiO_4^{4-}), showing their relation to a tetrahedron.

Box 3.2 *What is a covalent bond?*

Although the crystals of minerals are held together by the forces of attraction between positive and negative ions, many chemical compounds, and the complex ions we discussed earlier, are held together by covalent bonds. In a covalent bond, electrons are not acquired or given away, but shared. This type of bond is surprisingly strong, and as a result the complex ions that are held together by covalent bonds are able to survive even the extreme temperature conditions found in magma. We have seen above that a basic building block of the silicate minerals has a central silicon atom *covalently* bonded to four oxygen atoms.

If carbon forms its four covalent bonds to other atoms of carbon, and this process is repeated, with each of the attached carbon atoms also forming four carbon-to-carbon bonds, the structure of diamond, the hardest of all known minerals, is built up (Figure 3.4).

Since each oxygen atom has the capability of forming two bonds, each corner of the silicate tetrahedron (Figure 3.3) can form one additional bond. The silicon-to-oxygen covalent bond is very strong on a molecular scale, even stronger than the carbon-to-carbon covalent bond. So silicate tetrahedra can be linked together through each oxygen atom forming another oxygen-to-silicon bond. If this process of silicon-to-oxygen bonding is developed to build up a continuous 3-D array, the structure of quartz is the result (Figure 3.4).

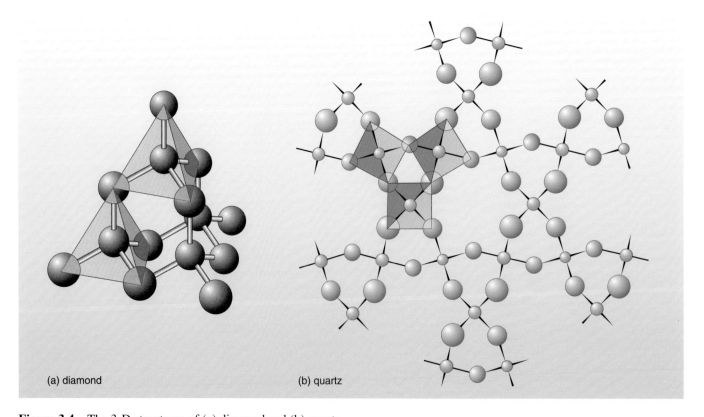

(a) diamond   (b) quartz

Figure 3.4   The 3-D structures of (a) diamond and (b) quartz.

The similarity of the quartz and the diamond structures is clearly visible. It is not therefore surprising to find that quartz is one of the most durable of natural materials.

There are other ways in which silicate tetrahedra may be linked together in common rock-forming silicate minerals, and we shall now explore these using the DVD *Air/Earth*.

Activity 3.4   Silicate structures at a molecular level

Open the section *Virtual Crystals* on the DVD *Air/Earth*, and, under Menu, select quartz from the drop-down menu at the top of the opening screen.

The screen display is typical for the whole range of minerals covered, and shows a frame on the left giving some general information about the structure, and a menu on the right. For each of the silicate rocks the first two items on this menu are the same, 'Single SiO_4 group' and '... represented as a tetrahedron'. Click on these buttons and they generate animated 'ball and stick' and geometrical representations, respectively, of the silicate tetrahedron in the left hand screen. You are also able to control rotation using the mouse. The next menu item shows how the tetrahedra link together, and the last item shows how the complete structure is built up, including the metallic ions when they are present.

The final button in the menu is Overview, and it sports a Movie icon. Click on this, and a continuous narrative takes you from the silicate tetrahedron to the complete structure. The movie often includes further information and illustration, and is probably the most useful part of each section.

Click on the menu items in turn to display all the information relating to quartz.

The drop-down menu shows which other minerals are also accessible in this program.

Select in turn olivine, pyroxene, amphibole, mica (biotite and muscovite) and feldspar, and explore each screen. For each mineral, note particularly: (a) the way the silicate tetrahedra are connected; (b) the overall structure that this leads to; and (c) the effect that the resultant crystal structure has on the character of the mineral, especially hardness and cleavage.

It is not necessary for this course to have a detailed knowledge of crystal structures.

Allow about 40 minutes for this activity, including Activity 3.5.

Activity 3.5 (optional) Exploring the properties of minerals

Properties of Minerals gives further general information about minerals. The most interesting of the sections for this course are those on the properties of shape, hardness and cleavage, and under 'colour' there are good illustrations of a selection of minerals. Browsing through this DVD section will revise and reinforce some of the concepts discussed in this section.

Description	Arrangement of tetrahedra	Mineral examples

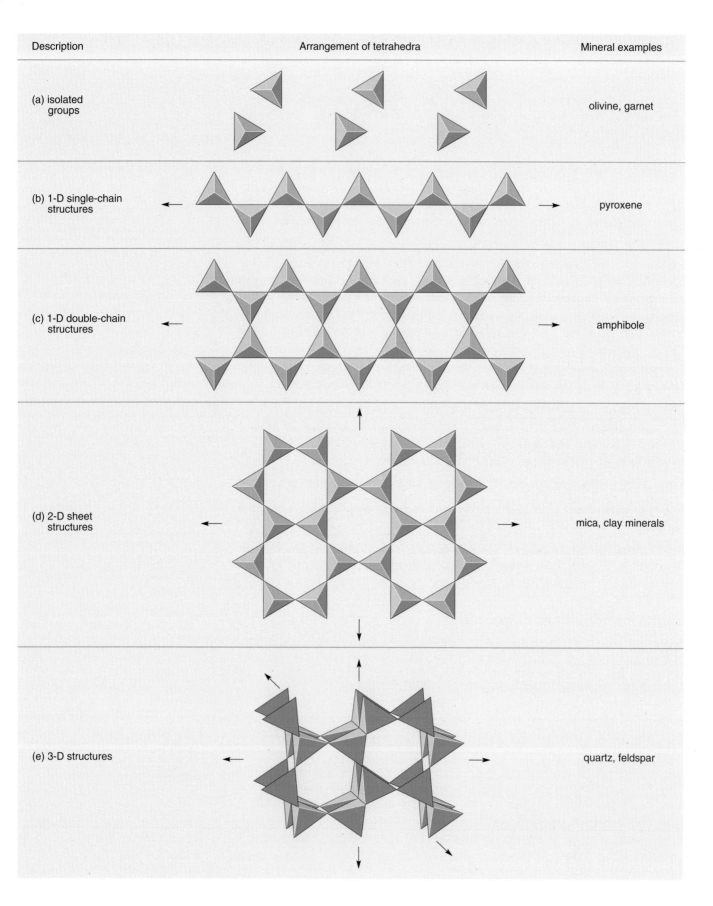

(a) isolated groups — olivine, garnet

(b) 1-D single-chain structures — pyroxene

(c) 1-D double-chain structures — amphibole

(d) 2-D sheet structures — mica, clay minerals

(e) 3-D structures — quartz, feldspar

◀ **Figure 3.5**   Different ways of linking silicate tetrahedra in rock-forming minerals.

The different ways in which silicate tetrahedra may be linked together are shown in Figure 3.5, which provides a brief summary of *Virtual Crystals*. The simplest form is found in the mineral olivine, where the SiO_4 tetrahedra are not linked, but each one is separate within the crystal structure (Figure 3.5a). Ions of silicate (SiO_4^{4-}) are balanced by twice as many ions of iron, Fe^{2+}, or magnesium, Mg^{2+}. If two corners of every silicate tetrahedron are shared with other tetrahedra, the form is an extended chain, in which an oxygen atom forms a bridge between every two silicate ions. The backbone of this structure is therefore $-O-SiO_2-O-SiO_2-O-SiO_2-O-SiO_2-$. Pyroxene has this structure, and forms single chains (Figure 3.5b), while amphibole forms double chains of silicate tetrahedra (Figure 3.5c). These are described as '1-D' structures in Figure 3.5.

In mica, the tetrahedra form sheets, called '2-D' structures in Figure 3.5, where they all share *three* oxygen atoms (Figure 3.5d). In the crystal, the sheets are stacked one on top of another, and this is reflected in the layered, platy structure of the mineral. The clay minerals have similar structures.

Lastly, as shown in Figure 3.5, the feldspars have 3-D structures, like quartz.

○   Which other elements, in addition to silicon and oxygen, are abundant in the silicate rock-forming minerals of the Earth's crust? Refer back to Table 1.1 for confirmation of your choice.

●   The other major elements are all metals; in order of abundance they are aluminium, iron, calcium, sodium, potassium and magnesium.

In the silicate tetrahedron, silicon is described as being *four co-ordinate*, that is, it has four nearest neighbours. Iron, magnesium and calcium are often found in minerals with six nearest oxygen neighbours, arranged at the corners of an **octahedron** (*octa* is the Greek word for eight and *hedra* for base). The six co-ordinate bonds form eight faces; you can verify this for yourself in Figure 3.6. As you have seen in Activity 3.4, mineral structures are often made up from these octahedra together with the silicate tetrahedra.

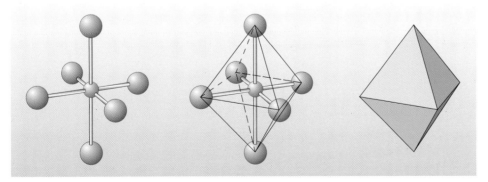

Figure 3.6   Six-coordinate iron, showing the octahedral arrangement of oxygen atoms.

Sodium and potassium, and many metallic ions that occur infrequently in mineral structures, usually occupy 'holes' or spaces in the crystal structure, left within the network of tetrahedra and octahedra. These spaces are clearly to be seen in the computer representations of structures. That accounts for all the common crustal elements except aluminium, the third most abundant element in the Earth's crust. Before we can discover where aluminium occurs in silicate structures, we shall have to look at the formulas that are given to the common rock-forming silicate minerals.

When you were carrying out Activity 3.3 you saw, in passing, the formulas of all these minerals, while you were looking for the elements that they contained. They might have looked quite formidable at the time, but now we are better able to understand how they arise. They are collected in Table 3.7.

Table 3.7 Formulas for the common rock-forming silicate minerals.

Mineral	Chemical formula
quartz	SiO_2
potassium feldspar (orthoclase)	$KAlSi_3O_8$
calcium/sodium feldspar (plagioclase)	$NaAlSi_3O_8 — CaAl_2Si_2O_8$
mica (muscovite)	$KAl_2(Si_3Al)O_{10}[OH,F]_2$
mica (biotite)	$K[Mg,Fe^{2+}]_3[Al,Fe^{3+}]Si_3O_{10}[OH,F]_2$
olivine	$Fe^{2+}_2SiO_4 — Mg_2SiO_4$
amphibole	$Ca_2[Fe^{2+},Mg]_4Al(Si_7Al)O_{22}[OH,F]_2$
pyroxene	$[Mg,Fe^{2+}]_2Si_2O_6$

Box 3.3 *Representing the chemical composition of minerals*

Table 3.7 shows the formulas used to describe a number of common minerals, but some of them look rather more complex than other chemical formulas that you might have encountered. There are two reasons for this. Firstly, the 'formula' of a mineral is not the formula of a molecule, like C_2H_5OH for ethanol. Instead it is the formula obtained by chemical analysis of the mineral. Just like the formula $CuSO_4.5H_2O$, which tells us the ratio of copper ions (Cu^{2+}), sulfate ions (SO_4^{2-}) and water molecules in the blue copper sulfate crystals, the formula of a mineral tells us the ratio of each metal ion to silicon and oxygen in the crystals. Secondly, unlike most chemical compounds, many minerals do not have an *exact* chemical composition. Although they have well defined structures it is possible to replace one metallic ion by another (or sometimes more than one), and still retain structural integrity.

Let's begin by looking at the formula of pyroxene, which is given as $[Mg,Fe^{2+}]_2Si_2O_6$. What this implies is that for every Si_2O_6 unit in the crystal there are two metal ions. Some of these are magnesium (Mg^{2+}), and some are iron in its Fe(II) or Fe^{2+} form. Pyroxene is a

ferromagnesian mineral, and in its crystal, iron and magnesium are interchangeable. This is only because iron (II) and magnesium have the same ionic charge of 2+, and a similar size. It is the presence of Fe(II) that gives the ferromagnesian (mafic) minerals their green colouration.

○ The so-called 'alkali feldspars' vary in composition between $KAlSi_3O_8$ and $NaAlSi_3O_8$. Suggest a formula for alkali feldspar to show this, and indicate which metal ions are interchangeable.

● A general formula would be $[K,Na]AlSi_3O_8$, and the interchangeable metal ions would be potassium (K^+) and sodium (Na^+).

In amphibole and the micas the silicate unit is not the only negatively charged ion included in the formula. Micas, like amphiboles, are termed *hydrous* minerals; that is, they include hydroxide (OH^-), or other anions in their structures, of which the commonest is fluoride (F^-). These anions arise from the water or fluorine gas dissolved in magma, but the ratio of hydroxide or fluoride may vary in different samples of mica.

○ Look at Table 3.7, and list the minerals of the Earth's crust where aluminium is to be found.

● Looking at the formulas, we see that aluminium is found in the micas, the feldspars and amphibole.

For this reason this group is better described as **aluminosilicate**, rather than just *silicate*, minerals.

We have already mentioned that feldspars are the most abundant crustal minerals. In the alkali feldspars, the ratio of aluminium to silicon is 1 : 3, but in the calcium-rich feldspar the ratio of aluminium is even higher at 1 : 1. As there is therefore more than one aluminium to every three silicon atoms in the feldspars overall, this accounts for a massive amount of aluminium in the crust.

Clay minerals also include significant amounts of aluminium, but as we shall see (Section 5.1.4), these aluminium atoms are each surrounded by six oxygen atoms in octahedral array (Figure 3.6).

But where do you think the aluminium might appear in other crystal structures? The term *aluminosilicate* may have provided a clue. Most aluminium is found as a replacement element in the silicate tetrahedra, but in a definite, rather than a random ratio. However, as the aluminium ion has a charge of 3+, compared with the formal charge of 4+ on silicon, there would not be enough positive charge to balance the negative charges on oxygen, unless one additional positive ion is provided for every silicon atom replaced. In the potassium-rich feldspar, with formula $KAlSi_3O_8$, one aluminium (Al^{3+}) and one potassium (K^+) together provide four positive charges; the three silicons (Si^{4+}) — remember these are only 'formal' positive charges — provide the other 12, to balance the 16 negative charges ($8 \times O^{2-}$) of the oxygens.

○ Satisfy yourself that the ionic charges on the elements in calcium-rich feldspar, $CaAl_2Si_2O_8$, add to zero.

● On the positive side we have calcium (2+), two aluminiums ($2 \times 3+$) and two silicon atoms ($2 \times 4+$); in total 16+. On the negative side we have eight oxygen atoms ($8 \times 2-$); in total 16−. The separate totals of the plus and minus ionic charges are equal, so their sum is zero.

Here each calcium ion (Ca^{2+}) accounts for *two* replaced silicons. Where alternative elements are possible, as shown by the square brackets in the formula of the alkali feldspar, $[K,Na]AlSi_3O_8$, the charge on either one only has to be counted once!

With the hydrous micas and clay minerals, there is an alternative way of neutralizing some of the excess positive charge on silicon. The hydrating ions, hydroxy (OH^-) and fluoride (F^-), both carry a negative charge, and so each one can neutralize one positive charge. For example in the formula of muscovite mica, $KAl_2(Si_3Al)O_{10}[OH,F]_2$, three silicons and three aluminiums contribute 21 positive charges, made up to 22 by the single potassium. These are balanced by 20 negative charges from oxygen, and two from either OH^- or F^-.

○ Look at the chemical formulas of quartz, mica, pyroxene and olivine in Table 3.7. What do you observe about the silicon : oxygen ratio in these minerals, and how can you explain the change in ratio? Note that with the micas, the iron or aluminium that replaces some of the silicon must be counted as well.

● In quartz the silicon : oxygen ratio is 1 : 2. As we go along the series we see that pyroxene has the ratio 2 : 6 (1 : 3), mica 3 : 10 (1 : 3.3), and olivine 1 : 4. As the silicate tetrahedra share more of their oxygen atoms in the crystal structure, the silicon : oxygen ratio falls.

Note that the more mafic minerals have the higher proportions of oxygen to silicon. This is because the amount of oxygen sharing by silicon tetrahedra increases from olivine, through pyroxene and amphibole, to mica and quartz. It is also interesting to note that the most compact crystal structures are found for olivine, pyroxene and amphibole, minerals with high melting temperatures, that crystallize under conditions of high pressure, deep within the Earth. The more open 3-D framework structures of quartz and feldspar are found in minerals that most frequently crystallize at lower pressures and temperatures nearer to the surface.

Question 3.1

The mineral augite has the formula $Ca[Mg,Fe^{2+}]Si_2O_6$.

(a) How do you think the silicate tetrahedra will be arranged in this mineral?

(b) Would you classify this mineral as a mafic or a felsic mineral?

(c) What is indicated by the part of the formula '$[Mg,Fe^{2+}]$'?

(d) What proportions of calcium, magnesium and iron (II) ions exist in this structure?

(e) Check that the oxidation level of iron must be 2+ or iron (II), not 3+ or iron (III), by balancing the positive and negative charges in the formula.

Question 3.2

Norite is an igneous rock consisting almost entirely of calcium-rich feldspar and pyroxene in about equal proportions; and syenite has the chemical composition (main components only; mass % as oxides) given below.

SiO_2, 62.5%; Al_2O_3, 17.6%; FeO/Fe_2O_3, 4.8%; MgO, 0.9%; CaO, 2.3%; Na_2O, 5.9%; K_2O, 5.2%

Using Figure 3.1 and Table 3.3 describe each of these rocks as felsic, mafic or intermediate.

Metamorphic rocks

In the brief survey of the major rock types in the earlier part of this book, we emphasized the different appearance of the metamorphic rocks in comparison to rocks of the igneous type. However, appearance is not the only significant difference between these two groups of crystalline rocks. They also differ in their *mineral* composition, although there is usually no significant change in their *chemical* composition, especially if it is igneous rocks that are metamorphosed. Metamorphism usually occurs in the solid state, without any melting, and so the nature of the minerals formed is principally dependent on those that were present in the original rock, as little or no transfer of material takes place. However, with sedimentary rocks metamorphic changes often do occur which involve the loss of water or gases, because of the high temperatures under which metamorphism occurs. The presence of water in the original rock may make metamorphic change more likely, and it may be incorporated into new hydrous minerals as well.

Each mineral has a limited range of conditions under which it is stable, and if temperature and pressure lie outside this range, minerals will be transformed into those that can better accommodate the applied stresses. Under severe conditions some small-scale migration of elements can also take place, and as a result, the pale-coloured felsic and the dark coloured mafic minerals may separate out into bands. This is the cause of the distinctive appearance of the group of rocks described as gneiss (Activity 2.3; Figure 4.1).

Figure 4.1 A hand specimen of gneiss.

Metamorphism can take place adjacent to an intruded magma body due to the high temperature of the magma, and this has the effect of baking the surrounding rocks. Unsurprisingly this process is termed **contact metamorphism**, and the area over which the effect is felt is called the *metamorphic aureole*. This feature is observed around the body of Dartmoor Granite in the Teign catchment, where all the rocks surrounding the granite show the effects of heat from the magma, often becoming hard and splintery. The metamorphic minerals found indicate that the magma was at a temperature of 600–700 °C. Contact metamorphism is always quite localized in its effect, extending to between 0.5 and 4 km from the boundary in the example of the Dartmoor Granite.

The more significant form of metamorphism, which extends over much larger areas, is termed **regional metamorphism**. This results when rocks are subject to an increase in pressure as well as temperature, both mainly related to depth of burial. Regional metamorphism may also be caused by tectonic activity, which also generates heat and pressure. High pressure has a significant effect on the nature of the metamorphic rock. Notably, those minerals like mica and amphibole that tend to form platy or elongate crystals tend to have this crystal habit accentuated; and this produces a characteristic 'layering' (Figure 4.2), usually obvious to the naked eye in a rock specimen, and in thin section under the microscope (Figure 4.3). It is characteristic of all regionally metamorphosed rocks, other than rocks like marble and quartzite that comprise only a single mineral.

Figure 4.2 Simplified diagram to show how platy minerals like mica (a), or elongate crystals like amphibole (b), change from a random distribution (c) to alignment in layers (d) under metamorphic conditions of heat and compression.

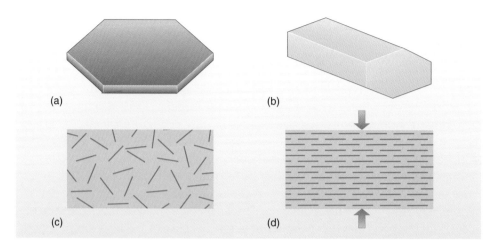

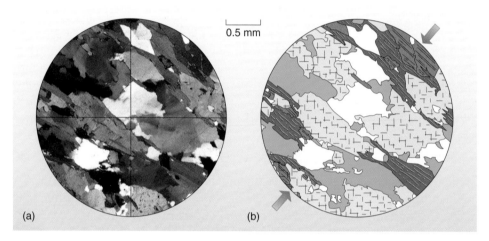

Figure 4.3 Appearance of metamorphic rock specimen (gneiss) under the microscope: (a) actual and (b) corresponding sketch.

The presence of these layers in slate enables it to be split into thin sheets, as it can be cleaved along planes of weakness (Figure 4.4). It is the layer of a platy mineral, often mica, on the cleavage surfaces that gives fresh slate its shiny appearance. Although this behaviour is described as '**slaty cleavage**', it is quite different from the cleavage shown by minerals. They do share one thing in common, however, in that both types of cleavage indicate weaknesses in the rock or mineral, and they are usually the sites at which physical and chemical weathering can begin.

The grain size of crystals in metamorphic rocks is determined by the intensity of the temperature and pressure conditions. Large grain size is usually the result of extremes of temperature and pressure on the original rock, which cause considerable change. Contrast this with igneous rocks, where the grain size is determined by how slowly the magma cools. Of course, the processes by which the minerals are formed are very different. Igneous minerals crystallize from a liquid melt, but the minerals of metamorphosed rock change without melting, to materials that are more stable under the prevailing conditions. Thus we can

Figure 4.4 Slate split into sheets along its cleavage planes.

conclude that the fine-grained minerals of slate have been subject to much less intense conditions, and so have undergone less change, than the large crystals found in gneiss. The contorted intermediate-sized crystals found in phyllites (Figure 4.5) and schists must have formed in conditions somewhere between the two extremes.

(a)

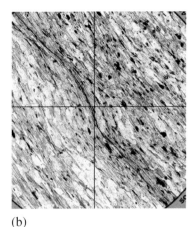

(b)

Figure 4.5 (a) Hand specimen and (b) thin section of phyllite.

Although the chemical composition of a metamorphic rock may be almost identical to the rock from which it was formed, the minerals it contains may be quite different. As an example of a metamorphic change, consider the equation below:

$$KAl_3Si_3O_{10}(OH)_2 + SiO_2 = KAlSi_3O_8 + Al_2SiO_5 + H_2O \qquad (4.1)$$

mica   quartz   feldspar   aluminium   water
                                      silicate

This reaction takes place above 550 °C at a pressure of 150 MPa. The aluminium silicate has a more compact crystal structure than the mica that it replaces.

Activity 4.1 The minerals of metamorphic rocks

Previously you examined igneous rocks of the *Rock Kit* to identify the minerals they contained. It is now time to do the same for metamorphic rocks. Tick the spaces in Table 4.1 when that mineral appears in the *Rock Kit* specimen, and write in the names of any other minerals.

Table 4.1 Minerals in metamorphic rocks (to be completed).

Rock type	No.	Mica	Feldspar	Quartz	Other minerals
gneiss	4				
schist	26				
phyllite	11				
slate	2				

Load the *Rock Kit* from the DVD *Air/Earth* and examine rock 4, the gneiss you looked at earlier. The differentiation of the mafic (dark) and felsic (pale) minerals is well shown in these samples. Using the various screens and the thin sections, indicate the minerals found in the gneiss in the appropriate column of Table 4.1.

Now do the same for the schist, rock 26. This is a rock you have not examined previously.

This is described as a *garnet mica schist*, and it is quite common to use the minerals prominent in a metamorphic rock in giving it a name. **Garnet**, a semi-precious stone, is a mineral regularly found after the metamorphism of mudstones and shales. It is well displayed in Figure 2.1a and as mineral X in the *Mineral Kit*, if you wish to examine it there. It can be even harder than quartz, and appears in a variety of colours.

○ Garnet is also a very dense mineral. Why do you think metamorphic minerals might have higher densities than the minerals of igneous rocks?

● Recall that metamorphism occurs at high pressures, and so results in crystals with tightly packed ions. Tightly packed crystal structures will probably be denser than more open structures.

Rock 11 is the phyllite you have seen before.

Examine the screens and add its characteristic minerals to Table 4.1 as well.

The mineral **chlorite**, which looks rather like a greenish mica, is also a characteristic metamorphic mineral.

Lastly look at rock 2, the slate, and again complete the columns of Table 4.1.

The closely spaced cleavage planes can be clearly seen in the samples shown. The original rock, before metamorphism, was probably a mudstone (shale) and the residual presence of clay minerals implies that it has been metamorphosed under fairly mild conditions. Indeed it is not uncommon to find slate as a product of contact metamorphism, as occurred in the Teign catchment in the neighbourhood of the Dartmoor Granite.

Check Table 4.2 on p. 308 for the items that you should have entered in Table 4.1 while carrying out this DVD activity.

Allow about 25 minutes for this activity.

4.1 Summary of Sections 3 and 4

1 Igneous rocks in the crust can be divided into coarse-grained (intrusive) and fine-grained (extrusive) groups; grain size is dependent on rate of cooling.

2 The pairs basalt/gabbro (mafic rocks), andesite/diorite (intermediate rocks), and rhyolite/granite (felsic rocks) each have similar chemical compositions, the first of the pair being the extrusive rock, the second the corresponding intrusive rock.

3 Mafic rocks contain a high proportion of mafic (ferromagnesian) minerals, olivine and pyroxene, and so high levels of iron, magnesium and calcium. Felsic rocks have a high proportion of felsic minerals, feldspars and quartz (silica), and so more sodium, potassium and silicon. Intermediate rocks have mainly feldspar and amphibole.

4 Each different silicate mineral has a characteristic melting range, which determines when they begin to melt and when they begin to crystallize from magma. As magma cools, mafic minerals crystallize first, then pyroxene and amphibole, then feldspars and mica, and finally quartz. As crystals form, metal and silicate ions are successively used up.

5 In their crystal structures, silicate minerals are built from SiO_4 tetrahedra, which can be free (olivine), can share two oxygen atoms in chains (pyroxene, amphibole), three oxygens in sheets (micas, clay minerals) or can share all four in giant 3-D structures (feldspars, quartz). The more mafic minerals therefore have a higher oxygen : silicon ratio.

6 Chemical formulas represent the proportions of each element in the mineral crystal. Positive metal ions (except aluminium) balance negative charges on silicate ions. In aluminosilicate minerals, aluminium (3+) replaces some silicon (4+), so one extra positive metal ion is needed for every silicon replaced.

7 Heat from cooling magma, or heat and pressure due to deep burial or tectonic activity, bring about contact and regional metamorphism, respectively. More intense metamorphism results in larger crystal size.

8 The chemical compositions of original and metamorphosed rock may not differ, but new, denser minerals are formed, which can better accommodate the applied stresses. In gneiss, pale felsic and dark mafic minerals may separate into bands. In schist and slate, this layering allows the rock to be cleaved into sheets.

9 Metamorphic minerals have closer packed crystal structures, and can be denser than the parent. If minerals acquire a 'platy' structure, like mica, this develops at right angles to the applied pressure.

5 Fragmentary rocks

Igneous and metamorphic rocks both fall in the general category of crystalline rocks, and magmatic and metamorphic processes are mainly responsible for the production of the huge variety of rock-forming minerals. However, as we noted earlier, the most obvious and abundant rocks visible at the surface of the Earth are formed from fragments of pre-existing rocks that have undergone the processes of sedimentation and lithification in the **rock cycle**. These are the **sedimentary** rocks.

There are two major groups of sedimentary rocks, those formed from fragments of rocks originally composed of silicate minerals, and those in which the principal minerals are carbonates, mostly calcium carbonate, $CaCO_3$, in the form of calcite. The first group is the larger and includes sandstones and shales; the second group includes limestone and the Chalk, in which the fragments are often skeletal debris from ocean dwelling plants and animals. However some limestones are derived chemically, by precipitation processes, rather than biologically.

As well as their fragmental nature, sedimentary rocks have another characteristic, in that the sediments from which they are derived are laid down in layers, or beds. This bedded structure relates to the environment of deposition of the fragments, and is usually identifiable in even small hand specimens. On a small scale this layered structure differs from the banding seen in metamorphic rocks because its mineral content remains consistent, even though the particle size may vary. On a large scale, interbedded rock formations of sandstones (quartz), shales (clay minerals) and limestones (calcium carbonate) are frequently encountered. The thickness of individual beds can vary from less than a millimetre in the mudstones (called **laminations**), to several metres in the **massive** sandstones and limestones.

5.1 Types and characteristics of silicate-based sedimentary rocks

5.1.1 Physical characteristics

The mineral composition of the silicate-based sedimentary rocks is very variable, and their descriptions tend to emphasize the size, size distribution and shape of the fragments. A chart showing a classification according to fragment size is given (Figure 5.1). The divisions on the chart represent successive halving of grain size, from 256 mm downwards.

The other characteristic of sedimentary particles is their shape. Figures 5.2 and 5.3 are charts to show degrees of sorting, and a classification of particle shapes, although the application of both is somewhat subjective.

Conglomerates (Figure 5.4a) have rounded, pebble-sized fragments (or larger) measuring from 2 mm upwards; if the rock is composed of angular fragments of this size, it is termed a **breccia**, like those of the New Red Sandstones at the coast of the Teign catchment. With such large fragments, the components are

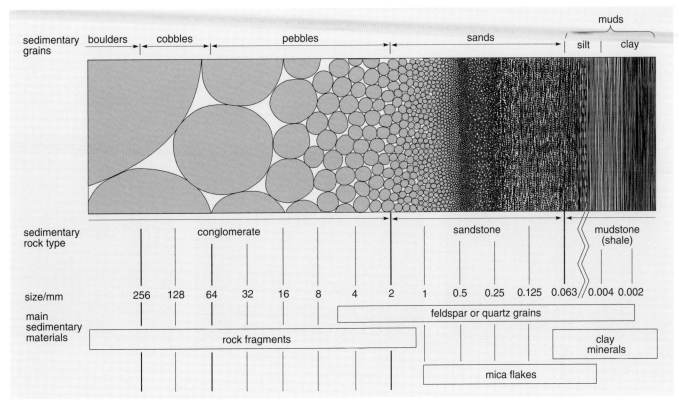

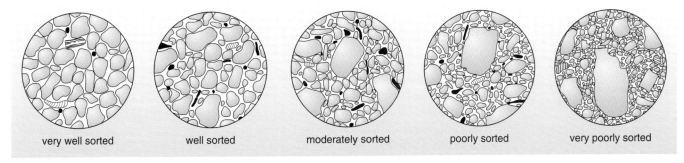

Figure 5.1 Sedimentary materials and rock types according to grain size.

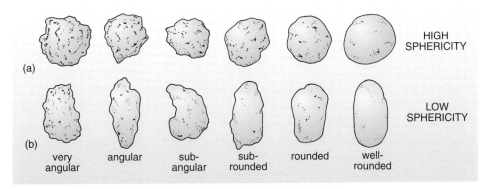

Figure 5.2 Cross-sections of sedimentary rocks to illustrate degree of sorting.

Figure 5.3 Classification of particle shapes.

pieces of rock rather than mineral grains, since the largest rock-derived quartz crystals are unlikely to have dimensions exceeding 5 mm. The rocks from which these fragments are derived are mainly igneous and metamorphic, as former sedimentary rocks are usually less durable. There is often a large variety of different sizes of fragment, with the spaces between the larger fragments being filled with finer material. As such the rock is said to be a **poorly sorted** sediment.

(a)

(b)

(c)

Figure 5.4   Hand specimens of (a) conglomerate, (b) sandstone and (c) shale.

If the particle size lies in the range 2 mm to 0.063 mm the general term **sandstone** (Figure 5.4b) is used, as this is the normal range for sand grains. At this size, most of the material of a well-sorted rock will be quartz grains, with varying amounts of feldspar crystals and mica flakes. Although Figure 5.1 indicates that the sizes of quartz and feldspar overlap with mica crystals, the thin flakes of mica occupy much less space than the quartz grains.

Feldspar and mica usually occur in greater abundance in immature sandstones, from sediments that accumulated near to their original rock source, where the breakdown of the minerals is incomplete. Feldspar and mica are both less stable in conditions that pertain at the Earth's surface than under the conditions found at depth where they crystallized, and so they tend to break down during the formation of sedimentary deposits (see Section 6 on weathering later in this block). Indeed the ratio of quartz to feldspar is a useful measure of the maturity of a sandstone, with a well-sorted mature sandstone being almost pure silica.

At the next lower size of particle it becomes difficult to discern individual mineral grains, even under the microscope. The mineral flakes that make up the mudstones are all below 0.063 mm; they are principally clay minerals, with some mica and quartz. The rocks are divided into the slightly grittier siltstones and the smoother claystones, again on particle size. They often include large quantities of organic matter, which accounts for their dark brown or black colouration. When the muds are compacted prior to lithification, the platy micas and clay minerals tend to become oriented in planes at right angles to the direction of applied pressure. This has a similar effect to the development of cleavage planes in slates, with the result that the rock becomes a **shale** (Figure 5.4c), easily split into thin sheets.

5.1.2 Cementation

Compaction of sediments would not by itself induce lithification; this is usually brought about by a process in which the particles are 'cemented' to each other. Rocks can be poorly cemented, when the grains crumble apart when gently abraded, or well cemented to the extent that they can be as resistant to fracture as crystalline rocks. The cement is often silica, but even with silica-based sandstones, the cementing mineral may frequently be calcite.

Silica cementation begins when silica dissolves in the aqueous fluid in the pores between the particles. The hard quartz grains are resistant to compaction, and so as pressure increases, silica dissolves at the point of contact, where the pressure is greatest — a process known as **pressure solution**. This silica can then be deposited from solution between the grains, cementing them together (Figure 5.5). The cement may grow onto the original crystal grain to form a continuous extended crystal, when it is said to be in *optical continuity* with the original grain.

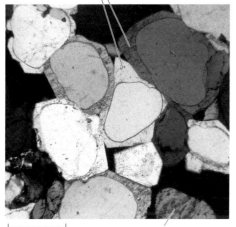

silica cement overgrowths in optical continuity with quartz grains

0.5 mm

feldspar grain (silica cement overgrowth is not in optical continuity with grain)

Figure 5.5   Cementation of quartz grains under the microscope. The 'ghost' outline of the original grains can easily be seen.

Coarse-grained sediments are usually better cemented, because fine-grained sediments are not so resistant to compaction, and so pressure solution is not as effective. However, quartz-rich muds can also become the cementing medium, either in mudstones, or in poorly sorted sandstones. If large and small crystals are present in a saturated solution, the large crystals will grow at the expense of the small crystals, cementing the mass together. Silica cement does not usually fill all the cavities between the grains, leaving many of the sandstones very porous. This has important environmental and commercial consequences, since it allows sandstones to act as **aquifers** (see Block 3 *Water*), and as reservoir rocks for oil and natural gas.

Since calcium carbonate can reach much higher concentrations in aqueous solution than can silica, calcite cement is even more readily precipitated than silica cement. Fine-grained carbonate mud is the usual source of calcite in pore fluids. Large calcite crystals are sometimes formed, enclosing several quartz grains in one crystal. As a result, calcite cemented sedimentary rocks are usually less porous than silica cemented ones.

Other minerals can also act as cement, usually in conjunction with silica. Iron (III) oxide can be precipitated out of aqueous iron (II) solutions under oxidizing conditions, but is rarely present in sufficient quantities to be the only cement. It is a thin layer of this mineral on quartz grains that gives red sandstones their characteristic colour. Sometimes clay minerals are found as cement, which indicates that they can form from solution as well as by direct breakdown of igneous rock minerals.

5.1.3 Minerals of silicate-based sedimentary rocks

The above discussion has indicated that the minerals of these sedimentary rocks are much less varied than the igneous and metamorphic rocks from which they originated, being limited essentially to quartz, mica and clay minerals, and a little feldspar. Table 5.1 shows an analysis of three typical silicate sedimentary rocks, and contrasts them with granite and gabbro. The low proportions of chemical elements other than silicon and aluminium points to the paucity of mineral content.

Table 5.1 Chemical analysis (% by mass) of typical silicate-based sedimentary rocks, in comparison with felsic and mafic igneous rocks (principal components only).

Element (measured as oxide)	Sedimentary			Igneous	
	Immature sandstone	Mature sandstone	Siltstone	Granite	Gabbro
SiO_2	78.1	99.1	58.1	70.8	49.0
Al_2O_3	11.8	0.4	16.4	14.6	18.2
iron oxides	1.2	0.1	6.6	3.4	9.2
MgO	0.2	0.0	1.5	0.9	7.6
CaO	0.2	0.3	0.2	2.0	11.2
Na_2O	2.5	0.0	1.3	3.5	2.6
K_2O	5.3	0.2	3.2	4.2	0.9

○ Compare the relative proportions of the elemental oxides in the three sedimentary and the two igneous rocks. Does the analysis of the sedimentary rocks resemble that of the granite or the gabbro more closely? Can you suggest an explanation?

● The chemical analyses of the sedimentary rocks show that they are closer in composition to granite than to gabbro. The high levels of iron and magnesium in gabbro are absent, and calcium also drops dramatically. There is more potassium than sodium in the sedimentary rocks and granite, whereas in gabbro the proportions are reversed. These observations fit the proposal that the minerals of the mafic rock gabbro — ferromagnesian minerals and calcium-rich feldspar — are less stable than those of granite, under conditions at the Earth's surface where sedimentary processes take place.

5.1.4 Clay minerals

The major constituents of the very fine-grained silica-based rocks are the **clay minerals**. A clay is, strictly speaking, a size classification, but not all particles that are small enough to be classified as clay are clay minerals. Very tiny particles of quartz are properly classified in this fraction, but are not clay *minerals*. Sand and silt are the solid unaltered particles left over after the *processes* of atmospheric weathering; whereas clay minerals are the *products* of this weathering. They are derived through chemical reactions taking place with the minerals formed in igneous and metamorphic processes. Individual clay particles cannot be seen with the naked eye; they are visible only under a microscope (Figure 5.6).

The clay minerals of the sedimentary rocks are **aluminosilicate clays**. They are formed by the partial decomposition of the aluminosilicate minerals, micas, feldspars and amphibole. Most of the soluble metallic ions (particularly sodium, potassium and calcium) are leached away, together with some silica, as the aluminosilicate tetrahedra break down. Much aluminium and some iron and magnesium are retained, and with the residual silicate ions they form sheet structures rather like the structure of micas. All the clay minerals are closely related, differing largely in the metallic ions that are incorporated into their structures. The three main family members are the aluminium-rich **kaolinite** (China clay), the potassium-rich **illite**, and the iron/magnesium-rich **montmorillonite**.

Figure 5.6 Particles of the clay mineral kaolinite under the electron microscope.

All aluminosilicate clays have the same basic structure. In a similar way to the rock minerals from which they are formed, they have repeating units of silicate tetrahedra, SiO_4^{4-} (Figure 5.7a), arrayed in sheets. In addition they have sheets made up from aluminium (Al^{3+}) ions, each surrounded by six oxide or hydroxide (OH^-) ions, arranged in the shape of an octahedron (Figure 5.7b). Alternatively an iron or magnesium ion may be included at the centre of the octahedron.

○ All aluminosilicate clay minerals consist basically of repeated arrangements of octahedral and tetrahedral groups. What then, do you think, imparts different characteristics to different clays, making them so distinctive?

● It is primarily the *arrangement* of these groups that gives clays their distinctive properties.

The arrangement of the tetrahedral and octahedral groups ranges from disordered or weakly ordered, to very regular, repeated patterns in the **layered silicate** clays. In these, the tetrahedral groups are linked in repeated units to form sheets called (not too surprisingly) **tetrahedral sheets** (Figure 5.7a). Similarly, the octahedral groups link together in **octahedral sheets** (Figure 5.7b).

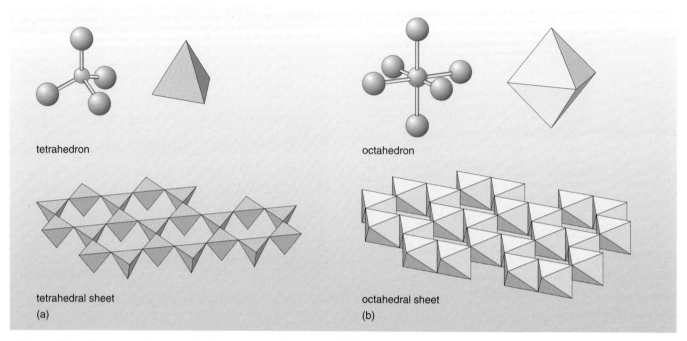

Figure 5.7   Structure of (a) tetrahedral groups and sheets and (b) octahedral groups and sheets.

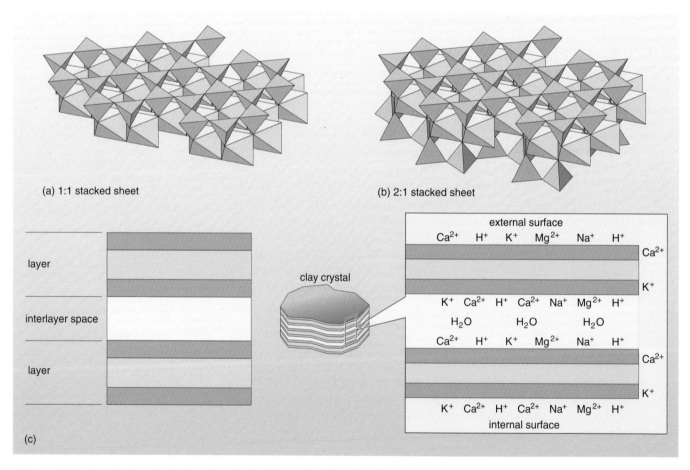

Figure 5.8   Stacked tetrahedral and octahedral layers in: (a) a 1 : 1 clay mineral and (b) a 2 : 1 clay mineral. (c) Overall structure of a 2 : 1 clay mineral.

The layered clays themselves fall into two groups. In the **1 : 1 clay minerals** like kaolinite, every (silicon–oxygen) tetrahedral sheet is stacked on an (aluminium/magnesium–oxygen) octahedral sheet (Figure 5.8a); whereas in the **2 : 1 clay minerals**, two tetrahedral sheets sandwich a single octahedral sheet (Figure 5.8b). These sheets are then stacked up on top of each other, separated by a layer containing cations balancing some of the negative charge, and sometimes water molecules (Figure 5.8c)

Figure 5.9 shows the bonding that holds the tetrahedral and octahedral sheets together. The lower tetrahedral layer has an outer surface of bonded oxygen atoms, whereas the surface of the upper octahedral layer consists of −OH groups. The hydroxyl (−OH) hydrogens of the octahedral sheets are linked to oxygens at the base of the tetrahedral sheets through **hydrogen bonds** (Box 5.1).

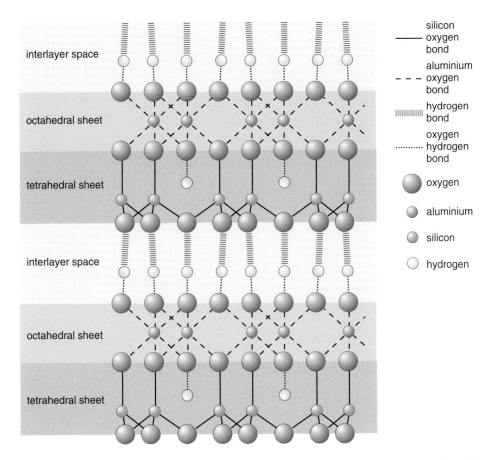

Figure 5.9 Overall structure of a 1 : 1 clay mineral (kaolinite). Note hydrogen bonds in the interlayer spaces.

Box 5.1 Hydrogen bonds

In the —OH group there is a covalent bond between hydrogen and oxygen. However, the electrons that are shared to form this bond are not evenly distributed between the two atoms. The oxygen atom attracts the electrons towards itself, and this makes it slightly negative (indicated as δ^-); the hydrogen atom becomes correspondingly slightly positive (indicated as δ^+). The charge difference is small, but sufficient to allow electrostatic attraction between the small positive charge on the hydrogen of one —OH group, and the small negative charge on the oxygen of another —OH group (Figure 5.10a). This attraction is

called a *hydrogen bond*. The attraction is much weaker than between two ions in a crystal structure, but if many hydrogen bonds are possible they can together exert a considerable force. For example, when an ionic material like potassium carbonate dissolves in water, it is the cluster of hydrogen bonds forming to the negatively charged carbonate anions (Figure 5.10b) that pulls them out of the crystal structure and into solution. Similarly, the positively charged potassium ions are surrounded by the slightly negatively charged oxygen atoms of many water molecules when in aqueous solution (Figure 5.10c).

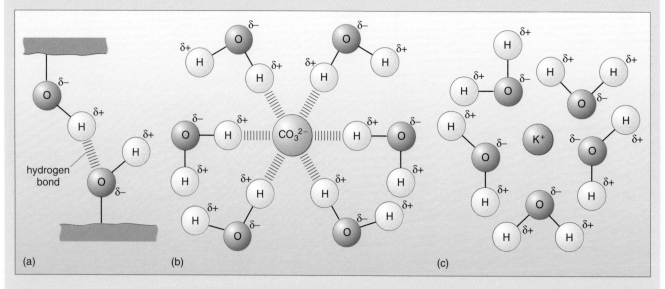

Figure 5.10 Hydrogen bonding (a) between two —OH groups and (b) between negative ions (e.g. CO_3^{2-}) and water molecules. In (c), water molecules surround positive ions (e.g. K^+) in solution. The 'surface' indicated in (a) represents a molecule of any size, or a crystal surface. The symbol δ^- or δ^+ indicates a 'partial negative' or 'partial positive' charge.

The most widespread 1:1 clay is kaolinite (china clay), where the hydrogen-bonded tetrahedral/octahedral sheets are stacked on top of each other (Figure 5.11a). This structure gives kaolinite the formula $Al_2Si_2O_5(OH)_4$.

○ Using the values for the formal charges on the ions given in Table 3.5 (p.180), calculate the overall charge on kaolinite.

● The positive charges are $2 \times 3+$ for Al, $2 \times 4+$ for Si, and $4+$ for the four hydrogen atoms, to give 18+ in all. There are 9 oxygen ions, each with two negative charges, making a total of 18−. Overall, kaolinite has zero charge.

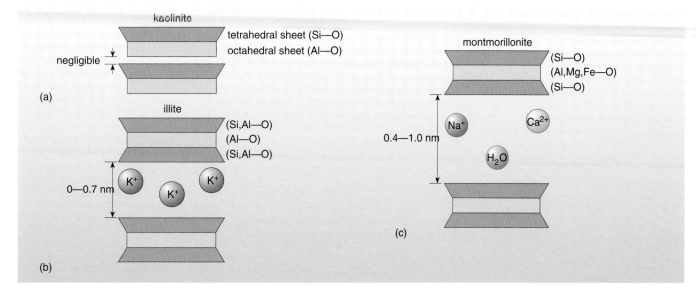

Figure 5.11 Diagrammatic structure of the three major aluminosilicate clay minerals. (a) kaolinite, showing alternating tetrahedral and octahedral sheets and stacked arrangement of 1 : 1 sheets, (b) illite, showing 2 : 1 sheets and interlayer K^+ ions, and (c) montmorillonite, showing 2 : 1 sheets and interlayer ions.

In the 2 : 1 clay structure, the oxygen atoms of the aluminium octahedra are now bonded above and below to silicate tetrahedra, with the result that there are fewer –OH groups in the octahedra than in the 1 : 1 clays, and these are located in the interior of the triple layer. In addition, some 2 : 1 clays show substitution of ions for others, such as aluminium for silicon in the tetrahedral layers, and iron or magnesium for aluminium in the octahedral layers. We shall cover this in more detail in Section 8.

The 2 : 1 clay illite has one in every four silicon atoms (at the centre of a tetrahedron) replaced by an aluminium atom, giving it the formula $Al_2(AlSi_3)O_{10}(OH)_2$.

○ Aluminium ions have a formal charge of 3+ and silicon a formal charge of 4+. What effect will this have on the overall charge of illite?

● For every formula unit of illite there will be one negative charge on the structure.

You can confirm this, if you like, by adding up positive and negative charges in the illite framework formula, and you should find that there is one excess negative charge. The negatively-charged aluminosilicate layers in illite are held together by their attraction, above and below, to positively charged potassium ions (Figure 5.11b). There is one K^+ ion for every negative charge on the layer, making the formula of illite $K^+[Al_2(AlSi_3)O_{10}(OH)_2]^-$. The strong negative charge causes the K^+ ions to be very tightly held, and the space between the layers to be small. This means that K^+ ions are very firmly held in illite clays.

The structure and formula of montmorillonite is much more complex. There is also exchange between Fe^{2+} or Mg^{2+} and aluminium. Since aluminium ions normally carry a charge of 3+, every time one of these exchanges takes place a further negative charge is added to the montmorillonite framework. Cations such as sodium (Na$^+$), magnesium (Mg^{2+}) and calcium (Ca^{2+}) reside between the layers, holding them together and cancelling out the negative charges. However, the larger space and weaker charge allows more free exchange of interlayer ions with the outside solution. It also allows water to enter the space, causing the clays to swell (Figure 5.11c).

Activity 5.1 Structures of clay minerals

The structures of kaolinite, illite and montmorillonite are shown in *Virtual Crystals* in the DVD *Air/Earth*. This takes you from a silicate tetrahedron and an aluminium octahedron to the layered structures of the clays. The 3-D structures can be rotated to show the structure in detail. Select each clay mineral in turn and explore the screens. The two micas, aluminium-rich *muscovite* and iron/magnesium–rich *biotite* are also shown for comparison.

Allow about 15 minutes for this activity.

The metallic ions are held relatively loosely in the gaps of montmorillonite, and may be exchanged for other ions from solutions in contact with the clay. In addition, if much water is trapped in the space between the layers of the sandwich, it causes the clays to swell, and also has the effect of lubricating the plates so they can slide over each other. This gives clay the properties that are so valuable commercially, but which are also responsible for subsidence damage to property and for landslips. Clay minerals are essential constituents of soils, where their water holding and ion exchange properties play a large part in determining the character of the soil, as you will learn later in this block.

Kaolinite is the end product of weathering and is commonly found in those **residual clays** still associated with the minerals from which they were formed. However, clay minerals can undergo alteration processes after formation, either during transport, or after deposition, or during lithification. Clay particles are usually transported in fresh water, but most **sedimentary clays** are deposited from seawater. Seawater is more alkaline (higher pH), and has a higher concentration of metal ions than fresh water, and under these conditions kaolinite is transformed to illite or montmorillonite. During lithification, montmorillonite becomes less stable, and the major clay mineral of mudstones is illite.

5.2 Sedimentary carbonate rocks

The other major group of sedimentary rocks is the limestones. Any rock having more than a 50% calcium carbonate content is usually placed in this group, which comprises about 25% of all sedimentary rocks.

The calcium carbonate of limestones can be generated chemically or biologically, and as in the silicate-based rocks, the fragments that make up the rock can vary in size from a few micrometres to several centimetres. The finest particles generate a carbonate mud, consisting of fine flakes, comparable in size to clay minerals. These fragments were largely indistinguishable, even under the microscope, before the development of electron microscopy. They may be precipitated directly from seawater saturated with calcium hydrogen carbonate (also known as calcium *bicarbonate*), but the major source is the minute skeletal plates of marine phytoplankton (coccolithophores) or zooplankton (foraminifera). The Chalk that forms the 'white cliffs of Dover' is a type of limestone composed almost entirely of the carbonate skeletons of countless numbers of these tiny creatures (Figure 5.12).

Lithification of carbonate mud will give a rock with a low porosity, similar to the non-porous shales formed from clay-sized particles. This mud also has an important role to play as a cementing medium.

Many limestones consist essentially of fossil debris, for example, the remains of corals and crinoids (a rare marine animal) and shell fragments, bound together by a clear crystalline calcite cement. The major fossil type present may be used in describing the limestone; for example, a shelly limestone (Figure 5.13a), or a crinoidal limestone (Figure 5.13b). As with the silica cement of sandstones, continuous crystals may be formed by calcite precipitating from solution onto the surfaces of fossil material from crinoids and sea-urchins. More usually, the calcite forms separate crystals, either within the pore space or enclosing grains. Limestones may also be porous, like silicate-based rocks, if the cement does not completely fill the pore spaces, but more frequently the cement occupies almost all the voids.

The most interesting products of direct precipitation from solution are the oolitic sediments. **Ooids** (or ooliths) are formed by the gradual precipitation of calcium carbonate from warm seawater onto a nucleus, which may be a sand grain or a shell fragment, often less than a tenth of a millimetre across (Figure 5.14). The nucleus slowly builds up a coating of calcite, developing a spherical shape as it is rolled back and forth by wave action. A cross-section of an ooid shows a series of concentric layers, which develop until it is about 0.5 mm in diameter. Lithification produces an **oolitic limestone** (Figure 5.13c), of which Portland Stone and Cotswold Stone are prominent examples.

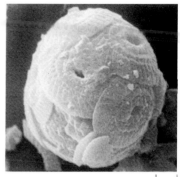

2 μm

2 μm

Figure 5.12 Skeletons of marine plankton from which chalk is formed (note the scale!).

(a) (b) (c)

Figure 5.13 (a) Shelly, (b) crinoidal and (c) oolitic limestones.

(a)

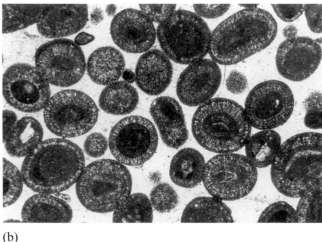

(b)

Figure 5.14 (a) Modern ooids and (b) thin section of oolitic limestone.

The precipitation of calcium carbonate is governed by the equilibrium represented by the equation below. When an equation (like 5.1) represents an equilibrium (that is when all components on both sides are present, and the reaction can proceed in either direction depending on the reaction conditions), we use the equilibrium arrow $\rightleftharpoons$ instead of the equals sign (=).

$$CaCO_3(s) + H_2O(l) + CO_2(aq) \rightleftharpoons Ca^{2+}(aq) + 2HCO_3^{-}(aq) \qquad (5.1)$$

calcium carbonate   carbon dioxide   calcium ions   hydrogen carbonate ions

---

Box 5.2   State labels in equations

Looking at Equation 5.1, you may have been puzzled by the letters that appeared in brackets after each item in the equation. These are simply labels to indicate the state of matter: solid (s), liquid (l), or gas (g) representing the condition of the preceding item. The additional label (aq) really has two meanings, both related to solution in water (*aqueous* solution). If the preceding species is ionic, (aq) indicates that it is a *solvated* ion surrounded by water molecules (Figure 5.10b and c); if it is non-ionic, it is simply a non-ionized material dissolved in water.

In this block you will meet many equations describing the interaction of aqueous ions (especially the positively charged hydrogen ion) with solid minerals, to yield ions in solution and precipitated solids. Also $SiO_2(aq)$ is used to indicate 'soluble silica', sometimes represented as silicic acid, $Si(OH)_4$. There are also equations where gases are labelled (aq) because it is most probable that dissolved gas rather than the gaseous material is the active agent.

As seawater warms up, some of the carbon dioxide is lost, with the result that the equilibrium moves to the left-hand side to generate more carbon dioxide in solution. When this happens, some calcium carbonate is precipitated at the same time.

○ Why should the water warming up lead to the loss of CO_2?

● Unlike most solids, gases are *less* soluble in liquids at higher temperatures.

The process of ooid formation can be observed at the present time in the warm shallow seas off the Bahamas.

5.2.1 Carbonate minerals

The common carbonate mineral is calcite, which forms soft, colourless rhombohedral crystals. Calcite is included among the minerals in the *Mineral Kit* (see Activity 5.2, Part 2). There is another crystalline form of calcium carbonate, also having the formula $CaCO_3$, called **aragonite**. Interestingly, many ocean creatures convert the ions they first extract from seawater into aragonite in order to form their shells or skeletal parts. However, aragonite is less stable than calcite, and most of the biogenic material found in fossils has reverted to calcite.

Another alteration process seen in limestones is the replacement of calcium ions by magnesium ions in the crystal structure of calcite to form **dolomite**, $CaMg(CO_3)_2$. It is not easy to distinguish dolomite from calcite, nor dolomitic limestone from all-calcium limestone.

Activity 5.2   Sedimentary rocks and minerals

Part 1   Silicate-based rocks

Open the *Rock Kit* and display rock 29, the conglomerate. Look at the various screens and fill in the columns in Table 5.2. 'Contents' will be rock fragments, or feldspar crystals, for example; and 'bedding/lamination' will require a 'yes' or 'no'. 'Cement' will be either calcite or silica. For the columns 'roundness' and 'sorting' you will have to make an estimate, based on the charts in Figures 5.2 and 5.3.

Now work through the series — micaceous sandstone (rock 20), coarse red sandstone (rock 10), and mudstone (shale; rock 15) — and for each rock type fill in the columns in Table 5.2. To estimate roundness and sorting in these rocks, it is usually easiest to use the microscope view of the thin section. It can be quite difficult to make these estimates from a hand specimen, especially if the rock is well cemented.

When you have completed your survey of the screens, compare your completed table with Table 5.5 on p. 308.

There are some other silicate-based sedimentary rocks in the *Rock Kit*, and if you have time you may like to explore these screens, noting down the rock characteristics as you do so. Rock 7 is a fine/medium grained 'white' sandstone; rock 12 is called a greywacke, a poorly sorted sedimentary rock with quartz, feldspar and rock fragments; rock 18 contains both quartz and feldspar, and is called an arkose; rock 27 is another mudstone.

Table 5.2   Characteristics of silicate-based sedimentary rocks (to be completed).

Rock type	No.	Contents	Cement	Roundness	Sorting	Bedding/lamination
conglomerate	29					
micaceous sandstone	20					
coarse red sandstone	10					
mudstone (shale)	15					

Part 2 Carbonate rocks and minerals

There are three limestones in the *Rock Kit* collection (Rocks 21, 22 and 24). You should look at the screens for each and note relevant data in Table 5.3. Then think about how you might describe these three limestones. Check your entries with Table 5.6 on p.309.

Table 5.3 Characteristics of carbonate sedimentary rocks (to be completed).

Rock type	No.	Muddy matrix	Crystalline matrix	Major fossil	Ooids
Limestone 1	21				
Limestone 2	22				
Limestone 3	24				

Finally in this section, we shall look at calcite on the DVD *Air/Earth*, in the *Mineral Kit*, *Properties of Minerals* and *Virtual Crystals*.

Two varieties of calcite are displayed in the *Mineral Kit*, the normal opaque mineral, and the more unusual clear variety called Iceland spar. Notice the hardness value of only three, suggesting that calcite is a very soft mineral. Cracks on all three rhombohedral faces show the planes of weakness of the mineral. If you were unsure about the shape of a rhombus, the Iceland spar crystal shows it perfectly.

Properties of Minerals follows up this preliminary view. Begin under Colour on the file tabs, and select Calcite (the bottom sample of the left hand row) for an enlarged view of the crystal. The Hardness tab leads to two useful buttons. How hard is hard relates the hardness of calcite to other minerals, and Mineral scratch test shows a copper coin scratching calcite. More examples under the Cleavage tab illustrates the fracturing of calcite with two movie sequences, and the Acid test tab shows the reaction of calcite with dilute hydrochloric acid.

The crystal structure of calcite is examined in the DVD section *Virtual Crystals*. Access the screen Layers of CO_3 groups and Ca, and rotate the structure using the mouse to see how the microscopic crystal lattice corresponds to the actual shape of the crystal. Note that the CO_3^{2-} ions occur separately in the crystal structure, rather like the silicate tetrahedra of olivine.

Allow about 40 minutes for this activity.

5.3   Ore minerals: Oxides and sulfides

An ore is a mineral containing a sufficiently high proportion of an economically important element to justify its extraction and processing. Most primary ore minerals are found in association with igneous rock because ores are principally formed as a result of magmatic processes. They are principally **oxides** and **sulfides** of the metallic elements.

When minerals like olivine and pyroxene crystallize early from hot basaltic magma they are sometimes accompanied by some dense non-silicate minerals containing iron, chromium and titanium. These minerals also sink to the bottom of the magma chamber. When the magma solidifies, layers of ore are found at the base of the magma chamber. Much of the world's chromium is extracted from ores that have accumulated in this way.

○ An important ore of titanium is ilmenite, $FeTiO_3$. What type of chemical compound is this ore?

● It is an *oxide* of the metals iron and titanium. Oxides are characterized by the presence of only oxygen and metallic elements in the formula.

Another way in which minerals can separate from sulfur-rich basaltic magma is as liquid sulfides. Molten iron sulfide forms blobs of liquid that also settle to the bottom of magma chambers, like drops of water in oil. This liquid, which is even denser than the first crystallizing silicates, also dissolves some metals more effectively than does magma, in particular copper and nickel. The liquid cools and solidifies to form layers of copper, iron and nickel sulfides (Figure 5.15).

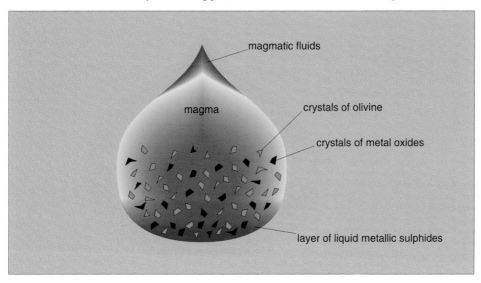

Figure 5.15   Formation of layers of ore by separation from magma.

While elements like chromium, nickel and titanium are segregating out, the magma is consequently being relatively enriched in other metals, including tin and tungsten. Eventually some of these metal rich magmas can crystallize as mineral veins, which can also be valuable sources of rare metals and gem-quality crystals.

At the high pressures associated with depth, magma contains a considerable amount of dissolved water, but as the anhydrous minerals, like olivine and pyroxene, crystallize, water separates and moves to the top of the magma chamber. Water at high temperature and pressure is an effective scavenger of metal ions from the magma. It also extracts metals from the surrounding rocks as it comes into contact with them, bringing about the process of alteration. As this hot, aqueous **hydrothermal fluid** approaches the Earth's surface it cools, gases may be given off, and it infiltrates cracks in the colder surrounding rocks. All these processes may initiate the formation of crystalline minerals, which are deposited from solution usually at quite shallow depths (Figure 5.16).

○ The major ore of copper is chalcopyrite, $CuFeS_2$. What type of chemical compound is this ore?

● It is a *sulfide* of the metals copper and iron.

Some important ore deposits result from percolating groundwater, or seawater, which can also become hydrothermal fluid at depth, or in the vicinity of a cooling body of magma (Figure 5.16). This fluid is especially effective in leaching out and concentrating metallic sulfides. Some rocks, including the mudstones and clays, readily yield up their metal ions and sulfur to percolating water. The mining of tin, copper, lead and zinc in the Teign catchment was of metal ores formed this way, where the cooling granite provided the source of heat.

Figure 5.16   Processes of ore formation from hydrothermal fluids, showing circulating seawater, circulating groundwater, and magmatic hydrothermal fluid.

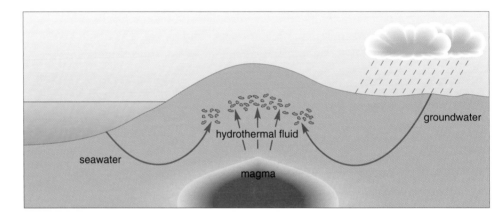

Question 5.1

Fill in the blanks in Table 5.4 with the correct descriptions from the lists below.

Table 5.4   Properties of rocks and minerals.

	Rock or mineral	Type	Texture/occurrence	Appearance	Other properties
1	shale				
2	conglomerate				
3	aluminium-rich mica (muscovite)				
4	slate				
5	shelly limestone				
6	kaolinite				
7	amphibole				

Types: 1. metamorphic rock; 2. sedimentary rock; 3. clay mineral; 4. felsic mineral; 5. mafic mineral; 6. carbonate rock; 7. sedimentary rock.

Texture/occurrence: 1. component of granites; 2. broken fragments of fossil shells; 3. very fine particles < 0.004 mm; 4. present in diorite and andesite; 5. pebbles of various sizes and rock types; 6. fine-grained, thin layers; 7. may retain signs of the original layered structure.

Appearance: 1. various coloured components; 2. dark green to black, shiny; 3. pale grey, rough surface; 4. dark grey; 5. tiny white plates under electron microscope; 6. shiny silvery surface; 7. dark colour.

Other properties: 1. cemented by calcite; 2. soft, splits easily into thin sheets; 3. soft, crumbles easily; 4. fragments set in finer-grained matrix, like concrete; 5. layered structure with tetrahedral and octahedral sheets in each layer; 6. include OH^- or F^- ions within their structure; 7. hard, formed from mudstones by heat and pressure.

Question 5.2

What is the difference between a 1 : 1 and a 2 : 1 aluminosilicate clay mineral?

5.4 Summary of Section 5

1 Silicate-based sedimentary rocks — conglomerates (breccias), sandstones and mudstones (shales) — are classified according to size, sorting and fragment shape.

2 Their main constituents are quartz, a little feldspar, and aluminosilicate clay minerals cemented by silica or calcite; their overall composition is closer to granite than to more mafic rocks.

3 The three main clay minerals are kaolinite (aluminium-rich), illite (potassium-rich) and montmorillonite (iron/magnesium-rich). They have a layered structure of 1 : 1 or 2 : 1 sheets of silicate tetrahedra and aluminium (or magnesium) octahedra, held together by hydrogen bonds, or metal ions between the layers.

4 Limestones consist mainly of calcium carbonate, generated by precipitation (oolitic limestone) or from fossil fragments, cemented by calcite.

5 Most ore minerals are oxides or sulfides forming within magma, or from hydrothermal fluids.

6 Weathering of rocks and minerals

(God) stood, and measured the earth; and the everlasting mountains were scattered, the perpetual hills did bow.

[Bible, Old Testament. Habbakuk Chapter 3, verse 6]

On a human time-scale, mountains and the rocks from which they are formed might at first appear to be indestructible and everlasting, but over geological time they have only a transitory existence. Rocks, and the minerals of which they are composed, are subject to a continuous barrage of forces that are gradually breaking them down and dissolving them. These are the processes of weathering and erosion, which result from the interaction between rocks, water and the atmosphere, and which have been operating since the first formed crust of the Earth encountered the primitive atmosphere.

However, confirmation of the very slow rate of weathering from fresh rock may be seen in any country churchyard. Gravestones are usually only 'moderately weathered', even after as long as 300 years, except in regions of regular severe weather conditions. The usual observation is that the lettering is worn, and may be illegible. It is nevertheless interesting to observe the comparative rates of weathering of gravestones made from different rock types, when it is clearly seen that igneous rocks (such as granite), and metamorphic rocks (such as slate, Figure 6.1a) weather much less readily than the fragmentary sandstones (Figure 6.1b).

Although some limestones are found to weather better than sandstones, monuments in limestone, or marble, are usually found only inside the church. We shall discover why in the next few sections.

(a)

(b)

Figure 6.1 Weathering of gravestones: (a) slate and (b) sandstone.

6.1 Physical weathering

On any walk in the hills, you must have been aware of the early stages of weathering as you encountered broken rocks surrounding a rock outcrop, or a scree slope tumbling down a mountain side (Figure 6.2).

Figure 6.2   A scree slope, southern Colorado, USA.

These familiar features result from the forces of **physical weathering**, which break down massive rocks into smaller and smaller pieces, making them more susceptible to further disintegration. Physical weathering is a mechanical process, which separates rock particles. Although the rock disintegrates, the chemical components of the rock maintain their integrity throughout. It is accompanied, and completed, by erosion and abrasion in transport until fragment by fragment, crystal by crystal, the rock is separated into its component parts.

The onset of physical weathering is often initiated by the process of **jointing** (Figure 6.3). Most rock formation takes place under conditions of intense pressure deep within the crust, as rocks crystallize or sediments are buried. Pressure increases rapidly at depth, with every 10 m of seawater, or 3–4 m of rock adding a further 100 kPa; that is, over 300 MPa at 10 km depth.

When these rocks are exposed at the surface, by uplift or erosion, pressure release is accompanied by expansion; a three-metre cube of freshly quarried granite can expand by up to three millimetres in each direction. This expansion can cause the rocks to break and split, forming joints to an extent and in a direction often related to the forces of uplift, or to pre-existing planes of weakness. In sedimentary rocks, joints are usually perpendicular to bedding (Figure 6.3a). Some sandstones and granites survive as massive blocks, but many rocks show a fracture pattern of roughly parallel and fairly evenly spaced near-horizontal and near-vertical open joints.

Figure 6.3 (a) Jointing in sandstone, and. (b) columnar hexagonal jointing in the basalt of the Giant's Causeway, N Ireland.

Massive rocks can display an alternative form of physical weathering, where fracture takes place parallel to rock surfaces. The rocks shed concentric shells or layers, as the outermost portion of the rock expands when the pressure load is relieved, and breaks away from the underlying body of rock. The layers can be from a few millimetres to about a metre thick. This process is called **sheeting** (Figure 6.4a), or on a smaller scale, the highly descriptive term **onion skin weathering** is used (Figure 6.4b).

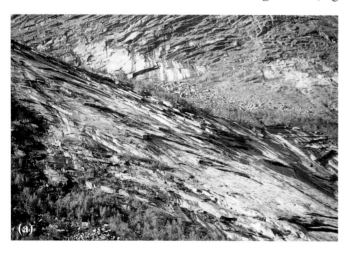

Figure 6.4 (a) Sheeting and (b) onion skin weathering in massive rocks.

Joint and sheet formation do not contribute greatly to the overall disintegration of rocks, but are an essential first step, allowing other agents of weathering — water, air and plant roots — access to cracks and fresh rock surfaces.

Once a crack has developed in a rock it may be forced further apart by the growth of crystals within it, for example by the infiltration of water, which then freezes. A unique property of water is that it expands on freezing, and any given mass of water increases in volume by about 9%. Consequently, as water in a confined space freezes, tremendous pressure builds up to a maximum of over 2×10^8 Pa, like the pressure of two cars resting on your fingernail! This level of pressure is not achieved in narrow open cracks, mainly because the ice-plug formed at the mouth of the crack is pushed out as the remaining water freezes, but a substantial force is nevertheless generated. So in a locality where melt water permeates rocks by day, and then refreezes at night, cracks are constantly subject to these powerful forces. This process, variously named as **freeze–thaw weathering**, frost shattering, or frost wedging, is responsible for a considerable amount of the broken rock to be seen on mountain tops (Figure 6.5a). Its effect is also familiar in decaying brickwork (Figure 6.5b).

Figure 6.5　(a) Shattered large boulder and (b) old brickwork and chimney pot deteriorating through freeze–thaw weathering.

○　Suggest a way in which frost shattering can bring about physical weathering in porous silicate-based sedimentary rocks.

●　If the pores become filled with water, and this then freezes, the expansion can generate forces that overcome the ability of the silica or calcite cement to hold the grains together.

The formation of other crystalline material has also been implicated in physical weathering, but is of lesser significance. Gypsum (calcium sulfate) is formed when rainwater containing dissolved sulfur dioxide accumulates in cracks in limestone. The sulfuric acid in the rainwater reacts with the calcium carbonate to liberate carbon dioxide and give calcium sulfate. The formation of these crystals in the confined space can cause flaking from the rock surface. This is another damaging effect of air pollution by sulfurous gases and the consequent acid rain.

Figure 6.6   Coastal cliff weathered by sea spray.

Similarly, evaporating sea spray can cause cracks in seashore rocks to widen, through the formation of sodium chloride crystals in them (Figure 6.6).

In hot, dry desert areas physical weathering is commonly observed (Figure 6.7); this has been ascribed to the effect of expansion and contraction, as the rocks heat up during the day and cool during the night. Interestingly, attempts to reproduce this form of rock shattering in the laboratory have failed to produce results. It is difficult, however, to conceive of an alternative mechanism under these climatic conditions. This may also be the process that causes physical 'weathering' to take place on the Moon.

Plants are also implicated in physical weathering, as their roots readily penetrate cracks and joints that have already been opened by other processes. However, the forces generated by a growing root are considerably less than the forces of frost shattering. The major effect of plants is therefore more chemical than physical, as metabolic processes produce organic acids and carbon dioxide, both of which act to degrade minerals in the rock. It is surprising how quickly plants colonize freshly exposed rock surfaces (Figure 6.8a).

Figure 6.7   Rock shattering under hot, dry desert conditions.

Figure 6.8   Bioweathering of fresh rock: (a) Plant growing out of the 'solid' rock of a recent lava flow and (b) lichen growing on the surface of an igneous rock.

Physical weathering is mainly responsible for the breakdown of massive rock into boulders or rock fragments. Further reduction in size, into rock particles or individual mineral grains is mainly due to chemical weathering, which is discussed below. A further significant factor, however, is the **abrasion** that occurs in transport.

Whenever rock fragments move, either through rock fall, or as a result of fluvial or glacial transport, they knock and grind against each other and are worn down. This is why the hardness of minerals is important, since the softer minerals will be worn away more readily.

○   Recall which are the hardest and the softest of the common rock-forming minerals (Table 3.2).

●   The hardest are quartz, feldspars and olivine; the softest are amphibole, pyroxene and the micas.

Amphibole, pyroxene and mica are also minerals with planes of weakness, and so tend to break up during transport; while those without, like quartz, survive. In temperate latitudes, rivers carry large amounts of sediment, which can wear away bedrock. In the tropics the river load is much less, because the more intense chemical weathering encountered there tends to break down the minerals in the rock fragments before they enter the river channels. The action of the sea in rounding pebbles (Figure 6.9a) and wearing away seashore cliffs (Figure 6.9b), and the erosion caused by sand-laden winds, are other examples of the

217

abrasion action of physical weathering. Strange shapes are sometimes formed, because softer layers of rock are eroded more easily than harder layers (Figure 6.10a). Geological structures are often emphasized for this same reason, as in Figure 6.10b, where the softer mudstone layers have been cut back making the harder limestone more prominent.

Figure 6.9   (a) Rounded seashore pebbles and (b) undercut cliff and arch, Malta.

Figure 6.10   Differential erosion: (a) dramatic rock shapes caused by wind erosion (Hoodoos, Alberta, Canada) and (b) sedimentary layering picked out by differential weathering.

6.2 Chemical weathering

The conditions of high pressure and temperature under which rocks are formed are very different from those encountered at the surface of the Earth. We know that rocks may undergo metamorphism when these severe conditions are applied, with the chemical composition of their minerals changing to give compounds that are more stable under the new regime. Similarly, when rocks are newly exposed at the Earth's surface to a pressure of only 100 kPa, and temperatures between −50 and 100 °C, many of the minerals they contain are unstable and can 'metamorphose' to compounds with lower density (or greater volume). This is a major driving force for **chemical weathering**, which is also greatly facilitated by the ubiquitous presence of water. Chemical weathering differs from physical weathering, as it involves the decomposition of the rock; the actual breakdown of the chemical structures of the minerals of which it is composed. Water is always implicated, but there are other agents of chemical weathering, principally carbon dioxide and oxygen. Each of these chemical agents contributes in specific ways to the overall weathering process.

Water is unique in that it exists at the planet's surface in solid, liquid and gaseous form. There is three times as much water in the upper five kilometres of the Earth's crust, if you include the oceans, as everything else put together; six times as much water as feldspar, the most abundant mineral of the crust.

Water is more effective as a solvent than any other fluid, so it is not surprising to find that all water contains dissolved substances. The purest natural form of water is freshly fallen snow, but it already contains ions from the atmosphere and as it melts and soaks into the ground its mineral content may increase by more than seven times as it dissolves minerals from the soil. Terrestrial water always contains carbon dioxide, with a much higher concentration being present in water in contact with soil than in rainwater. Silicic acid (soluble silica, $Si(OH)_4$) is also always present, but in general the mineral content of water relates directly to the rocks with which it comes into contact.

Rainwater dissolves carbon dioxide from the air as it falls, and so is always slightly acidic, but the concentration is much higher in soil water and groundwater. The **respiration** of soil bacteria and plant roots (the process by which an organism breaks down the organic materials in food into carbon dioxide, water and energy) and the decay of plant material both provide additional carbon dioxide. Consequently the concentration of carbon dioxide in the air in the pore spaces of the soil is much higher than the normal concentration in the atmosphere. Plant rootlets also release organic 'plant acids' into soil water, and these contribute to the weathering of rocks and minerals. The action of growing (and decaying) plants on breaking down minerals in rocks is called **bioweathering**. The plant benefits from the liberation of metal ions from the minerals, which it may then absorb as nutrients. The rate of chemical weathering is therefore related to the amount of biological activity in the soil.

Chemical weathering begins at the rock surface, but can penetrate quite deeply. The larger the surface area of the rock the more is available to weathering, and so as rock fragments become smaller they weather more rapidly. A permeable rock is more susceptible to weathering for this same reason. The processes of chemical weathering give a fourfold result: some of the minerals undergo chemical reaction to produce new materials; part of the rock material dissolves and is washed away; some of the minerals resist weathering and remain largely unaffected; and the rock disintegrates to participate in the subsequent processes of erosion, transport or soil formation (Figure 6.11).

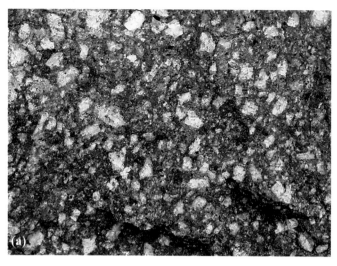 

Figure 6.11 (a) Fresh and (b) heavily weathered specimens of igneous rock.

6.2.1 Chemical weathering of carbonate rocks

Carbon dioxide dissolves in water (Equation 6.1), to generate a low concentration of hydrogen ions in solution. As this is an equilibrium, the higher the concentration of carbon dioxide, the more will dissolve, and so the concentration of hydrogen ions will increase.

$$H_2O(l) + CO_2(g) \rightleftharpoons CO_2(aq) \rightleftharpoons H^+(aq) + HCO_3^-(aq) \tag{6.1}$$
$$\text{hydrogen carbonate ion}$$

Box 6.1 Acidity, hydrogen ions, moles and pH

Acidic properties in an aqueous solution are related to the concentration of hydrogen ions, H⁺, in that solution. This concentration can vary enormously from very strong acid to very strong alkali, but there will *always be some* hydrogen ions present in water, because of the equilibrium in Equation 6.2.

$$H_2O(l) \rightleftharpoons H^+(aq) + OH^-(aq) \tag{6.2}$$

Concentrations are expressed in **moles** per litre ($mol\,l^{-1}$), where a mole is about 6.02×10^{23} particles of any chemical species, i.e. atoms, molecules or ions. One **formula mass**, i.e. the mass in grams of the sum of the relative atomic masses, called the **relative molecular mass (M_r)** of all the atoms in the formula, contains 6.02×10^{23} particles, or *one mole* of the species represented by that formula. For water, H_2O, with a formula mass of 18.0 [2×1.01 (since the **relative atomic mass (A_r)** of hydrogen = 1.01) plus 16.0, (the A_r of oxygen)], there will be 6.02×10^{23} water molecules in 18.0 grams of water. A mole of hydrogen ions is about 6×10^{23} hydrogen ions, or 1.01 gram of hydrogen ions, but the number of hydrogen ions in a litre of pure water is very much less than that. It is in fact near to 6×10^{16}, or $1 \times 10^{-7}\,mol\,l^{-1}$.

○ How many molecules of carbon dioxide are there in one mole, and what will be their mass? [Carbon has $M_r = 12.0$; oxygen has $M_r = 16.0$]

● *Like all moles,* a mole of carbon dioxide contains about 6×10^{23} particles (molecules), and they have a mass of $(1 \times 12.0) + (2 \times 16.0) = 44.0$ g.

These huge numbers are rather inconvenient to use, and so the **pH scale** was developed to provide an easier way to express acidity. pH is defined as 'minus the log to base 10 of the hydrogen ion concentration', but you do not need to remember that. What it means is, for example, that a concentration of 10^{-9} moles of hydrogen ions per litre is pH 9, or an H⁺ concentration of $10^{-2}\,mol\,l^{-1}$ is a pH of 2. Just remember that (i) water is taken as neutral, and that its pH value is 7; and (ii) that any solution with a pH of less than 7 is acid, and any solution with a pH more than 7 is alkaline. Also (iii) the greater the pH value, the fewer hydrogen ions present, and so the less acid will be the solution; and vice versa.

○ Which is the most acidic of the solutions with these pH values: pH 4.40, pH −1.00, pH 8.90? Which solutions are acid and which alkaline?

● The solution with pH −1.00 is the most acidic, about the same concentration as concentrated hydrochloric acid. The solution with pH 4.40 is also acidic, as 4.40 is *less* than 7; but the solution with pH 8.90 is alkaline, as 8.90 is *greater* than 7.

Using the equilibrium from Equation 6.2, we can establish a constant called the **ionic product** of water (K_w):

$$K_w = [H^+(aq)] \, [OH^-(aq)]$$

where the quantities in square brackets are concentrations in $mol \, l^{-1}$.
In a neutral solution the amount of H^+ is equal to the amount of OH^-. If a neutral solution contains $1.00 \times 10^{-7} \, mol \, l^{-1}$ of hydrogen ions, the value of K_w must be $(1.00 \times 10^{-7} \, mol \, l^{-1})^2 = 1.00 \times 10^{-14} \, mol^2 \, l^{-2}$. Figure 6.12 relates pH to the concentrations in the equation as a diagram.

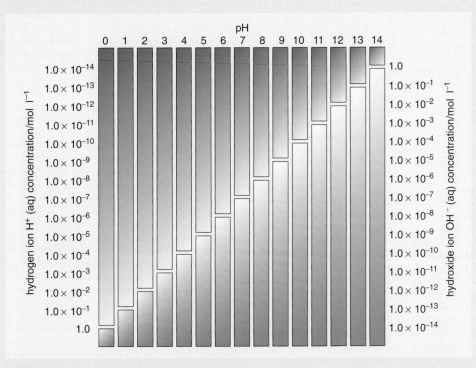

Figure 6.12 Relationship between pH and the concentrations of H^+ and OH^- in aqueous solution.

○ What is the pH of a solution which has (a) a hydrogen ion concentration of $1.00 \times 10^{-5} \, mol \, l^{-1}$; (b) a hydroxide ion concentration of $1.00 \times 10^{-4} \, mol \, l^{-1}$?

● (a) The first solution has pH of 5.00; and (b) the second solution has a pH of 10.0 (i.e. $14 - 4$).

○ Is a solution with pH of 6.00 more or less acidic than one of pH 6.50?

● Remember the larger the pH value, the lower the concentration of hydrogen ions. So an acid with a pH 6.50 is weaker than an acid with pH 6.00, but stronger than one of pH 7.00.

The interaction of an aqueous solution of carbon dioxide with minerals is termed *carbonation*, and not surprisingly it is a very common process. Such a solution is a very weak acid, more than six times weaker than acetic (ethanoic) acid. A saturated solution of carbon dioxide in equilibrium with the normal concentration in the atmosphere has a concentration of about $36\,\mathrm{mmol\,l^{-1}}$, but most of this is present as $CO_2(aq)$. Less than 2% of the CO_2 molecules generate hydrogen ions, giving a pH of only 5.6.

The most notable effect of solutions of carbon dioxide on the landscape is their dissolution of the carbonate rocks, limestone and the Chalk. It is responsible for the magnificent underground cave systems found in limestone regions (Figure 6.13), and the 'limestone pavement' feature of the landscape (Figure 6.14). You may think that the ornament of limestone pavement (that is the deep weathered fissures) have been caused by the action of rainwater on bare rock surfaces, but it is believed that almost all this weathering took place beneath a soil cover, which has since been removed.

○   Why do you think that this is more probable?

●   As stated above, water saturating soil that is resting on the limestone surface has a higher level of acidity than rainwater falling on the surface of bare rock. The soil water will also stay in contact with the limestone for longer, because rainwater will run off or evaporate.

Figure 6.13 Cave system in limestone, showing the formation of stalactites and stalagmites, Derbyshire, England.

Figure 6.14 Limestone pavement in North Yorkshire, England,

Limestone is almost insoluble in pure water, because calcium carbonate has an ionic product of only 4.5×10^{-9} $(mol\,l^{-1})^2$. This means that the concentration of a saturated solution is less than $7\,mg\,l^{-1}$. However, the presence of hydrogen ions gives rise to the reaction in Equation 6.3.

$$CaCO_3(s) + H^+(aq) + HCO_3^-(aq) \rightleftharpoons Ca^{2+}(aq) + 2HCO_3^-(aq) \qquad (6.3)$$

Box 6.2 Ionic product

Rather than express the solubilities of sparingly soluble ionic materials in terms of grams per $100\,cm^3$ or moles per litre, the ionic product may be used. For an ionic solid, AB, which separates into ions A^+ and B^-, in contact with its saturated aqueous solution, the ionic product, K_i is given by the equation:

$$K_i = [A] \times [B]\text{(written as [A] [B])}$$

where the quantities in the square brackets, [A] and [B], are the concentrations of the ions A^+ and B^- in $mol\,l^{-1}$. If the material gives rise to more than two ions, the concentrations of all the ions are multiplied together. This means, for example, that the ionic product for iron (III) hydroxide, $Fe(OH)_3$, is given by the equation:

$$K_i = [Fe^{3+}] [OH^-] [OH^-] [OH^-] = [Fe^{3+}] [OH^-]^3$$
$$= 2 \times 10^{-39}\,(mol\,l^{-1})^4$$

The units of each component also have to be multiplied together, which is why the ionic product for iron (III) hydroxide will have the units $(mol\,l^{-1})^4$.

Ionic products for some other common minerals are given in Table 6.1 in $(mol\,l^{-1})^2$.

Table 6.1 Ionic products for some common minerals $(mol\,l^{-1})^2$.

Mineral	calcite	magnesite	gypsum	barite	pyrite	galena	sphalerite
formula	$CaCO_3$	$MgCO_3$	$CaSO_4$	$BaSO_4$	FeS_2	PbS	ZnS
ions	Ca^{2+} CO_3^{2-}	Mg^{2+} CO_3^{2-}	Ca^{2+} SO_4^{2-}	Ba^{2+} SO_4^{2-}	Fe^{2+} S_2^{2-}	Pb^{2+} S^{2-}	Zn^{2+} S^{2-}
ionic product	4.5×10^{-9}	3.5×10^{-8}	2×10^{-5}	1×10^{-10}	6×10^{-18}	1×10^{-28}	2×10^{-24}

Calcium hydrogen carbonate, $Ca(HCO_3)_2$, is about 30 times more soluble than calcium carbonate and so the weathering effect on the land surface is considerable, reducing the height of limestone hills by an estimated 200–550 mm over a period of 1000 years.

○ Estimate the mass of calcium carbonate that could be dissolved by 1000 litres of pure water, then estimate the amount dissolved by the same amount of rainwater (the M_r of $CaCO_3$ is 100 g mol^{-1}).

● Ionic product of calcium carbonate = $[Ca^{2+}] [CO_3^{2-}] = 4.50 \times 10^{-9}$ 1 (mol l^{-1})^2.

So the concentration of either ion is $\sqrt{4.5 \times 10^{-9}}$ mol l^{-1} = 6.71×10^{-5} mol l^{-1}.

The mass of calcium carbonate dissolved in pure water is therefore 6.71×10^{-3} g l^{-1}.

Since the amount that can be dissolved by rainwater is 30 times as much, one litre would dissolve $30 \times 6.71 \times 10^{-3}$ g; and 1000 litres, 30×6.71 g = 201 g.

As the calcium carbonate dissolves, the insoluble impurities like clay minerals and quartz sand previously incorporated into the limestone are left on the surface to be added into the soil. If the limestone included iron oxide, this too is insoluble, and accounts for the reddish colour of the soils often found on top of limestone.

Calcium carbonate dissolving in weakly acidic water gives the water 'temporary hardness', so called because heating the water causes the carbonate to precipitate out, resulting in 'fur' in kettles and pipes. The natural equivalent of this is the formation of reprecipitated carbonate rocks (Figure 6.15), due to slow evaporation of hydrogen carbonate rich waters. The stalactites and stalagmites visible in Figure 6.13 were also formed this way.

Question 6.1

Use Equation 6.3 to explain why calcium carbonate is reprecipitated when a saturated solution of calcium hydrogen carbonate evaporates.

Figure 6.15   Reprecipitated limestone, Pamukkale, Turkey.

6.2.2 Chemical weathering of silicate rocks

The various minerals present in silicate rocks may be placed in a series that shows their increasing stability against weathering (Figure 6.16). It is interesting to note that this series is the inverse of their readiness to crystallize from hot magma during igneous rock formation (Figure 3.2).

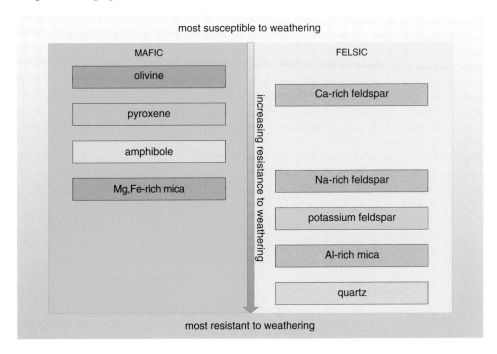

Figure 6.16 Relative chemical weathering of minerals derived from magma.

Olivine, a major mantle mineral that crystallizes at depth under conditions of high temperature and pressure, weathers very easily. Among the feldspars, calcium-rich feldspar, which crystallizes at a higher temperature than sodium-rich feldspar, is the more readily weathered, and the lower-melting potassium feldspar, is more resistant still. Micas are also quite resistant, but the mica that can often be seen 'glinting' on the surfaces of sandstones and shales is usually the more resistant aluminium-rich rather than the dark coloured more mafic variety. The least readily weathered mineral is quartz, silicon dioxide (SiO_2), which is the last to crystallize from cooling magma, and is virtually insoluble and unaffected by weakly acidic solutions. As it is also hard and resists abrasion, it is not very susceptible to physical weathering either, and so tends to 'weather out' as individual grains and accumulate as seashore and desert sand, while other minerals disintegrate. The trend is therefore for the mafic *minerals* to weather more easily than the felsic *minerals*, and consequently the mafic *rocks* weather more readily than the felsic *rocks*.

The weak acid formed by the solution of carbon dioxide in rainwater is also the initiator of the chemical reaction between silicate rocks and water, the process known as **hydrolysis**, that is breakdown by water, directly from the Greek *hydro*, water and *lysis*, splitting. The hydrogen ion is the true reagent in reactions that generate 'silicic acid', for example the hydrolysis of olivine:

$$Mg_2SiO_4(s) + 4H^+(aq) \rightleftharpoons 2Mg^{2+}(aq) + Si(OH)_4(aq)$$

olivine   'soluble silica' (silicic acid)

○ This reaction removes the hydrogen ions from solution, and in a weak acid there are not many there in the first place. So what will happen when the hydrogen ions are used up? (*Hint*: where did the hydrogen ions come from in the first place?)

● Equation 6.1 shows a system in equilibrium, so when the hydrogen ions on the right hand side of the equation are used up, more carbon dioxide will dissolve to generate more hydrogen ions, and maintain the equilibrium position.

'Silicic acid' is so weak an acid that it is better to regard it as shown in the equation:

$$Si(OH)_4(aq) \rightleftharpoons SiO_2(aq) + 2H_2O(l)$$

For this reason it is sometimes described as 'soluble silica', but this does not imply that we should assume that quartz will readily dissolve in the presence of water! For water in contact with quartz the mass ratio $Si(OH)_4 : SiO_2$ is only $1 \times 10^{-4} : 1$.

The weathering reaction occurring most frequently at the Earth's surface is the hydrolysis of the potassium/sodium/calcium aluminosilicate minerals, the feldspars. When feldspar is exposed to an acidic aqueous solution of carbon dioxide it reacts, releasing metal ions and dissolved silica into solution and leaving behind a solid, kaolinite. Remember that this is a clay mineral with a regular 1:1 aluminosilicate sheet structure. The clay may form a barrier to further penetration of water through the soil to the unweathered rock beneath. The hydrolysis of potassium-rich feldspar under acid conditions is shown below:

$$2KAlSi_3O_8(s) + 2H^+(aq) + 9H_2O(l) = Al_2Si_2O_5(OH)_4(s) + 4Si(OH)_4(aq) + 2K^+(aq) \quad (6.4)$$

potassium feldspar   kaolinite

Recall that organic acids and carbon dioxide released by plant roots, and microbial action (like decomposing vegetation), can also considerably increase the acidity of the solution in contact with rocks. Human activity may also contribute through the generation of acid rain, so the rate of chemical weathering may even be related to the severity of air pollution in the region. Each of these acids will act upon silicate minerals in a similar way. The positive soluble metal ions balance the hydrogen carbonate and other negative ions generated in the ionization of the acids that initiated the weathering reactions.

○ From Equation 6.4, which of the three elements in this weathering reaction go into solution, and which are retained in the clay mineral?

● Potassium is *entirely lost* to the solution as K+ ions; aluminium is *absent* from the solution, but present in the newly-formed clay mineral; and *some* silica is present in *both* the new mineral *and* the solution.

Kaolinite is one of several clay minerals that are the normal products of feldspar weathering. Sodium and calcium feldspars are hydrolysed even more easily under these weakly acidic conditions. Kaolinite has the simplest formula, and is the end product of relatively strong chemical weathering in which all the soluble ions have been washed away. Sodium-rich feldspar weathers similarly to release sodium ions instead of potassium ions into solution, but the overall result is the same. Both feldspars produce kaolinite, and the more easily-dissolved sodium and potassium ions are released from the original mineral. Kaolinite is a commonly occurring clay, renowned for its water-adsorbing properties. It is the major ingredient in many anti-diarrhoea products, because a given amount can absorb large quantities of water in the gut.

The more intense the weathering, the more calcium, magnesium, sodium and potassium and, finally, silicon are lost.

○ Thinking back to Section 5.1.4, would you consider 2:1 or 1:1 clays to be the more weathered?

● The 2:1 clays have two silicon layers and one aluminium layer, but the 1:1 clays have only one layer of each, a ratio of (1 Si:1 Al). Therefore 2:1 clays are more silica-rich, since illite and montmorillonite have not lost as much 'soluble silica' as kaolinite. Thus 2:1 clays must be less strongly weathered.

Under less intense weathering, feldspars can also yield the other two principal clay minerals, illite and montmorillonite, but both are easier to obtain from the weathering of more mafic minerals (Figure 6.17). In confirmation of the answer to the question above, both illite and montmorillonite are 2:1 clays. Recall too that illite retains some potassium ions held between the aluminosilicate sheets.

$$3KAlSi_3O_8(s) + 2H^+(aq) + 12H_2O(l) = KAl_2(AlSi_3)O_{10}(OH)_2(s) + 2K^+(aq) + 6Si(OH)_4(aq)$$
potassium feldspar   illite

You are not expected to remember this equation showing the formation of illite, but do note that this hydrolysis both releases potassium into solution and retains some in the clay mineral.

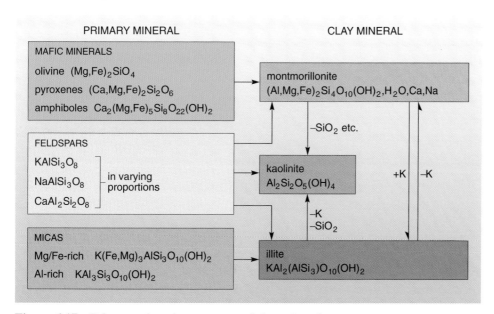

Figure 6.17 Primary minerals as sources of clay minerals.

Figure 6.17 shows the relationship between the primary rock-forming minerals and the clays produced from them by weathering. If the weathered mineral is ferromagnesian, the initial product is usually montmorillonite, which retains some magnesium and iron in its aluminosilicate sheet structure, and calcium and sodium ions, and water, in the layers between the sheets. Micas first weather to illite.

In general, when any **primary mineral** (a mineral formed directly from magma) weathers to produce a clay mineral, the new mineral is usually deficient in elements such as potassium, sodium, calcium, magnesium, and silicon, which more easily weather out of the original mineral than do aluminium or iron.

○ From Equation 6.4 note the initial ratios of Al : Si in the original rock-forming mineral, and in the clay mineral that is produced from it. How does the ratio change after weathering?

● In the potassium-rich feldspar the ratio is (1 Al : 3 Si), but in kaolinite the ratio is (2 Al : 2 Si). Thus, the clay has been enriched in aluminium in comparison with the original mineral, but depleted in silicon, because some silicon has gone into solution.

Once in solution, potassium ions are available as an essential plant nutrient. Potassium is readily absorbed out of solution, but it may also be incorporated into clay minerals by ion exchange, which may cause problems of nutrient deficiency for plants on certain clay soils. Sodium, calcium and hydrogen carbonate ions are abundant in river water as a result of weathering, and sodium ions accumulate in the sea.

○ What happens to the calcium and hydrogen carbonate ions when they reach the sea?

● They are utilized by many marine organisms to produce skeletal parts and shells.

Figure 6.18 Silica skeletons of freshwater diatoms.

Soluble silica is extracted from solution by diatoms in the fresh water, again for the construction of their intricate skeletons (Figure 6.18), which accounts for the lower concentration of silica in seawater—much silica never reaches the sea. Silica that does reach the sea may be extracted by marine organisms.

Accompanying hydrolysis there is frequently **hydration**, which is the part-physical, part-chemical absorption of water by minerals. The water molecules are incorporated unchanged into the structure of the mineral, and may be driven off by heating. The layered structure of the aluminosilicate clays allows water to penetrate between the layers, causing them to swell (Figure 5.11c). The champion of swelling clays, sodium montmorillonite, is able to absorb up to 15 times its own volume of water.

Oxidation is a further weathering process of importance, especially for minerals that include iron (II) ions in their chemical structure. Oxidation often accompanies hydrolysis in the weathering of rocks exposed to the atmosphere. You are probably aware that iron needs both air (oxygen) and water in order to form rust. The same is true for iron in minerals, but there is enough oxygen dissolved in rainwater or groundwater to bring about oxidation of ions of iron (II) (the soluble form of iron) to iron (III) (the highly insoluble form). The mafic mineral olivine, as we have noted earlier, weathers very easily under acid oxidizing conditions (Equation 6.5).

$$4(Fe,Mg)_2SiO_4(s) + O_2(aq) + 8H^+(aq) + 10H_2O(l) =$$
olivine
$$2Fe_2O_3.3H_2O(s) + 4Mg^{2+}(aq) + 4Si(OH)_4(aq) \quad (6.5)$$
hydrated iron (III) oxide

Weathering out of iron, and its oxidation to the highly insoluble, and exceptionally stable, red-brown iron oxide, is responsible for the characteristic iron stain observable on many rock surfaces. There are some organic processes that can reduce iron (III) to the more soluble iron (II) ions, but these are not common in the Earth's oxidizing atmosphere.

Box 6.3 Oxidation and reduction

The simplest definition of oxidation is the *addition* of oxygen to a chemical compound, and this works satisfactorily for many oxidations like the combustion of fuels, and respiration:

$$CH_4(g) + 2O_2(g) = CO_2(g) + 2H_2O(l)$$

$$(CH_2O)_n(s) + nO_2(g) = nCO_2(g) + nH_2O(l)$$

Similarly reduction may be considered as the opposite of oxidation, and so the *removal* of oxygen from a compound during a reaction. Again there are many examples where this definition holds good, as in the smelting of ores. Note that an oxidation is always accompanied by a reduction, just as here, where the iron oxide is reduced but the carbon is oxidized. In becoming oxidized itself, the carbon is acting as a 'reducing agent':

$$2Fe_2O_3(s) + 3C(s) = 4Fe(s) + 3CO_2(g)$$

This also allows us to think of *oxidizing conditions*, such as the surface of the Earth exposed to the oxygen of the atmosphere, and *reducing conditions* where oxygen is absent (described as **anaerobic**), such as the muddy bottom of stagnant ponds.

However, it may not always be easy to see what has been oxidized and what has been reduced. In Equation 6.5, for example, has the iron been oxidized? If we look at another simple oxidation we can then answer this question. Copper is oxidized in air to copper oxide:

$$2Cu(s) + O_2(g) = 2CuO(s)$$

This is clearly an oxidation since oxygen has been added, but during this oxidation the copper *atoms* of the metal have been converted into copper *ions*, Cu^{2+}. In the oxidation process the copper has lost electrons. So to check whether an oxidation has taken place we have to find out whether electrons have been gained or lost during the course of the reaction.

○ When copper is exposed to the air it forms copper carbonate, $CuCO_3$, the green deposit you see on copper roofs. Similarly, silver is often converted to black silver sulfide, Ag_2S. Have these metals been oxidized?

● In both examples the metal atoms have been converted to metallic ions, Cu^{2+} and Ag^+, *and have lost electrons*, so oxidations have taken place. The second reaction is an example of an oxidation that does not involve oxygen.

Look at Equation 6.5. In the iron form of olivine, formula Fe_2SiO_4, the silicate ion has four negative charges (SiO_4^{4-}) so each ion of iron must be Fe^{2+} if the charges are to balance. In the iron oxide, Fe_2O_3, however, there are three oxygen atoms (six negative charges) so the two ions of iron must now be Fe^{3+}, to provide the balancing six positive charges. Iron has lost electrons in going from Fe^{2+} to Fe^{3+}, *and this shows it has been oxidized*. We say that iron has gone from the Fe(II) oxidation state to the Fe(III) oxidation state. The Roman numerals in brackets, which correspond to the number of plus charges on the ion, are called *oxidation numbers*.

6.2.3 Chemical weathering of metallic ores

When sulfide ores are exposed to the oxidizing action of air, water and certain bacteria, sulfurous and sulfuric acids are produced:

$$FeS_2(s) + H_2O(l) + 3O_2(aq) = Fe^{2+}(aq) + SO_4^{2-}(aq) + H_2SO_3(aq)$$
pyrite   sulfurous acid

then:

$$2H_2SO_3(aq) + O_2(aq) = 2H_2SO_4(aq)$$
sulfurous acid   sulfuric acid

This increases the acidity of surface and groundwater, and considerably enhances their ability to leach metals from surrounding ores:

$$2H_2SO_4(aq) + CuFeS_2(s) = Fe^{2+}(aq) + Cu^{2+}(aq) + 2SO_4^{2-}(aq) + 2H_2S(g) \quad (6.6)$$
chalcopyrite   hydrogen sulfide

Oxide ores are less prone to weathering than sulfide ores; they are in effect oxidized already! Weathering of iron ores is characterized by the production of ochre, the red-brown hydrated oxide of iron, which is highly insoluble and can be found blanketing river and stream beds (Figure 6.19).

Figure 6.19 A stream bed blanketed with ochre.

Once again oxidation of the soluble sulfate of the Fe(II) ion to the insoluble hydrated oxide of the Fe(III) ion has occurred:

$$4Fe(II)SO_4(aq) + 10H_2O(l) + O_2(aq) = 2Fe(III)_2O_3 \cdot 3H_2O(s) + 4H_2SO_4(aq)$$

6.3 Weathering and climate

Since weathering results from the direct interaction of the atmosphere and the hydrosphere on the lithosphere, it is to be expected that climate has a significant effect. Different types and different intensities of weathering are observed in tropical, temperate and arid environments. The differences can be related directly to temperature and precipitation, and indirectly to the nature of the resulting vegetation. As chemical reactions generally occur more rapidly at higher temperatures, more intense weathering is observed in tropical regions, especially as levels of precipitation are also high in these climates. Figure 6.20 shows the correlation between the thickness and composition of the weathering residue, and the variation in rainfall, evaporation and temperature, on a section from the Arctic Circle to the tropics. It is clear that the intensity of weathering decreases with depth.

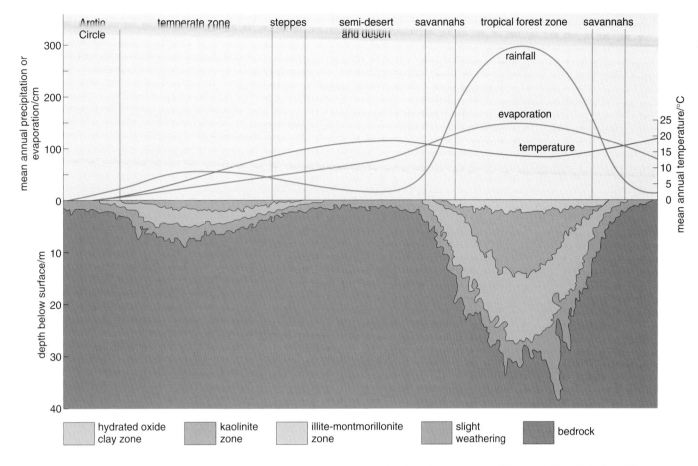

hydrated oxide clay zone | kaolinite zone | illite-montmorillonite zone | slight weathering | bedrock

Many of the breakdown products of chemical and physical weathering become components of the layer of soil that blankets much of the Earth's continental surfaces. This is particularly true of the clay minerals, which form a significant proportion of many types of soil.

Figure 6.20 Relationship between weathering, climate and latitude (highly generalized).

Figure 6.20 suggests that in the temperate zone, the aluminium-rich 1 : 1 clay kaolinite dominates the more intensively weathered surface zone, while the more silicon-rich 2 : 1 clays illite and montmorillonite are found in a deeper zone of milder weathering. However in the forested tropical latitudes the surface deposits after weathering are predominately different materials. Under these hot, wet conditions, as well as the aluminosilicate clay minerals, another important group of clay minerals is found.

6.3.1 Hydrated oxide clays

Hydrated oxide clays are oxides of iron, aluminium and manganese that are formed, as their name suggests, by the reaction of these elements, in ionic form, with water and oxygen. They play an important role in nutrient availability, contaminant transformations and enhancement of soil structure. These clays are a frequent component of many soils, but they dominate in highly weathered soils (with low organic content), especially in the tropics, under climatic conditions of high temperatures and high rates of precipitation.

With heavy rainfall and fairly low soil pH, not only are the more soluble cations such as Ca^{2+}, Mg^{2+}, Na^+ and K^+ leached out of the soils, but SiO_4^{4-} as well, leaving only resistant aluminium, iron and manganese compounds. With continued leaching and high acidity, or under reducing conditions (low oxygen, often due to waterlogging), the ionic forms of these elements will also be leached from the surface soil. Deeper in the soil, where pH generally increases and water flow slows, or if oxygen levels increase, the dissolved Fe, Al and Mn precipitate, forming oxides and hydroxides:

$$4Fe^{2+}(aq) + O_2(aq) + 6H_2O(l) \rightleftharpoons 4FeO(OH)(s) + 8H^+(aq) \tag{6.7}$$
$$\text{goethite}$$

$$4Fe^{2+}(aq) + O_2(aq) + 4H_2O(l) \rightleftharpoons 2Fe_2O_3(s) + 8H^+(aq) \tag{6.8}$$
$$\text{hematite}$$

$$Al^{3+}(aq) + 3H_2O(l) \rightleftharpoons Al(OH)_3(s) + 3H^+(aq) \tag{6.9}$$
$$\text{gibbsite}$$

$$2Mn^{2+}(aq) + 2H_2O(l) + O_2(aq) \rightleftharpoons 2MnO_2(s) + 4H^+(aq) \tag{6.10}$$
$$\text{birnessite}$$

Hydrated oxide clays can be strongly to weakly crystalline. They form regular sheet or layered structures composed of a metal ion (Fe^{3+}, Al^{3+}, Mn^{4+} or Mn^{2+}) bound to O^{2-} (oxygen) and/or OH^- (hydroxyl) groups in an octahedron. Since they are composed of only octahedral groups, hydrated oxide clays are structurally simpler than aluminosilicate clays, which are composed of both octahedral and tetrahedral groups (Figure 6.21).

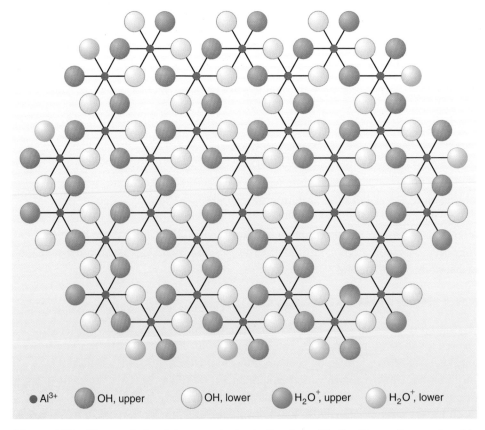

● Al^{3+} ⬤ OH, upper ◯ OH, lower ⬤ H_2O^+, upper ◯ H_2O^+, lower

Figure 6.21 Sheets of aluminium octahedra in the clay gibbsite. Depending on the pH, unshared OH groups at edges may be protonated to form H_2O^+.

The most common aluminium oxide clay is **gibbsite** ($Al(OH)_3$), in which aluminium ions form octahedra by sharing OH^- groups. These octahedra combine to form sheets that are stacked on top of each other and held in place by hydrogen bonds (Figure 6.21). As you will see later in this block, the formation and dissolution of gibbsite is thought to play a major role in maintaining a constant soil pH in soils affected by acid rain.

Of the iron oxide minerals in soils, **goethite** ($FeO(OH)$) is the most common. The Fe^{3+} ions share O^{2-} and OH^- ions and are arranged in octahedra. The octahedra are arranged in double rows. Goethite appears as needle-shaped crystals when examined with an electron microscope. A less common iron oxide clay is hematite (Fe_2O_3), in which the Fe^{3+} ions share O^{2-} with other Fe^{3+} ions in octahedra. Hematite is the major component of common rust. Under moist conditions it can be hydrated to goethite:

$$Fe_2O_3 + H_2O \rightarrow 2FeO(OH) \tag{6.11}$$
$$\text{hematite} \qquad\qquad \text{goethite}$$

Manganese oxides are less abundant than the iron and aluminium oxides, yet they are present in virtually all soils. Though present in only small amounts, manganese oxides have an importance that is much greater than their low abundance would suggest. These minerals play a key role in the storage, release, and chemical changes of many organic and inorganic soil components, including contaminants such as arsenic, chromium and many organic pesticides. Many contaminants, both organic and inorganic, are also oxidized when they come into contact with Mn oxides. The Mn^{4+} ion, Mn(IV), is reduced during this process, as it gains two electrons and is converted to Mn^{2+}, Mn(II) (see Box 6.3).

All these oxide minerals are formed through hydrolysis reactions. The reactions of Fe^{2+} and Mn^{2+} are also oxidation reactions, since the ions lose electrons, and generate iron and manganese in the Fe(III) and Mn(IV) oxidation states respectively.

○ What is common to reactions in Equations 6.7–6.10?

● They all produce H^+ ions.

Through the production of hydrogen ions, formation of hydrous oxide clays from hydrolysis or oxidation of metal ions can contribute to soil acidification.

Hydrous oxide minerals may also be formed under conditions of intense weathering from the partial dissolution of other minerals, for example, gibbsite can form from kaolinite:

$$Al_2Si_2O_5(OH)_4(s) + 5H_2O(l) \rightarrow 2Al(OH)_3(s) + 2H_4SiO_4(aq) \tag{6.12}$$
$$\text{kaolinite} \qquad\qquad\qquad \text{gibbsite} \qquad\quad \text{silicic acid}$$

or hematite can form from the mineral olivine:

$$4(Fe,Mg)_2SiO_4 + O_2 + 8H^+ + 4H_2O \rightarrow 2Fe_2O_3 + 4Mg^{2+} + 4H_4SiO_4(aq) \tag{6.13}$$
$$\text{olivine} \qquad\qquad\qquad\qquad\qquad \text{hematite} \qquad\qquad \text{silicic acid}$$

When deep weathering takes place in humid tropical environments containing felsic rocks, with their high proportions of feldspars, kaolinite can be formed. Subsequent leaching of silica from the kaolinite can give fairly pure deposits of **bauxite**, mainly gibbsite $Al(OH)_3$, a major source material for aluminium production (Equation 6.12).

The deep weathering of mafic rocks produces **laterite** soils, from the Latin *later*, a brick, because since ancient times these soils have been moistened and sun-baked to provide a rock-hard building material. This property makes them rather less useful for agriculture! The mafic minerals are broken down by the movement of water through the rock, forming kaolinite together with insoluble hydrated oxide clays. Laterites develop a layered structure, where the iron tends to accumulate as the top layer and the deeper layers become richer in clay. The weathering therefore acts to concentrate ore minerals.

○   How do you think limestones weather in tropical areas, rich in vegetation?

●   Carbonate rocks are intensely weathered and dissolved in these conditions, as acids from living and decomposing plants are in abundant supply. The normal features of a limestone landscape will be exaggerated.

Seasonal variation has more of an effect in the humid mid-latitude climate. Freeze–thaw fragmentation takes place in the cold seasons, while at the same time plants and microbes have a lower demand for nutrients for growth.

In desert regions, whether hot or cold, the shortage of water, and the limited range and cover by vegetation (or even microbes), do not allow chemical processes to operate to any great extent (Figure 6.20). Instead, physical weathering predominates, resulting in a landscape of angular cliffs and peaks littered with rock fragments. Weathering is slow, and often rocks retain their original appearance, unstained by iron oxide. Under arid conditions, crystalline rocks are less resistant to weathering than are limestones, the reverse of the behaviour observed in humid climates. Fragmentary rocks may crumble through wind erosion. In hot arid conditions, groundwater is drawn upwards by capillary action, and dissolved salts may crystallize as water evaporates to form crusts at or just below the surface.

Question 6.2

A sample of the intermediate rock diorite contains sodium feldspar, pyroxene, mica and a little quartz. Use Figures 3.2 and 6.16 to decide what will be the relative susceptibility of these minerals to chemical weathering, and why. How does weathering susceptibility relate to their melting temperatures? How will weathering begin, and what will be the principal products of the chemical weathering that follows?

Question 6.3

Why are hydrated oxide clays much more likely to be formed in tropical latitudes? How are they formed under these climatic conditions?

6.4 Summary of Section 6

1   Interaction between atmosphere, water and rock-forming minerals results in physical and chemical weathering. Physical weathering is a mechanical process, which breaks rocks into smaller fragments through frost shattering, crystal formation, heating and cooling, root growth, and abrasion during transport.

2   Chemical weathering occurs because minerals formed at depth are unstable at the Earth's surface, in the presence of oxygen and aqueous acid (especially CO_2 in water). The result for silicate rocks is disintegration, formation of new minerals, an aqueous solution of metal ions and 'silicic acid' (sometimes described as 'soluble silica'), and residual resistant material. Calcium carbonate in limestone dissolves to form calcium and hydrogen carbonate ions. Mafic silicate minerals weather more readily than do felsic minerals, the inverse of their readiness to crystallize from cooling magma.

3   When a primary mineral weathers to a clay mineral, it becomes deficient in potassium, sodium, calcium, magnesium and silicon, which weather out more easily than do aluminium or iron. All clays are enriched in aluminium in comparison with the original mineral, but depleted in silicon. Feldspars principally weather to kaolinite. Less intense weathering of feldspar, and the weathering of mafic minerals, produces illite and montmorillonite. Some metal ions, and water, are incorporated into the less weathered clays.

4   Oxide ores are less readily weathered than sulfide ores, which weather to sulfuric acid.

5   Aluminosilicate clays are the most common clay constituents of soils in temperate regions. The highly weathered soils of the tropics are dominated by the hydrated oxide clays, oxides and hydroxides of iron, aluminium or manganese that are formed by hydrolysis and oxidation reactions.

6   Weathering depends on climate, and in the tropics some hydrated oxide clays form by the further breakdown of aluminosilicate clays to hydrated aluminium oxide (gibbsite). Mafic minerals are also oxidized, with soluble Fe(II) becoming insoluble Fe(III), and precipitating as hydrated iron(III) oxides (goethite and hematite). Acid or reducing conditions cause Fe(II) and Al(III) ions to dissolve; they reprecipitate when conditions become oxidizing and more alkaline.

Learning outcomes for Sections 2–6

After working through these sections you should be able to:

1   List the principal elements that make up the Earth's crust. (*Question 1.1*)

2   Specify the major divisions of rock classification — igneous, fragmentary (sedimentary) and metamorphic, and describe how each rock type is formed. (*Questions 2.1 and 2.2; Activities 2.1 and 2.4*)

3   Summarize the distinguishing characteristics of igneous, fragmentary (sedimentary) and metamorphic rocks, and place a given rock in one of these classes. (*Questions 2.1 and 2.2; Activities 2.1 and 2.4*)

4   Know the major rock forming silicate minerals, describe their structures at a molecular level, and account for the place of metal ions (including aluminium) in the crystal structure. (*Question 3.1; Activity 3.3*)

5   Understand what is conveyed by the formula of a silicate mineral, and check the balance of the formal positive and negative charges on the ions. (*Question 3.1*)

6   Place the major silicate-based rock-forming minerals in order of melting, in order of formation from magma, and in order of susceptibility to weathering. (*Question 6.2*)

7   Using a table of chemical composition, or mineral composition, identify mafic, felsic and intermediate igneous rocks, and recognize intrusive and extrusive igneous rocks from their grain size. (*Questions 3.2 and 5.1; Activities 2.1 and 3.2*)

8   Describe the different types of metamorphic rock — gneiss, schist and slate, and relate their nature, appearance and grain size to the conditions of metamorphism. (*Question 5.1; Activities 2.3, 2.4 and 4.1*)

9   Describe the main types of fragmentary rock — conglomerate, sandstone, mudstone (shale) — in relation to fragment size and composition. (*Question 5.1; Activities 2.1 and 5.1*)

10 Describe the different types of limestone and explain the origins of their constituents. (*Question 5.1*)

11 Outline the general structure of aluminosilicate clays, and (in general terms) their formation from other silicate minerals. (*Question 5.2*)

12 Describe how ores are formed by separation from magma, and from hydrothermal fluids.

13 Explain the processes, agencies and results of physical weathering and bioweathering. (*Question 6.2*)

14 Explain how water can become acid thus enhancing chemical weathering, and describe the dissolution and reprecipitation of calcium carbonate during weathering. (*Questions 6.1 and 6.2*)

15 Outline the weathering process for silicate-based rocks and sulfide ores, and describe the changes experienced by metallic elements (potassium, iron, aluminium) and silica. (*Question 6.2*)

16 Outline the formation of hydrated oxide clays, and the conditions necessary for their formation. (*Question 6.3*)

17 Explain why the course of weathering is dependent on climate and latitude, and outline the differences observed. (*Question 6.3*)

Introduction to soils

In the last section, we saw how different types of rock are weathered by the action of physical, chemical and biological forces. These processes result initially in rock fragments, loose unconsolidated material, ranging in size from boulders to quite small particles. There are also individual mineral crystals, especially quartz, left over from the chemical disintegration of igneous or metamorphic rocks, but also other resistant minerals, like iron or aluminium oxides, and clay particles formed in the chemical transformation of the original silicate minerals. In addition there are soluble positive and negative ions in groundwater in contact with the rock debris. The nature of the underlying rock influences the composition of this mixture, and to some extent determines its properties as it matures to become a soil. But soil is 'biologic', as well as 'geologic'. There are still many more steps to take in the transformation of weathered rock into soil.

From the popular perspective, soil is rather unglamorous. We grow our crops in it, and it sticks to our shoes when we go for walks. When asked to describe their environmental significance, most people will probably come up with the fact that soils are necessary for growing things like food and trees (which isn't strictly true!). Some may say that soil can erode, or that soils can become contaminated, and that this contamination can move up the food chain or leach into groundwater.

All these things occur with soils, and much more. In fact, on the land surface of the Earth, soil is the most critical interface between organic and the inorganic environments. Some of the many functions of soil include:

- *Water storage*: Soils provide a long-term steady supply of water for vegetation and other biota, as well as storage of water during rainfall events, thereby buffering storm surges in rivers.

- *Habitat*: Soils are the home for enormous numbers of organisms, including organisms that break down the remains of other organisms.

- *Recycling*: Carbon and oxygen are recycled primarily through terrestrial vegetation and microbes in soil.

- *Nutrient storage and retention*: Nitrogen, phosphorus, potassium, calcium and other critical elements are released, stored and recycled through soils.

- *Provision of water, nutrients and anchorage for many plants and animals.*

- *Filtration and purification of toxins.*

Having touched on the subject briefly at the end of Section 6, we shall now go on to delve into this world of soils. We shall cover their component parts, the similarities and differences among types of soils, and their development over time and in a single place. We shall also consider some of the major processes that are mediated in soils and which can affect local and regional ecosystems, as well as the global biosphere.

8 What is soil?

8.1 Common threads and different features

We all have some image in mind when we say 'soil' — perhaps your garden, or a farm, or a woodland. Equally, bog peat is soil, the lichens you may find clinging to rocks are growing on primitive soils: peatlands and rice paddies, sand dunes and prairies are all associated with unique soils. Soils may be black, orange, red, brown, white or many colours in between, they may have the consistency of squelchy ooze or dust, they may be exceedingly fertile or hardly able to support life. There is nearly as wide a variety of soils as there are habitats on Earth. Obviously it is not colour, or fertility, or moisture that link soils together. So what does?

Nearly all soils have at least two properties in common:

Their basic composition. Soils contain at least some of each of the following:

- inorganic matter (weathered rock, other minerals)
- decayed organic matter
- living organisms
- water
- air.

Their basic structure. Soils are fundamentally a loose and porous material. Even if baked into a hard pat, a soil is never completely solid like a rock.

From these two points, their basic composition and their overall structure, we will develop all the major physical and chemical properties of soils that make them vital to the biosphere (humans included).

A good working definition of soil is:

> Soils are naturally occurring, unconsolidated materials consisting of mineral and organic components that are potentially capable of supporting plant growth.

Briefly, then, soils are composed of solid inorganic material, organic matter, water and air (Figure 8.1). Soils cover much of the Earth's land surface, and may exceed several metres in thickness. Areas considered not to have soil are, for example, recent dumps of fill (e.g. rocks and rubble), bedrock, shifting sand and areas permanently covered by more than 1 m of water.

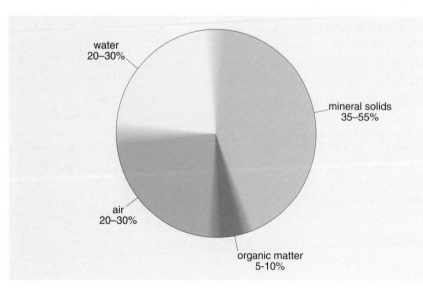

Figure 8.1 Pie chart showing percentages of mineral, organic, air and water in an 'average' soil. (Soils may have proportions out of these ranges.)

water 20–30%

mineral solids 35–55%

air 20–30%

organic matter 5-10%

Much of the variety we find between soils is due to different proportions of the main 'ingredients' (Figure 8.2). For instance, peaty soils have a high proportion of organic matter (soils with organic matter content > 30% are classified as organic soils), soils in rice fields, estuaries or wetlands have a high water content, soils in mountainous regions tend to have more inorganic material and can be very stony and dry, soils in tropical regions are clay-rich (high inorganic matter) but organic matter-poor.

Figure 8.2　An extract from Nature's recipe book.

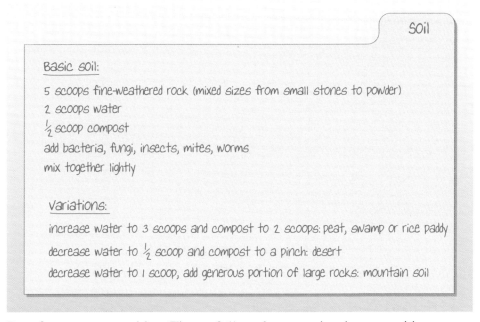

Soil

Basic soil:

5 scoops fine-weathered rock (mixed sizes from small stones to powder)

2 scoops water

$\frac{1}{2}$ scoop compost

add bacteria, fungi, insects, mites, worms

mix together lightly

Variations:

increase water to 3 scoops and compost to 2 scoops: peat, swamp or rice paddy

decrease water to $\frac{1}{2}$ scoop and compost to a pinch: desert

decrease water to 1 scoop, add generous portion of large rocks: mountain soil

But of course, composition (Figure. 8.1), and even variety in composition (Figure 8.2), isn't by any means the whole story.

○　What is missing from Figures 8.1 and 8.2?

●　One thing that is missing is the *structure* of the soil.

Figures 8.1 and 8.2 no more really tell us how to build a soil than a description of all the chemicals that make up a human tell us how to build one. It is the basic material, *together with* the structure and organization of the material that actually becomes the functioning soil. For instance, most of the organic matter in a soil tends to be on the top, because that is where the vegetation is. Roots and biota also tend to be concentrated in the upper layer. For this reason, much of the cycling of water and nutrients occurs in shallow upper soil layers. Soil particles may be clumped together, which may allow water to flow through more freely. This percolating water can dissolve some chemicals from the upper (organic-rich) layers, and move them into lower layers. In addition to all this, organisms such as earthworms, and the roots of plants, can riddle the soil with burrows and fissures, which affect how fast water flows through the soil.

We will discuss all these factors, and many others, in the remainder of the book. But first it would be useful to go outside and have a look at what we have been talking about so far (Activity 8.1).

Activity 8.1   Soil — up close and personal

Take a sample of soil from your garden or another local area. Please check first that no one will object and *it is not a protected area.* Use a garden trowel or a spade and carefully dig up a sample of soil, making sure you are not just digging up surface compost or wood chips. First examine the soil 'as it is' in the trowel or spade. Do you see any differences between the upper and lower parts, for instance, in the colour of the soil or the concentration of roots? (Can you think of any reasons why you may *not* see any differences?) Do you feel sand or grit if you rub it between your fingers? If moistened, does it stick together or fall apart? Now, spread some of it out on a sheet of unlined white paper and pick through it. List what you find, noting, for instance:

- colour
- particles in clumps or aggregates
- roots of plants
- pieces of dead root, stems, leaves
- stones, gravel
- sand
- organisms.

If you wish, try another area, such as a beach, a lakeside, or look at the soil exposed in a road cutting, and compare with your first soil.

Allow about 30 minutes for this activity.

In the remainder of Section 8 we will examine each of these components — solid inorganic material, organic material, water and air — in more detail.

Table 8.1   Size distributions of gravel, sand, silt and clay for soils (International System).

Particle	Size(mm)
gravel	> 2.0
sand	2.0–0.02
silt	0.02–0.002
clay	< 0.002

8.2 Mineral components of soil: An overview

The mineral components of soil are divided by size class into four categories: gravel, sand, silt and clay (Table 8.1).

There are two things to note about Table 8.1. First, the different categories are based on size alone. That means that the actual material that falls into each category can be as varied as any inorganic solids on Earth. For instance, particles of quartz, lead, gold, brick, bone or clamshell would all, if between 2 mm and 0.02 mm in size, be considered sand.

The other matter to note in Table 8.1 is that the 'International' system for soil particle sizes is used. There are different classification systems in use for soil particle sizes, and you will find out later in this book that there are also different systems in use to classify soils. This is because soil science has developed independently in a number of countries over the last two centuries, and different countries have naturally emphasized the soils of most importance to them. So, in your further reading (and certainly if you become a soil science professional), you may come across slightly different systems and classification schemes in use. It is therefore often necessary for soil scientists to describe which system they are using when publishing information on the properties of soils.

The proportion of sand, silt and clay in a sample of soil is called its **texture**. The proportion of gravel is generally not considered in the texture, but, as with other special qualities of particular soils, is noted if it makes up an important component of the soil.

Do you recall from Block 1 measuring the texture of the soil at Dunsford Wood? There are several different ways of measuring texture from soils in the lab, but all of the common methods rely on the basic theory that, when shaken in water, larger particles will settle out quickly (sand settling within a minute), and smaller particles will settle out slowly (fine clay particles taking hours or days to settle). In this way the different size fractions are separated. By measuring the mass of material remaining in solution at pre-determined time intervals, the relative proportions of sand, silt and clay in the original sample can be calculated (Figure 8.3).

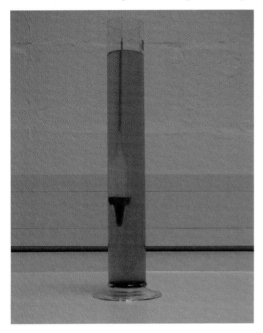

Figure 8.3 Particle size determination in the laboratory using a density method.

The overall texture of a particular soil is described by the size fraction that most dominates the physical and chemical characteristics of the soil. The next size fraction often modifies this name. For instance, a *silty clay* is a clay-dominated soil with sufficient silt to modify its properties; a *clayey sand* is a sandy soil but with a significant proportion of clay. If no fraction is dominant, the soil is described as a **loam**. These descriptions are shown in Figure 8.4a. The diagram consists of a triangle, with corners representing 100% sand, 100% silt, and 100% clay, and the sides representing different proportions of each. To determine a soil texture, we need to know the proportions of at least two of these main texture classes.

To read the diagram, find the proportion of one size fraction (e.g. silt) on the corresponding side of the triangle. Draw a straight line from the percentage value, following the direction of the guides, through the diagram. Do the same for the second fraction (e.g. clay). The intersection of these two lines at a point marks the texture of the soil. A line drawn from the third texture class will go through that point of intersection, and can be used as a check that you have done it correctly. If the point of intersection falls on a borderline, then the soil texture is described as between the two (or three) textures.

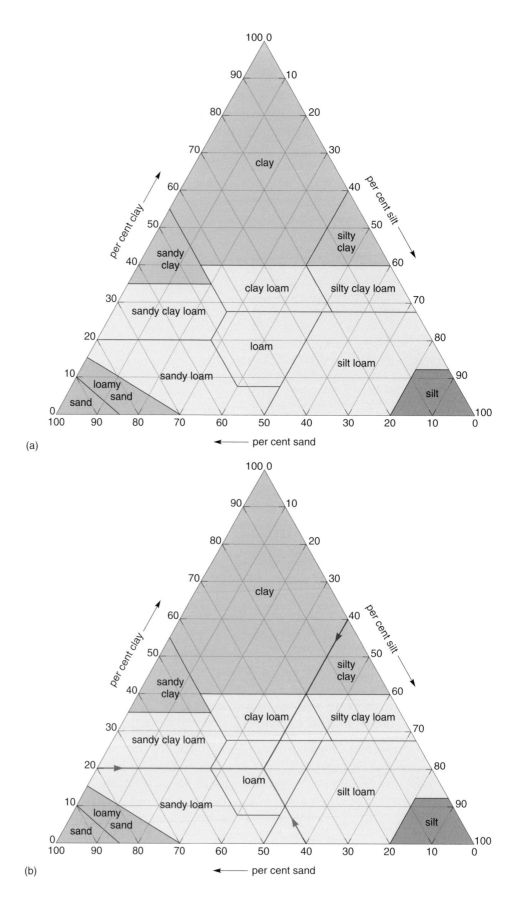

(a)

(b)

Figure 8.4   (a) Soil texture diagram. (b) How to read the texture diagram.

Let's go through an example (Figure 8.4b) and then you can try a few on your own. Assume that you measure the texture of a soil as 40% sand, 40% silt and 20% clay, and you wish to determine the texture class. First, the percentage of silt in the sample (40% in this example) is found on the 'per cent silt' axis, and a line drawn downwards along the direction of the guides — see the red line on Figure 8.4b. Then the same is done for the percentage of clay (20%); here it is drawn across to intersect the red line — see the blue line. In this example, the lines intersect in the zone marked 'loam', and the soil is classified as a loam. The line from 'per cent sand' (40%) should intersect the other two and can be used as a check (green line).

Question 8.1

Using the diagram in Figure 8.4a, how would you describe a soil that is (i) 20% sand, 30% silt and 50% clay, and a soil that is (ii) 20% sand, 70% silt and 10% clay?

Describing a soil as 'sandy' or 'clayey' tells more about its physical characteristics than its chemistry or fertility. For instance, you may, from experience, expect a soil with a high clay content to hold more water than a very sandy soil, but you cannot necessarily say which soil is more fertile. In addition, the structure of a soil (whether soil particles are clumped or dispersed) acts with texture to determine important hydrologic factors like drainage or the diffusion of oxygen. Nevertheless, soil texture can give us a good preliminary estimate of many important soil characteristics. It gives us a feel for the material the soil formed from, what kind of plants may grow best there, how well it will retain nutrients, how well it may conduct water and air, and much else.

For this reason, soil texture is one of the first qualities that soil scientists determine when considering the origin or biology of a soil, or what crops to recommend for planting there. Although it is properly done in the lab, with some experience a very good estimate of texture can be made by rubbing moist soil between the fingers — sand feels gritty, silt is variously described as smooth, silky, or soapy, clay sticks and can form balls and tubes. Such rapid field tests are useful for making quick surveys of the soils of an area. After years of practice, many soil scientists develop their own individual methods of determining texture. Some soil scientists 'listen to' or 'taste' soil, especially for fine sand; others chew it, although we don't recommend this technique!

You probably already have an intuitive feel for how texture affects some physical properties, such as water-holding capacity, of a soil. In the next two sections we will explore these attributes explicitly and, in some cases, quantitatively. We will focus on the question: what is it about each texture class that imparts particular physical attributes to a soil?

8.3 Sand and silt: What's left from the old rock

As mentioned above, the particles that make up the sand or silt fraction of a soil can be composed chemically of anything inorganic as long as they are in the proper size fraction. However, because some solid material (e.g. that which comprises silicate or carbonate rocks) is far more likely to be found on the

Earth's surface than others (e.g. piles of gold, pieces of rust, bones or bricks), sand and silt are usually derived from rocks.

Recall from earlier in this block that sand and silt are usually the residue of rock that has been physically weathered, that is, broken physically into smaller bits by the action of wind, water or biology. Often they have experienced some chemical weathering as well, leading to the leaching of ions from less resistant minerals such as carbonates, micas and feldspars. For this reason, sand and silt are often composed primarily of grains of quartz that remain after other minerals have washed away. Soil water and stream water also include some soluble silica ($Si(OH)_4$) from the release of silicate tetrahedra into solution during weathering.

In areas of more moderate weathering, the sand and silt fraction may also contain micas and feldspars, and if weathering is very mild, and if the bedrock is dominated by carbonate, much of the sand and silt fraction may be remnants of carbonate rock. In contrast, as discussed in Section 6, some areas, especially in the tropics, may have such intense weathering that the silicate tetrahedra from aluminosilicate minerals themselves dissolve, and the sand and silt fractions are very low.

After the more soluble minerals have partially dissolved and ionic components have leached away, the remaining minerals in sands and silts are not chemically changed. The chemical composition of the sand and silt minerals thus reflects both the material that weathered to form the sand and silt (the **parent material**) and the strength of chemical weathering (as influenced by rainfall and temperature) in the environment. But the important points to remember are (i) that the remaining minerals in the sand and silt fraction are *unchanged* from their original state; and (ii) they are usually dominated by particles of quartz (in areas of fairly strong weathering), micas and feldspars (in moderate or mild weathering areas), or carbonates (in mild weathering areas, where there is carbonate rock).

Sand and silt impart many important qualities to soils, particularly in relation to drainage and water retention. Their contribution is mainly to the soil's physical properties rather than the soil chemistry. The *chemical* powerhouses of the soil are contained in the smallest — but by no means least important — texture fraction, the clays.

8.4 Clay minerals in the soil

Recall from earlier in the book that although a *clay* is, strictly speaking, a size classification, not all particles that are small enough to be classified as clay are clay minerals. The aluminosilicate clay minerals that you studied in Section 5, and the hydrated oxide clays that you met in Section 6, are the products of chemical weathering. If you wish to refresh what you learned earlier about the clay minerals, it might be useful to look back at the summaries of these two sections.

Together with organic matter, clay minerals play the central role in imparting the characteristic physical and chemical properties to soils that determine properties like fertility and the cycling of carbon, nitrogen and phosphorus, as well as the loss or retention of water, nutrients, cations, anions, pesticides, heavy metals and other toxins. They have an important role in determining nutrient availability, in bringing about the transformation of contaminants and in enhancing soil structure.

Because they are strongly influenced by the duration and intensity of weathering, the proportion of different minerals in the clay fraction can give a good indication of the development of a soil, or of the dominant climate to which the soil has been exposed (Table 8.2).

Table 8.2 Soil development stage as indicated by the dominant minerals in the clay fraction.

Development stage	Dominant minerals in the clay fraction
early	gypsum and other soluble salts, calcite, olivine
intermediate	micas, feldspars, quartz, 2 : 1 clay minerals (e.g. illite, montmorillonite)
advanced	kaolinite, gibbsite, goethite, hematite

8.5 Sand, silt and clay: How important are they?

Texture is one of the most important variables determining fundamental soil properties such as fertility, water holding capacity and susceptibility to erosion (Table 8.3). Differences in many of these properties among soils can be attributed to the strong dependence of texture on soil mineralogy. Quartz, feldspars and micas usually occur in the sand and coarse silt fractions, while the much more reactive oxides and clay minerals dominate the clay and fine silt fractions. The importance of texture as a fundamental soil property is further emphasized by its relative permanence. Soil texture can change only over very long periods of time, through erosion, mineral weathering, or movement of particles through the soil profile.

Table 8.3 Effect of texture on several soil properties in an unstructured soil.

Property	Textural class		
	Clay	Silt	Sand
water-holding capacity	high	moderate	low
drainage rate	slow (unless cracked)	moderate	fast
water erosion susceptibility	moderate	high	low
wind erosion vulnerability	low	high	moderate
cohesion, stickiness, shrink–swell	high	moderate	low
inherent fertility	high	moderate	low
ease of pollutant leaching	low (unless cracked)	moderate	high
ease of compaction	high	moderate	low

Different proportions of sand, silt and clay can produce soils of dramatically different qualities (Table 8.3). For example, clays have high water-holding capacity and nutrient retention, but can form sticky, compacted soils that are hard to cultivate. Too much clay can lead to poor soil aeration, puddling and cracking when dry. On the other hand, sandy and silty soils are well aerated and not prone to compaction, but may have poor water and nutrient retention and erode easily.

Pollutants are more easily leached from sandy or silty soils than clayey soils. As with most things in life, a balance is usually best in a soil, and the most desirable soils for agriculture, forestry and building are usually loams.

○ From a plant's point of view, would a sandy or a clay soil be better in a dry climate?

● The clay soil would be better because it would be more likely to remain moist during dry periods.

○ What problem might plants experience with clay soils in very wet conditions?

● The soil may become waterlogged, which could mean that the plants roots die from lack of oxygen.

Question 8.2

The following plots (Figure 8.5) shows the proportions of clay and sand in soil samples taken at 10 cm depth at six different locations near the headwaters of the Mississippi River in northern Minnesota, USA.

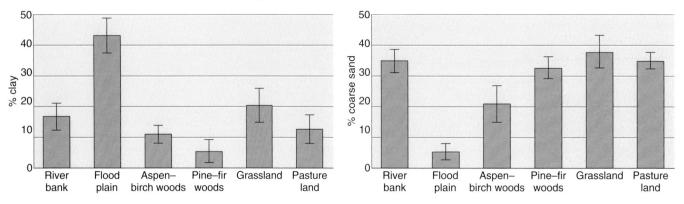

Figure 8.5 Proportions of clay and sand in soils samples taken from six locations in Itasca County, Minnesota. Values represent the mean of three samples; error bars show variation in these samples.

(a) Based on the mean textural analyses alone, which of the soils would you expect to have the highest: (i) water-holding capacity; (ii) fertility; (iii) drainage rates; (iv) vulnerability to erosion; and (v) vulnerability to compaction?

(b) Give two reasons for treating the above conclusions with caution.

8.6 Organic matter

In the context of soil studies, organic matter is matter that is or was living. The chemical composition of organic matter, be it earthworm or humus, is primarily carbon, hydrogen and oxygen, with lesser amounts of nitrogen and phosphorus. Together with clay minerals, organic matter regulates the cycling of these and other major elements in terrestrial ecosystems.

8.6.1 Litter and life

The non-living organic component of soil is primarily formed from fresh or partially decomposed plant material such as roots, stems, leaves and woody material, as well as animal faeces and remains. On soil where such material is available, it ranges in size from the large woody fragments of fallen trees to highly decomposed remnants of plants and animals, in which any semblance of the original organism has gone.

The fresh litter provides the basis of carbon and nitrogen for nearly the entire soil biota, primarily microbes. The more easily digestible fractions of vegetation are consumed first, such as the relatively nitrogen-rich leaves, and the more resistant material is decomposed more slowly. This carpet of partially decomposed leaves, wood and other pieces of vegetation provides habitats for most of the animals that live in the soil. The process of decomposition of this organic matter releases large amounts of carbon dioxide, which increases soil acidity and enhances weathering.

The common living organisms in soil include bacteria, fungi, algae, protoctists (known also as protists or protozoa), and invertebrate animals such as earthworms, snails, millipedes, centipedes, mites, and springtails (Figure 8.6). A single gram of moderately fertile soil typically contains millions of bacteria; a hectare (100 m × 100 m) of productive pasture can contain ten million earthworms! Strictly speaking, organisms that live in the soil are not 'soil' (they are soil organisms), any more than dolphins or jellyfish are the 'sea'. However, they are so tightly integrated with both the inorganic and the non-living organic part of the soil (most are microbes) that it is sensible to consider them all together.

Figure 8.6   Some soil detritivores: (a) the larva of a ground beetle and (b) an earthworm.

The living organisms in soil break down vegetation, cycle nutrients and other elements, maintain soil structure and maintain soil fertility. Starting from the moment a piece of vegetation falls on the surface of a soil, its carbon, nitrogen and other elements are gradually returned to the atmosphere, biosphere, surface waters and soils by a broad range of organisms.

The invertebrate animals in soils are mostly **detritivores** (*detritus*, broken-down material; *vores*, feeding on). These shred the vegetation into smaller pieces and consume the most easily digested parts, leaving the material in many fine particles. Earthworms, centipedes, mites, nematodes, insect larvae and springtails are all important detritivores in the soil. They can work their way through large amounts of soil: it has been estimated that a single earthworm can ingest each day an amount of soil nearly ten times its own weight.

The casts or faecal pellets produced by these organisms are generally enriched in nitrogen and phosphorus relative to the soil as a whole, and are thus a rich source of nutrients for further decomposition by bacteria and fungi.

The larger organisms, such as earthworms, also impart important structure to the soil. As they burrow, earthworms mix lower layers of soil with upper layers, thus bringing nutrients up to the rooting zone and mixing them with organic carbon. Furthermore, earthworm tunnels provide conduits for air and water, greatly enhancing drainage, reducing compaction and bringing oxygen to lower soil layers.

Detritivores break down particles into many smaller units, creating a much greater surface area for attack by microscopic algae, fungi and bacteria, the **decomposers**. These organisms are the ultimate recycling factory, breaking down vegetation, dead animals and other decomposer organisms into their constituent elements. Like the detritivores (and other aerobic organisms, including ourselves), bacteria and fungi gain energy and nutrients from the breakdown of organic matter into CO_2 and water. The carbon in vegetation is released as CO_2, as shown in this simplified overall reaction:

$$(CH_2O)_n + nO_2 \rightarrow nCO_2 + nH_2O \tag{8.1}$$
carbohydrate

The detritivores and decomposers that make up most of the community of living organisms in the soil are not readily apparent, but their importance is massive, as a simple back-of-the-envelope calculation will show.

○ A forest floor of an oak forest is blanketed to a depth of 20 cm by fallen leaves each autumn. If none of those leaves decomposed or blew away, calculate how high the leaves would be in 50 years.

● The leaves would stack to up to 1000 cm (10 m) high.

In fact, the availability of fallen leaves in autumn triggers an enormous increase in rates of decomposition by soil fauna and microbes. A large amount of CO_2 is consequently released — so large, in fact, that an autumn peak of CO_2 can be measured in the atmosphere (Figure 8.7). Note that the late summer concentration of CO_2 shown in Figure 8.7, which was generated from measurements of atmospheric CO_2 in stations around the world, reaches its highest level in the northern latitudes, where there is more land than in southern latitudes. Figure 8.7 illustrates the low points in atmospheric CO_2 concentration that are seen in the northern hemisphere's spring and summer, when photosynthesis in growing plants locks up CO_2 in carbohydrate production. This is followed by peaks in autumn when CO_2 is released by the actions of billions of detritivores and decomposers chomping their way through organic litter.

Given enough time, pretty much any organic matter can be broken down by some combination of detritivores and microbial decomposers. But since any soil is a snapshot in time, it contains fresh material, partly decomposed material and highly resistant material, all being worked on to a greater or lesser extent by the soil biota.

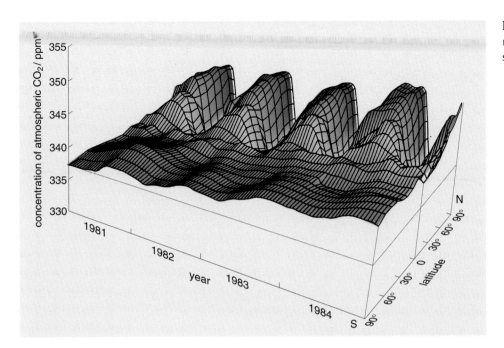

8.6.2 Humus

After several years of decomposition, most of what remains of the once-living matter is a dark-coloured material in which the structure of the original material is lost through extensive decomposition and chemical transformation. This is **humus**. Humus is a catch-all term for a group of dark, highly degraded organic compounds that resist further decay. The structure of humus is poorly understood, although under the microscope it can be seen to consist of intertwined chains of complex organic compounds forming an amorphous particle. Much of it adheres strongly to clay minerals in the soil. A great variety of chemical compounds have been identified in humus, including complex sugars (polysaccharides), lipids, waxes, lignin, cellulose and organic acids. About half the nitrogen in humus is from amino acids, originally part of the proteins of living vegetation.

Figure 8.8 shows a hypothetical chemical structure of a major component of humus, a humic acid. Rather than try to untangle the chemistry, just note a few things about the structure. First, broadly, humic acid (and humus in general) is composed of chains of carbon rings linked together. Second, and most importantly, attached to these rings are various chemical groups, shown as RCOOH, ROH, etc. These chemical groups are of primary importance to soils because it is mainly these groups that participate in chemical reactions in the soil. They are known as **functional groups.** Functional groups are groups of atoms that, when bonded together in a chemical compound, usually behave in a particular way, i.e. have a particular chemical function.

Several functional groups have been consistently identified in humus (Figure 8.8). Particularly important are the **phenolic** hydroxyl group (—OH), in which a hydroxyl group is attached to a benzene-like ring of carbon and hydrogen atoms, and the **carboxylic acid** group (—COOH), in which one oxygen is bonded to the carbon via a single bond and the other is bonded by a double bond. These groups participate in chemical reactions with the surrounding solution, for instance they can release hydrogen ions and become negatively charged, and, depending upon their charge, bind or release metal ions or nutrient ions into solution.

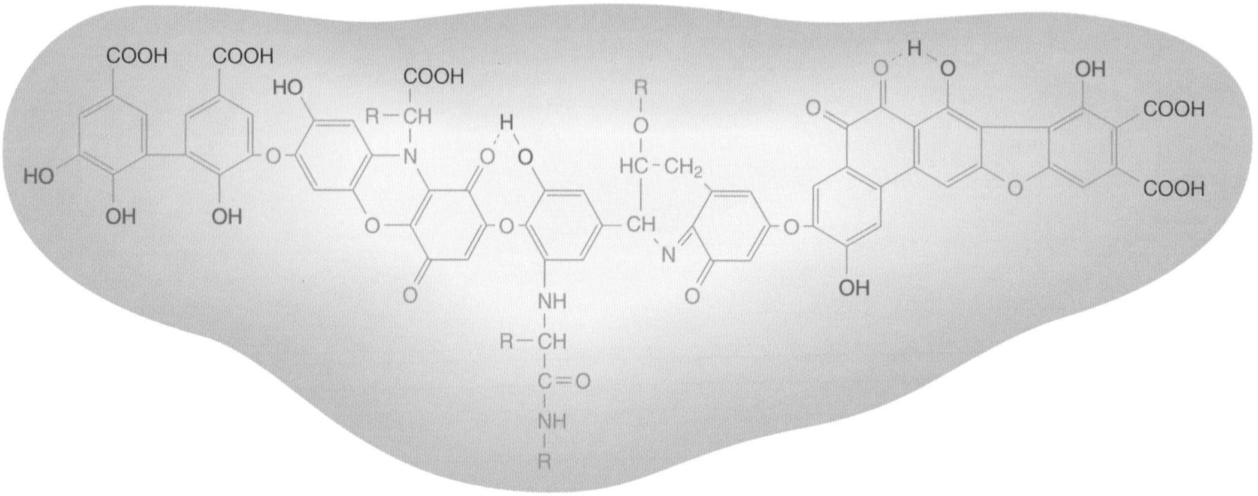

Figure 8.8 Structure of an idealized humic acid, showing functional groups.

In addition to its important chemical qualities, humus is also highly important for water retention in a soil. On a mass basis, humus can hold about five times more water than can clay minerals. In addition, the dark colours characteristic of humus-rich soils at the surface absorb more sunlight than those of a lighter colour and therefore give rise to warmer soils. However, humus may also hold pollutants, both organic and inorganic, and prevent or retard their movement from the soil to groundwaters.

8.6.3 Properties of soil organic matter

The physical and chemical qualities of soil organic matter, and humus in particular, can strongly influence plant growth, soil fertility, and overall ecosystem functioning. Among these qualities are:

Nutrient supply. Soil organic matter is the primary source of nitrogen, and a major source of phosphorus and sulphur to plants. In addition, organic matter provides energy and carbon for soil organisms, and its decomposition releases beneficial vitamins and amino acids.

Ion adsorption and desorption. Organic matter plays a major chemical role in the soil by adsorbing and desorbing cations, many of which are used as nutrients by plants. Thus organic matter-rich soils can hold large amounts of these necessary cations in a form readily available to plants.

Structure. Fungal strands and fine roots bind soil particles together, reducing compaction and allowing water and air to pass through to lower layers. Humus acts as a cement between particles of clay, silt and sand, binding them together. Organic matter-rich soils are generally less vulnerable to erosion, and allow for easy root penetration and seedling emergence.

Water-holding capacity. In addition to the water-retaining properties of humus, organic matter also enhances water holding efficiency indirectly, through its effect on soil structure. The structure of organic matter-rich soils allows easier water infiltration and reduced run-off.

The amount of organic matter in a soil sample is estimated by a delightfully simple method: burning it. The soil is first dried in a low-temperature oven (typically 105 °C) and its mass measured by weighing it on a sensitive balance. Then it is placed in a furnace at a temperature of about 400 °C for about 16 hours. It is then removed and weighed again, and its new mass recorded. The new mass is always less than the original mass, and the difference, (old mass − new mass)/old mass, is the **loss on ignition** or LOI. Soil that is suspected to have significant amounts of $CaCO_3$ is pre-treated with HCl to remove any $CaCO_3$ present before initial weighing, as $CaCO_3$ would also burn in a high-temperature oven, releasing CO_2 and giving false high values. Another method for determining organic matter in soils involves treating the soil with a strong oxidizing agent, which releases the organic matter into solution.

Burning removes the organic matter from the solid phase, converting it to gaseous CO_2 and water vapour. The net chemical equation is the same as that for the breakdown of organic matter by living organisms (Equation 8.1). Since all the organic matter is therefore lost to the atmosphere, the mass remaining is the non-organic fraction.

8.6.4 Factors influencing organic matter content

The upper layer of a well-drained mineral soil (the zone where most plant roots are found) usually contains 1–8% organic matter. The amount present in any given soil depends on a variety of factors. Organic matter content is greatest in cool, moist regions and lowest in warm, dry areas such as deserts. Poorly drained soils, such as those in depressions, often have a large amount of organic matter due to greater moisture from run-off and the resulting increase in plant growth. At the other extreme, very wet soils tend to have high organic matter content because the breakdown of organic matter slows greatly when there are low levels of oxygen in the soil. Vegetation with extensive rooting systems, such as grasses, provides more soil organic matter than do forests, which contribute organic matter mainly through above-ground leaf litter.

Cultivation greatly reduces the amount of organic matter. This reduction is due to removal of plant material by harvesting or weeding, and to increased microbial decomposition caused by greater soil aeration. When a virgin soil is first cultivated, there is a rapid decline in organic matter content, especially in tropical ecosystems. However, the organic matter content eventually stabilizes at a new, lower level (Figure 8.9).

There are several management practices that may be implemented to increase soil organic matter content. The most important of these are to increase the amount of crop residue returned to the soil, adopt cultivation strategies that reduce the amount of tillage required, and incorporate into the crop rotation plants, such as many grasses, that have extensive rooting systems.

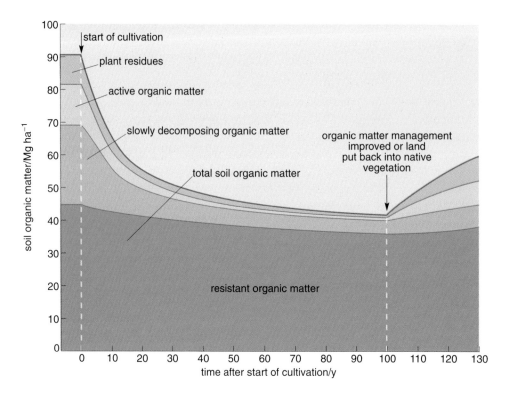

Figure 8.9   Organic matter content as a function of soil management.

Question 8.3

A typical forest loam soil has an LOI of about 1–4%, grassland soils generally range around 2–8%, and peatland soils have an LOI of around 95%. What is the order of these soils in terms of the amount of organic matter they contain? Can you come up with some possible reasons for that order?

8.7 Chemical properties of clays and humus

Clays and humus do not exist in isolation. In particular, these important components of the soil interact strongly with dissolved ions in the **soil solution**. The soil solution is the term used for the aqueous solution of ions and dissolved gases that is in contact with the soil.

Although we have been considering them separately, frequently clays and humus are closely associated, forming essentially a single complex. Clays and humus selectively adsorb and desorb nutrient ions such as Ca^{2+}, Mg^{2+}, K^+, ammonium (NH_4^+) and phosphate (PO_4^{3-}), thus regulating the concentration of these ions in the soil solution and largely determining their availability to plants and their release into the wider environment. Clays and humus also adsorb and desorb water.

The structure of the clay mineral or humus molecule strongly influences these characteristics. In this section we will examine the main reasons behind this far-reaching physical and chemical activity of clay minerals and humus.

8.7.1 Permanent charge of aluminosilicate clays

Perhaps the most important reason for the chemical importance of clays, as opposed to the primarily physical importance of sands and silts, is the fact that clays possess a high net negative electrical charge. Recall from earlier in this block that some of this charge, particularly on clays derived from micas, is acquired at the time of the formation of the parent mineral. This negative charge is further enhanced in soils through exchange of ions in the clay matrix with ions dissolved in the soil water, in an ongoing process called isomorphous substitution.

Isomorphous substitution (*iso*, same; *morph*, shape) occurs when ions of a similar shape and size (Figure 8.10) are swapped between the soil solution and clay sheets. The result is that some of the structural ions, such as aluminium and silicon in the clay matrix, are substituted for ions of a similar size. In particular, Al^{3+} in solution (ionic radius 0.051 nm) often substitutes for another small ion, Si^{4+} (ionic radius 0.042 nm) in tetrahedral sheets (Figure 8.11). Similarly, the intermediate-sized ions Mg^{2+} (0.066 nm) and Fe^{3+} (0.064 nm) can just fit into the more open octahedral sheets and substitute for Al^{3+}. In contrast, large ions such as K^+ (0.133 nm) or Ca^{2+} (0.99 nm) cannot undergo isomorphous substitution with Al^{3+} or Si^{4+} because they are too big to fit into the lattice. (Recall from earlier that the charge of +4 on the Si^{4+} ion is a formal charge.)

○ What is the consequence of an ion with charge $n-1$ (e.g. Al^{3+}) substituting for an ion with charge n (e.g. Si^{4+}) in a clay mineral?

● As the clay mineral as a whole has lost one positive charge, it becomes more negatively charged.

The more 'open' the clay is to the soil solution, the more exposed the clay lattice is to isomorphous substitution.

○ Recall the structures of aluminosilicate minerals discussed in Section 5.1.4 and in the DVD *Virtual Crystals*. Which one, of 1 : 1 clays or 2 : 1 clays, has a more open structure?

● The 2 : 1 aluminosilicate clays have a more open structure.

The initial substitution at a clay's formation, and the ongoing isomorphous substitution, results in most 2 : 1 aluminosilicate clays being negatively charged (Figure 8.11). The negative charge that results from isomorphous substitution in 2 : 1 clays is called a **permanent charge** because the charge does not change if the pH of the soil solution changes. The negative charge on clays is responsible for many of the important physical and chemical properties of soils, including the retention against leaching of vital cations such as calcium, magnesium, potassium and ammonium. We will come back to this important concept later in the block.

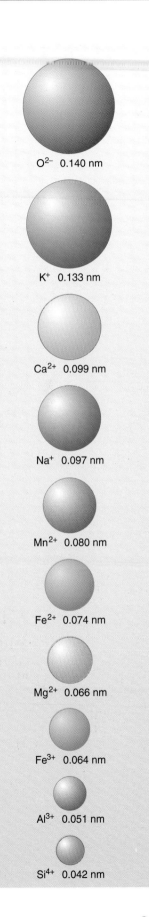

O^{2-} 0.140 nm

K^+ 0.133 nm

Ca^{2+} 0.099 nm

Na^+ 0.097 nm

Mn^{2+} 0.080 nm

Fe^{2+} 0.074 nm

Mg^{2+} 0.066 nm

Fe^{3+} 0.064 nm

Al^{3+} 0.051 nm

Si^{4+} 0.042 nm

Figure 8.10   The radii of ions commonly found in soil minerals.

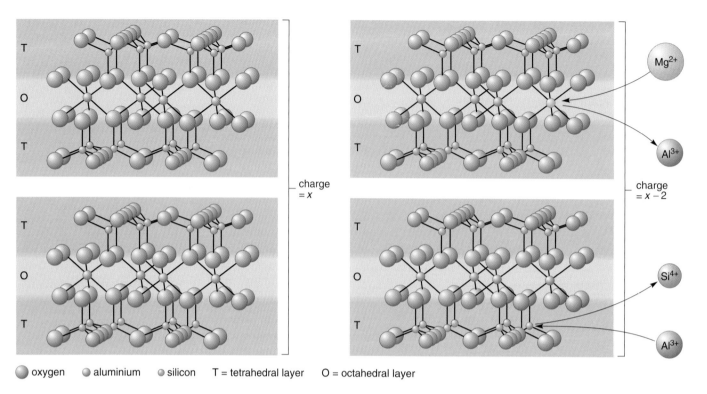

oxygen   aluminium   silicon   T = tetrahedral layer   O = octahedral layer

Figure 8.11   Isomorphous substitution leading to permanent charge in 2 : 1 clays.

Unlike the permanent charge on clay minerals, which is fixed at the time of the mineral's formation, the negative charge on clays through isomorphous substitution continues to increase to some extent as a soil develops over time. The extent of substitution and the type of ions involved are related to factors such as the nature of the parent material and the chemical composition of the soil water, the latter of which is in turn related the activity of biota and the climate. For that reason, the amount of isomorphous substitution, and the type of ions substituting, can tell us something about the environment that the clay has been exposed to.

Isomorphous substitution is not limited to aluminosilicate clays. For example, substitution of Al^{3+} for Fe^{3+} in the octahedra of goethite is common, especially in highly weathered soils, where there is an abundance of Al^{3+} in solution. There is no change in charge, but there are likely to be changes in morphology and a weakening of the crystal structure. These imperfections are areas where the mineral is more vulnerable to solvation, thus highly substituted clays tend to be more easily degraded.

8.7.2 pH-dependent charge

Humus and certain aluminosilicate clays may be strongly or weakly negatively charged, and even neutral, depending on the pH of the soil solution. At lower pH values, but still within the range of many soils, hydrous oxide clays may become positively charged. This is called a **pH-dependent charge**.

pH-dependent charge arises because the broken surfaces and edges of clays, and reactive surfaces of humus, can interact chemically with the soil solution. In clays, these edges contain hydroxyl (−OH) groups that are either bonded to Al^{3+}

cations (at the edges of octahedral sheets) or to Si^{4+} cations (at the edges of tetrahedral sheets). The corresponding groups in humus are the phenolic (−OH) and carboxylic acid (−COOH) groups, which occur in soil organic matter (Figure 8.8).

Let's consider humus first. Figure 8.12a shows a simplified diagram of the phenolic and carboxylic acid groups attached to humus. Recall that the distribution of charge is not even on the −OH group, but that the oxygen atom tends to be partially negatively charged and the hydrogen atom tends to be partially positively charged (Box 5.1). This slight difference in charge density allows the −OH group to lose or gain a proton (H$^+$) depending partly upon the chemical environment of the soil solution, and partly upon the chemical characteristics of the compound.

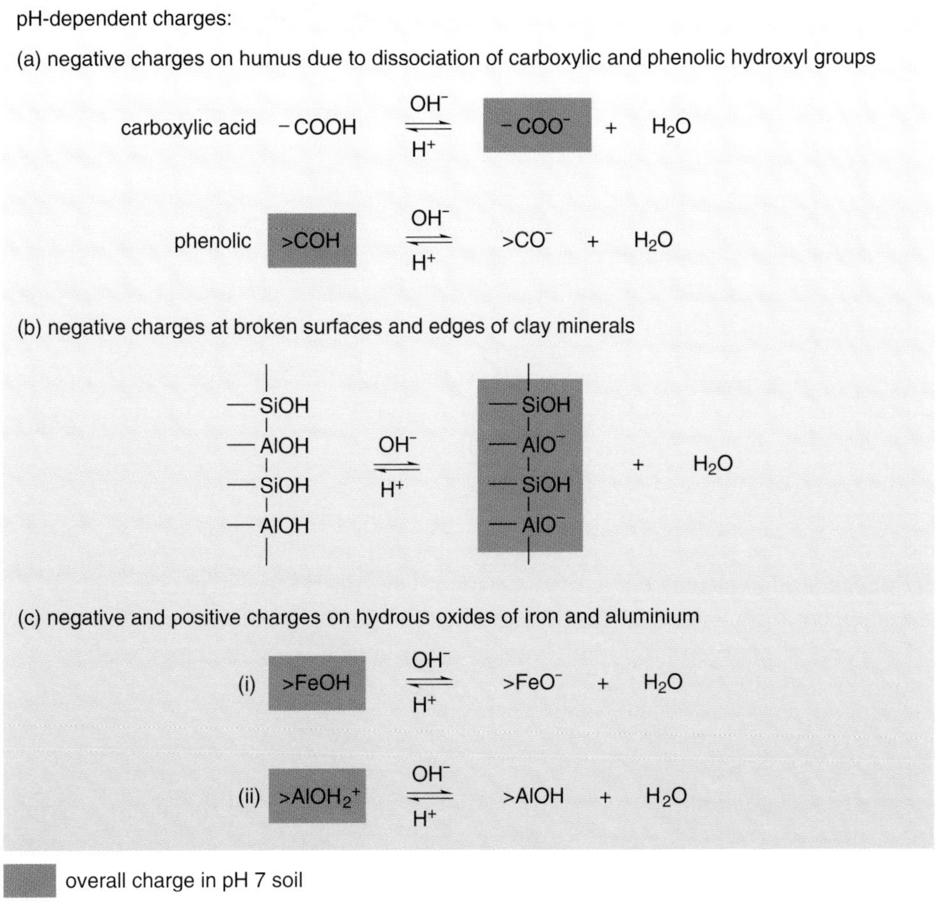

pH-dependent charges:

(a) negative charges on humus due to dissociation of carboxylic and phenolic hydroxyl groups

(b) negative charges at broken surfaces and edges of clay minerals

(c) negative and positive charges on hydrous oxides of iron and aluminium

overall charge in pH 7 soil

Figure 8.12 Generation of pH-dependent charge on the edges of clay minerals and humus. The shaded areas show the dominant form at a soil solution pH of 7.

If the soil solution has a low concentration of H$^+$ (high concentration of OH$^−$) the hydrogen atom at the end of a carboxylic or phenolic group will be attracted into the soil solution (right-hand direction of Figure 8.12a). It will react with hydroxide ions (OH$^−$) in the soil solution to give H_2O, leaving that site on the humus with a negative charge. The carboxylic acid groups of humus begin to lose, or **dissociate**, the proton on the −OH group when the solution reaches a pH of around 4.5, and are essentially fully dissociated at a pH above 7. Phenolic −OH groups dissociate at higher pH (8–10). Thus, at the pH of most soils, the carboxylic acid functional groups of humus tend to be dissociated into R−COO$^−$, whereas the phenolic groups are undissociated. Figure 8.12 shows the dominant form of the functional groups at pH 7. The dissociation of carboxylic acid groups on humus gives the humus compound a net negative charge of approximately three moles of charge per kilogram humus at pH 7—by far the highest charge per kg of material of all of the soil reactive components.

If the soil solution is more acid than pH 4.5 (for carboxylic acid groups) or pH 8 (for phenolic groups) there is enough H^+ in the solution for the compound to remain undissociated (Figure 8.12a, shaded area). In theory, at lower pH values, the undissociated functional groups themselves will become **protonated** (from H^+, a proton), and thus positively charged. But for humus this occurs under such extremely acid conditions (pH < 2) that it is not seen under natural soil conditions (even under conditions of extreme acid rain).

pH-dependent charge can also be generated at the edges of clay minerals, where unshared hydroxyl groups are exposed to the soil solution. If there is a high enough concentration of OH^- in the solution, these hydroxyl groups, primarily in the aluminium octahedra and along the edges of the clay minerals, can lose protons to form water (Figure 8.12b, to the right).

Like the carboxylic acid and phenolic groups of humus, there is no precise pH value at which all of the hydroxyl groups dissociate; it is a function of the type of clay and other environmental conditions such as the temperature. However, in all but the most acid soils considerable numbers of the hydroxyl groups are dissociated, contributing to pH-dependent negative charge (Figure 8.12b, shaded area).

Because it arises at broken edges of clays, the pH-dependent charge of aluminosilicate clays is greater on clays that have a less regular structure, or are more 'disordered'. Among the layered silicates, a 1 : 1 clay such as kaolinite, with a regular structure and forming large crystals, has fewer broken exposed surfaces, and thus shows less pH dependent charge, than a 2 : 1 clay such as montmorillonite, which has more substitution and disorder. Kaolinite has some pH-dependent negative charge over a wide range of pH values, but the amount of charge is not high—around 20–60 mmol charge per kg of kaolinite. Much more pH-dependent charge is found on weakly ordered or disordered clays such as imogolite and allophane, respectively, such as those formed from volcanic ash.

Iron and aluminium hydrous oxide clays also possess pH-dependent charge (Figure 8.12c). However, unlike humus and aluminosilicate clays, they are not usually negatively charged at the pH of most soils. Indeed, for the hydrous oxide clays, there may be substantial *positive* charge generated at the pH of many soils (pH < 7 for goethite, pH < 9 for gibbsite), through adsorption of protons from the solution (Figure 6.21). Especially in the acid tropical soils where these clays are often found, there is enough H^+ in solution to bind to, or protonate the hydroxide of the hydrous oxide clays, and thus render them positively charged.

Charge is expressed in units of **moles of charge** (mol_c) per mass of soil, commonly, $mol_c\,kg^{-1}$ dried soil or $mmol_c\,g^{-1}$ dry soil. The moles of charge are calculated by multiplying the number of moles of each ion by its charge. So, a solution containing one mole of Al^{3+} would have three moles of charge, but a solution containing one mole of Na^+ would have only one mole of charge.

0.7.3 Cation adsorption and exchange

Having described how charge is generated on clay minerals and humus, we can now explore the implications of this. One of the most important results of the charge generation on these particles, and a major property of a soil for determining its fertility, is its **cation exchange capacity** (CEC). Cation exchange capacity is the ability of a soil to adsorb and exchange cations with the surrounding soil solution.

○ Where do you think the cations in the soil solution originate?

● They originally come from weathered rock, often the bedrock in the valley.

Cations in soil solution can also originate in the atmosphere as, for example, dust in rainfall, sea spray (an especially important source of Na^+ and Mg^{2+}), and dissolved ammonia from waste products of livestock, which is protonated to ammonium (NH_4^+).

Clay and organic matter do not remain electrically charged, they attract these cations to restore electroneutrality. The negatively-charged clay or humus particles attract positively charged ions such as Al^{3+}, Ca^{2+}, Mg^{2+} and K^+ from the soil solution. These cations surround the particle (Figure 8.13). Some of the cations accumulate in the spaces between layers of clays, others in a shell or halo around the clay, humus, or clay–humus particle (Figure 8.14). This 'halo' is known as the **diffuse layer**. Sufficient cations accumulate to make the whole unit, particle plus diffuse layer, electrically neutral.

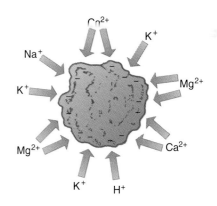

Figure 8.13 Overview of clay–humus particle with charge-balancing cations surrounding particle.

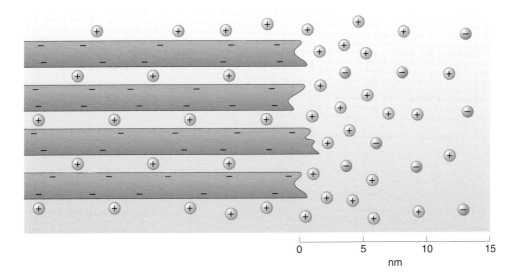

Figure 8.14 Cations in clay interlayer spaces, and ions in the diffuse layer. Note that the proportion of cations to anions decreases with distance from the particle surface, until they are equal.

In general, the distance of an ion from the surface of a particle (i.e., the width of the diffuse layer) is due to a balance between two opposing forces: (1) the diffusion of the ions *away* from the surface, due to movement from higher to lower concentrations in the surrounding solution; and (2) the electrostatic attraction of the positively charged ion *to* the negatively charged surface. For the first reason, exchangeable cations are held closer to the surface in more highly concentrated soil solutions than in dilute solutions. If the solution is highly concentrated, individual clay particles can be brought close enough to stick together and form aggregates.

So, whereas the solid particles are slightly negatively charged, and the soil solution surrounding that particle is slightly positively charged, the whole particle–solution unit is neutral. The charges balance themselves out extremely close to the surface of the particle, so that the dissolved ions in the soil solution as close as 10^{-8} m (10 nm) away from the surface are often in chemical balance.

Any positively charged ion can potentially neutralize the negative charge on a soil particle and thus contribute to a soil's CEC. But the ions that contribute most to CEC are not random; some ions are far more likely to be found close to the surface than others.

In *general*, the greater the charge on the cation, the more strongly attracted it is to the negatively charged surface. The result is that cations with a charge of three, such as Al^{3+}, have more affinity for the cation exchange surfaces than ions with a charge of two, such as Ca^{2+}. Ions with a charge of two in turn tend to have higher affinity than singly charged ions such as K^+. Na^+ is the weakest cation adsorbed because it readily forms a large hydration 'shell' of water, which keeps it well away from the surface. (Figure 5.10c shows a hydration shell around K^+.)

The net result is that, in theory, the cations that are commonly found in the soil solution are attracted to negatively charged surfaces in the order:

$$Al^{3+} > Ca^{2+}, Mg^{2+} > NH_4^+, K^+ > Na^+$$

However, the true cation selectivity of any soil will depend strongly on the nature of the cation exchanger, including its surface charge density, surface structure (edges and wedge sites, cracks, etc.) and the type of clay or organic material present. For instance, recall that some large ions, particularly K^+, are incorporated into the mineral lattice during formation of aluminosilicate minerals, particularly the potassium micas biotite and muscovite. Both NH_4^+ and K^+ can also become trapped under environmental conditions of wetting and drying, when the sheet layers of some clays expand and contract.

This physical entrapment of ions is only important for 2 : 1 clays, and the extent of it is related to the permanent charge on the clay and the distance between individual sheets. Illite has a large permanent negative charge, and potassium and ammonium ions trapped in the interlayers are strongly held by electrostatic forces. These ions neutralize some of the negative charge of the clay and exchange only slowly with the outside soil solution. Soils that are rich in illite clays, therefore, will have a strong selectivity for K^+ and NH_4^+. Figure 8.14 shows some interlayer ions that may or may not be easily exchanged with the soil solution.

With their single charge and large hydration shell of water, sodium ions are the most poorly adsorbed of the major cations in the soil solution. Saturating the soil solution with sodium, therefore, will cause dispersion of the clay–humus particles. For this reason, soils with a high concentration of sodium ions in the soil solution, called **sodic** soils, tend to be poorly aggregated and difficult to work with. On the other hand, saturating the soil solution with calcium or magnesium ions will cause a more compact diffuse layer and allow the particles to approach each other more closely, aiding in aggregate formation.

The hydrogen ion is a special case. Although it is only singly charged, H^+ can be strongly attracted to cation exchange sites and falls in the general series between Al^{3+} and Ca^{2+}. The reason for this is that H^+ is very small, with a very high charge per unit surface area, and can therefore closely approach the negative surface of the clay–humus particle. However, as soon as H^+ is adsorbed it begins to react with the surface, breaking up the structure and releasing metal cations such as Ca^{2+}, Mg^{2+} or, at high H^+ concentrations (pH < 5), Al^{3+} ions, which may then adsorb onto the newly exposed surfaces.

One outcome of the exchange series is that, *given equal molar concentrations in the initial soil solution*, a cation that is more attracted to a negatively charged surface tends to displace one that is less attracted to the surface.

○ In an equal molar ('equimolar') solution of Ca^{2+} and Na^+, which ion would tend to displace the other one on exchange sites, and why?

● Ca^{2+} would tend to displace Na^+ because it has a higher charge (two versus one).

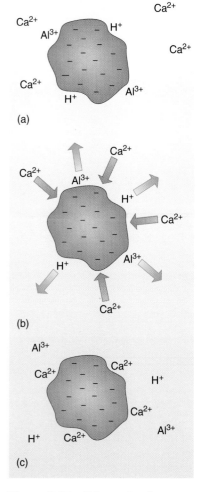

Note that the exchange series described above only gives the relative affinity of different cations for negatively charged surfaces, *all else being equal*. Other processes can, however, affect the actual adsorption of ions. For instance, the series applies to equimolar solutions. If a soil solution was strongly dominated by one cation, this may displace other cations from exchange sites due to a simple concentration gradient, swamping any effect due to relative affinity. Farmers and gardeners make use of this fact whenever they add lime to acid soils (Figure 8.15): the hydrogen and aluminium cations on exchange sites are displaced by Ca^{2+} (or, if dolomitic limestone is used, with Ca^{2+} and Mg^{2+}) by exposing them to an 'overwhelming' concentration of these ions:

$$\text{clay–humus}-2H^+ + Ca^{2+} \rightleftharpoons \text{clay–humus}-Ca^{2+} + 2H^+ \qquad (8.2)$$

$$\text{clay–humus}-Al^{3+} + Ca^{2+} + K^+ \rightleftharpoons \text{clay–humus}-K^+Ca^{2+} + Al^{3+} \qquad (8.3)$$

Al^{3+} released into solution rapidly hydrolyses to form hydrogen ions:

$$Al^{3+} + 3H_2O \rightleftharpoons Al(OH)_3 + 3H^+ \qquad (8.4)$$

These hydrogen ions, together with the hydrogen ions directly released from the clay–humus surface, are neutralized by reaction with more lime:

$$CaCO_3 + H^+ \rightleftharpoons Ca^{2+} + HCO_3^- \qquad (8.5)$$

Liming is also practised on large areas of forests in Scandinavia and central Europe, where soils have recently experienced accelerated acidification by a combination of acid rain, reforestation, and natural soil-forming processes.

In general, the capacity for cation exchange by a soil particle (clay or humus) is in part related to the amount of negative charge per area, or **charge density**, both permanent and pH-dependent, held by that particle. As we saw earlier, permanent charge is mainly a function of the extent to which the (2 : 1) clay has undergone isomorphous substitution, whereas pH-dependent charge is mainly a

Figure 8.15 Schematic of mass–balance effect of liming — flooding exchange sites with Ca^{2+}.

259

function of the disorder of the structure (which determines the extent to which free hydroxyl groups are exposed to the soil solution).

In addition to the charge density, the cation exchange capacity of a particle is related to the amount of surface area exposed to soil solution, which can retain or release adsorbed cations. Clays such as montmorillonite, in which the aluminosilicate layers are loosely held together (Figure 5.11c), have a higher surface area, and therefore a higher CEC than illite, in which the layers are very tightly held by electrostatic attraction (Figure 5.11b; Table 8.4). Despite their high negative charge and the large amount of potassium that balances that negative charge, those ions are held so tightly that they may be effectively non-exchangeable. Humus, with its highly disordered structure and large concentration of freely exposed phenolic and carboxylic acid groups, has the highest CEC of any soil constituent.

Table 8.4 Net charge available for cation exchange of soil constituents at pH 7.

Component	CEC (mmol$_c$ kg^{-1})	Main source of charge
humus	3000	pH-dependent
2 : 1 clays:		
montmorillonite	1000	permanent
illite	300	permanent
kaolinite	20–60	pH-dependent
hydrous oxide clays	5	pH-dependent

The fact that water molecules can freely enter the interlayer spaces causes soils dominated by montmorillonite-type clays to swell when wet and shrink when dry, giving them the common name of 'swelling clays'. The shrink–swell properties of these clays are also important considerations when constructing roads, bridges and other buildings, sometimes ignored to great cost, with subsidence becoming a homeowner's nightmare (Figure 8.16).

Figure 8.16 Cracks developing in a house built on swelling clays.

What about the adsorption of other metal ions? Metals like copper (Cu^{2+}), zinc (Zn^{2+}), lead (Pb^{2+}), iron (Fe^{3+}) and nickel (Ni^{2+}) can form strong bonds to oxygen and hydroxyl ions at the surfaces of clays and organic matter. These ions can be effectively adsorbed and do not generally undergo cation exchange. A positive aspect of this is that, in contaminated soils, potentially toxic metals are effectively fixed in the soil and are not available for plant uptake. However, a negative consequence is that clay-rich soils that are contaminated by heavy metals are often very difficult to reclaim. The metals are so strongly adsorbed that, unlike aluminium or hydrogen on acid soils, they cannot be displaced by even large concentrations of Ca^{2+}, Mg^{2+} and K^+ cations. Only strong complexing agents or concentrated acids will do the job, and then the soil is often more damaged than when it began. Clay soils that are contaminated by metals therefore often have to be hauled away and treated as toxic waste, which can be an enormously expensive undertaking.

Question 8.4

(a) Kaolinite is a common 1 : 1 clay that forms large, regularly ordered crystals. Would you expect its CEC to be high or low, and why?

(b) Do you agree or disagree with the following statement, and why?

'Since all negative charges have to be balanced by positive charges, the cation exchange capacity of a soil clay mineral can be determined by measuring the total amount of negative charge on that clay mineral'.

Question 8.5

In the spring of 1986, the Chernobyl nuclear reactor exploded, sending clouds of radioactive material all over Europe. Some of the material, such as radioactive iodide, decayed relatively quickly and is no longer in the environment. However, radioactive forms of caesium still persist. The caesium ion, Cs^+, has an ionic radius of 0.167 nm, and its persistent radioactive forms $^{137}Cs^+$ and $^{134}Cs^+$ have the same radius to this degree of precision.

Which of the major soil solution ions would you expect $^{137}Cs^+$ and $^{134}Cs^+$ to behave most similarly to? In which kinds of soils would you today expect radioactive caesium to persist?

Measuring cation exchange capacity

Cation exchange capacity is one of the most important chemical characteristics of a soil, and is one of the most commonly measured. There are several different standard methods of measuring CEC, applicable for different types of soil. These methods all rely on overwhelming the soil exchange sites with a high concentration of an ion and then measuring the amount of that ion that has been adsorbed.

One widely used method for measuring CEC is to leach or shake soil with a solution containing an ammonium salt, for example ammonium acetate (NH_4Ac), ammonium chloride (NH_4Cl) or ammonium nitrate (NH_4NO_3). The NH_4^+ ions replace all the exchangeable cations present. The displaced ions in the solution that has percolated through the soil are analysed, and CEC is calculated from the sum of all of these cations. This method also gives the composition of all of the adsorbed cations that make up the CEC. If only the total CEC is desired, the original NH_4^+-saturated soil can be leached with another ion such as KCl, and the displaced NH_4^+ that collects in the percolate solution can be measured.

CEC is expressed in the same units as charge: in units of moles of charge (mol_c) per mass of soil. By working through the two exercises in Box 8.1 you will learn how to calculate moles of charge and cation exchange capacity.

Box 8.1   *Moles of charge and cation exchange capacity*

Exercise 1:   *Moles of charge*

How many moles of positive charge are in a solution containing two moles of $CaCl_2$ dissolved in solution? How many moles of negative charge?

We need to first write the equation for the dissolution of $CaCl_2$ in solution:

$$CaCl_2(s) = Ca^{2+}(aq) + 2Cl^-(aq)$$

So, two moles of $CaCl_2$ dissolved in solution would give us double the above products:

$$2CaCl_2(s) = 2Ca^{2+}(aq) + 4Cl^-(aq)$$

Therefore, two moles of $CaCl_2$ dissociate to two moles of Ca^{2+} and four moles of Cl^-.

To determine the number of moles of positive charge, we simply multiply the number of moles of Ca^{2+}, which is two, by the charge on each ion, which is two. This gives us four moles of positive charge.

To determine the number of moles of negative charge, we multiply the number of moles of Cl^-, which is four, by the charge on each ion, which is one. This gives us four moles of negative charge.

This leads us to an important conclusion: *In an electrically neutral solution of ions, the number of positive charges always equals the number of negative charges.*

Exercise 2:   *Cation exchange capacity*

A sample of 10 g of soil is percolated with ammonium chloride in order to measure its CEC. A CEC percolate solution is collected and found to contain 2 mmol (millimoles, or 10^{-3} moles) of Mg^{2+} and 2 mmol of Na^+. How many moles of positive charge are in the solution? If this were all of the ions adsorbed on the soil surface, what is the CEC of the soil?

We first need to determine the total number of moles of positive charge that was displaced by the NH_4Cl.

Multiplying the number of moles of Mg^{2+} by its charge of 2 gives 4 mmol of positive charge. Multiplying the number of moles of Na^+ by its charge of 1 gives 2 mmol of positive charge. Therefore this solution has $4 + 2 = 6$ mmol of positive charge.

Thus the soil particles contain 6 mmol of exchangeable negative charge. In order to calculate the cation exchange capacity, we need to relate that amount of negative charge to the original amount of soil used. So, if we used x g of soil, the CEC is 6 mmol per x g.

Cation exchange capacity is usually standardized to moles of charge per kg of soil. So, in the above example, the cation exchange capacity is calculated as:

$$6 \text{ mmol charge}/10 \text{ g} \times 1000 \text{ g/kg} = 600 \text{ mmol charge/kg of soil}$$

Question 8.6

A sample of soil is collected from a grassland in southern England. Ten grams of that soil are leached with ammonium acetate, and the solution (100 ml in total) is collected. The solution is found to have the following *concentrations* (not charge) of ions: Ca^{2+} 12.5 mmol l^{-1}; Mg^{2+} 1.4 mmol l^{-1}; K^+ 0.105 mmol l^{-1}; Na^+ 0.24 mmol l^{-1}; H^+ 6.5 mmol l^{-1}.

An additional 33.8 mmol l^{-1} *charge* was determined from the solution, presumably mostly from various charged species of aluminium in the solution.

What is the cation exchange capacity, in cmol charge/kg of soil?

(NB a cmol, or *centi*mole, is 10^{-2} mol.)

Base saturation

The cations Ca^{2+}, Mg^{2+}, K^+ and Na^+ are informally known as **basic** (or base) **cations**. They are called basic cations not because they themselves are bases, but because the oxides of these elements form soluble hydroxides that dissociate in solution to generate hydroxide anions (OH^-), e.g.:

$$CaO + H_2O \rightleftharpoons Ca(OH)_2 \rightleftharpoons Ca^{2+} + 2(OH^-) \qquad (8.6)$$

Oxides such as CaO, MgO and K_2O are rarely found in soils (except perhaps for CaO in very dry soils), so basic cations do not *themselves* generally act as bases in soils. However, they do contribute to maintaining a constant pH in soils by exchanging readily for hydrogen ions during mineral weathering and cation exchange:

mineral weathering:

$$2KAlSi_3O_8 + 2H^+ + 9H_2O(l) \rightarrow 2K^+ + 4Si(OH)_4 + Al_2Si_2O_5(OH)_4 \qquad (8.7)$$
$$\text{orthoclase} \qquad\qquad\qquad \text{silicic acid} \quad \text{kaolinite(s)}$$

cation exchange:

$$\text{clay–humus}-Ca^{2+} + 2H^+ \rightarrow \text{clay–humus}-2H^+ + Ca^{2+} \qquad (8.8)$$

The presence of significant amounts of Ca^{2+}, Mg^{2+}, K^+ and Na^+ ions on soil exchange sites indicates that the soil is not experiencing intense acidification. These soils generally have high to neutral pH. Ca^{2+}, Mg^{2+} and K^+ are also important plant nutrients, and a soil with a high proportion of these cations on exchange sites is, in general, a fertile soil.

Because of their importance in soil fertility and pH, the ratio of the basic cations Ca^{2+}, Mg^{2+}, Na^+ and K^+ to the total CEC is an important parameter to measure in soils. This is called the **base saturation** (BS), and it is calculated as:

$$BS = \frac{(mol_c \text{ of } Ca^{2+} + Mg^{2+} + K^+ + Na^+)}{(\text{total exchangeable cation } mol_c)} \times 100$$

○ Which of the above ions would you generally expect to dominate the exchange sites in temperate regions, and why?

● One would generally expect to find more Ca^{2+} and Mg^{2+} on exchange sites than K^+ and Na^+, because Ca^{2+} and Mg^{2+} are the most abundant cations in the solution in most temperate region soils.

The other cations that generally make up the cation exchange capacity, H^+, $Al(OH)_2^+$, $Al(OH)^{2+}$ and Al^{3+} are known as **acidic cations**. Aluminium in particular is an important acidic cation in low pH soils. In aqueous solution, Al^{3+} ions take on a hydration shell of water:

$$Al^{3+} + 6H_2O = Al(H_2O)_6^{3+} \qquad (8.9)$$

The basic cations also acquire a hydration shell in water, and are more completely written $Ca(H_2O)_n{}^{2+}$, $Mg(H_2O)_n{}^{2+}$, $K(H_2O)_n{}^+$ or $Na(H_2O)_n{}^+$ (Figure 5.10c), although the surrounding water molecule is usually ignored in the formula. However, in Al^{3+} the water molecules forming the hydration shell can be partially hydrolysed:

$$[Al(H_2O)_6]^{3+} \rightarrow [Al(OH)(H_2O)_5]^{2+} + H^+ \tag{8.10}$$

The partial hydrolysis of Al^{3+} in water releases hydrogen ions. Aqueous Al^{3+} released into solution can thus act as an acid.

Due to cation exchange, once a cation has been released by weathering it is not immediately lost in the soil solution, but a certain proportion can be retained on the solid surfaces of the soil particles. Not only does this retard their loss in runoff, it allows cations to be available for plant uptake. Through their exchanges in the soil solution, exchangeable cations are relatively easy for plant roots to extract — at least a lot easier than getting them out of the rock — and thus the **fertility** of the soil (its ability to provide essential nutrients) is strongly influenced by the CEC. CEC from basic cations such as Ca^{2+}, Mg^{2+} and K^+ also helps maintain the soil pH at levels desirable for the health of plants. Organic matter is very important for its high CEC; this is especially important in sandy soils.

○ Why especially in sandy soils?

● These soils are by definition clay-poor, and clays are the other major source of CEC to soils.

The relation between the concentrations of cations on the exchange sites and in the soil solution is strongly related to the CEC, but is modified by physical factors such as the rate of water percolation, soil temperature, and the level of biological activity. For this reason, the response of soils (and ecosystems) to acid inputs can differ even if they have the same CEC and BS, although CEC and BS give a very good indication of how they will behave.

In contrast to weathering reactions, soil exchange reactions can occur rapidly and reversibly. Thus the base saturation of a soil is not static, but changes over time and space. Soil that is initially formed from weathering of primary minerals may have a high base saturation as more and more cations released from weathering are adsorbed on the clay and oxide surfaces. But as time goes on, water percolation, the action of CO_2-rich acid soil solution from respiration, and acid rain, remove more and more basic cations from these exchange sites, and BS declines.

Somewhat less straightforward is the fact that CEC *itself* declines in a soil as the soil ages over hundreds or thousands of years. One would think that, although the calcium, magnesium, sodium and potassium that are removed from exchange sites are replaced by either hydrogen or aluminium, the total cation exchange capacity (the ability of the soil to exchange cations) would remain constant. In fact, this is only true over the time-scales of human observation at one site.

However, as weathering continues over hundreds or thousands of years, the very nature of the clays changes. In regions where weathering is not intense, such as in temperate, boreal or desert areas, soil clays tend to be dominated by 2 : 1 minerals such as montmorillonite or illite. Prolonged leaching with acid, however, can remove so many basic cations that the clay mineral itself weathers to a 1 : 1 mineral such as kaolinite, for example:

$$\text{illite} \xrightarrow{\ -\text{K, } -\text{SiO}_2\ } \text{kaolinite} \qquad\qquad (8.11\text{a})$$

$$\text{montmorillonite} \xrightarrow{\ -\text{Mg}^{2+}, \ -\text{SiO}_2\ } \text{kaolinite} \qquad (8.11\text{b})$$

In this example, in addition to disintegration of the tetrahedral silicate layers, illite loses potassium, and montmorillonite loses magnesium and iron (this is also shown in Figure 6.17). Hydrogen ion weathering 'dissolves' some of the clay, thus reducing the CEC.

With even more weathering, the 1 : 1 clay minerals can be converted to oxides of aluminium and iron:

$$\underset{\text{kaolinite}}{\text{Al}_2\text{Si}_2\text{O}_5(\text{OH})_4(\text{s})} + 5\text{H}_2\text{O}(\text{l}) \rightarrow \underset{\text{gibbsite}}{2\text{Al}(\text{OH})_3(\text{s})} + \underset{\text{silicic acid}}{2\text{Si}(\text{OH})_4(\text{aq})} \quad \text{(see Equation 6.12)}$$

The 1 : 1 clay minerals have much lower CEC than 2 : 1 clay minerals and the iron oxides have even lower CEC (Table 8.4). (In fact they may have a substantial *anion* retention capacity, as we shall see in the next section.) So, not only does the *amount* of basic cations held on exchange sites become lower as soils weather, the *capacity* of those soils to hold and exchange cations declines. This means that the potential fertility and buffering capacity of the soil declines.

○ Why don't all soils, even in temperate zones, eventually become dominated by 1 : 1 clays and Al and Fe oxides?

● New 2 : 1 clay minerals are released from bedrock by weathering.

Question 8.7

In samples of soil collected from different environments in the Teign Valley in Devon, the LOI, pH and amounts of exchangeable ions were calculated, and the values entered into Table 8.5. Complete the table (overleaf) with the calculated cation exchange capacity and base saturation of the soils. Based on these values, which soils would be the most desirable from an agricultural perspective for growing crops? Speculate on the reasons for the differences in CEC and % base saturation you find.

Table 8.5 Chemical data for some soil samples in the Teign Valley, Devon, UK. (All exchangeable ions are given in mmol charge per 100 g soil.)

Soil series	Swanaford	Yarner Wood	Haldon Forest	Trusham 1	Trusham 2
Depth (cm)	8–23	0–28	0–13	0–28	0–23
Surface vegetation and environment	ash/willow, floodplain of River Teign	oak/bilberry	pine plantation	semi-natural meadow	agricultural pasture
LOI	8.1	3.5	3.4	7.0	5.2
pH (H_2O)	5.7	4.5	4.3	5.9	6.2
Exchangeable Ca^{2+} ($mmol_c$/100g soil)	12.0	1.4	1.2	25.0	32.8
Exchangeable Mg^{2+}	2.0	0.20	0.20	2.8	1.4
Exchangeable K^+	0.11	0.04	0.03	0.11	0.1
Exchangeable Na^+	0.20	0.05	0.05	0.24	0.13
Exchangeable H^+	2.0	0.51	1.1	0.7	0.03
Exchangeable Al^{3+}	16.2	7.5	9.1	16.4	1.02
CEC ($mmol_c$/100g soil)					
BS (%)					

8.7.4 Anion adsorption and exchange

All reactive soil components with pH-dependent charge can, in theory, become positively charged at a low enough pH. In practice, as you saw in Section 8.7.2, this is only important for the hydrous oxide clays, particularly gibbsite (at soil pH < 9) and goethite (at soil pH < 7):

$$-(Fe,Al)OH + H^+ = -(Fe,Al)-OH_2^+$$

(See also Figure 8.12c.) In a similar process to cation exchange, the charge on these particles can attract and hold ions from the soil solution. In this case, however, the surfaces are *positively* charged and it is the negatively charged ions such as PO_4^{3-} and SO_4^{2-} that are held and exchanged.

The capacity of a soil to retain anions is known as its **anion exchange capacity** (AEC). Soils with a high proportion of hydrous oxide clays, such as highly weathered tropical or semi-tropical soils, can have substantial anion exchange capacity. Many temperate soils can also have high AEC, especially under conditions of high weathering in which aluminium and iron form oxides through dissolution, leaching and re-precipitation.

Anions are held by soil minerals in two main ways. The first way is through simple electrostatic attraction, in a manner similar to cation adsorption:

$$\text{soil particle}-(Fe,Al)-OH_2^+ + A^- = \text{soil particle}-(Fe,Al)-OH_2^+(A)^- \qquad (8.12)$$
$$\text{anion 'A'}$$

The ions are not physically adsorbed to the particle, but are held in a loose shell or 'halo' around the soil particle. The adsorbed ions can exchange with other anions in solution:

$$\text{soil particle–(Fe,Al)–OH}_2{}^+(A)^- + B^- = \text{soil particle–(Fe,Al)–OH}_2{}^+(B)^- + A^- \quad (8.13)$$

anion 'B'   anion 'A'

Similar to cation exchange, the higher the charge on the anion, the stronger the attraction to anion exchange surfaces:

$$PO_4{}^{3-} > SO_4{}^{2-} \gg Cl^-, NO_3{}^-$$

The anion exchange capacity of hydrous aluminium and iron oxides is pH-dependent. Figure 8.17 summarizes the pH dependence of surface charge on these oxides, and in turn the capacity to attract anions or cations.

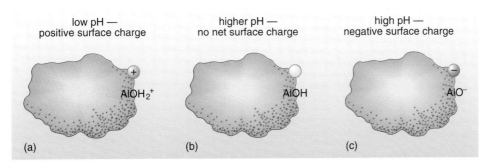

Figure 8.17 pH-dependency of anion adsorption. (a) low pH, (b) pH at which there is no surface charge and (c) high pH.

A second way for anions to be held by soil particles is by means of **complexation**, the formation of strong chemical bonds with the surface of soil particles. Anions bound in this way cannot exchange as readily with ions in solution as can those anions that are held more weakly, through simple electrostatic attraction. Ions that form complexes include phosphate, silicate, and probably sulfate and chloride. All these ions can also be held electrostatically.

Electroneutrality is maintained during ion exchange. This means that one mole of negative charge at the surface must be exchanged with exactly one mole of negative charge in solution. The anions released to solution by this exchange reaction are available for uptake by plants. Because soil particles become more positively charged as pH decreases, AEC is greatest at low pH.

Anion exchange influences the concentration of nutrients and pollutants in soil solution, and controls the leaching of these ions to groundwaters. The type of anion exchange is therefore important: tightly held ions such as phosphate are difficult for plants to extract from anion exchange complexes. For this reason, tropical soils with high anion exchange capacity may be strongly phosphate limited. The strength of anion retention is also important when trying to predict the effects of acid rain on soils: soils with significant anion retention capacity will retain $SO_4{}^{2-}$ from acid rain, thus reducing the amount of acid which leaches into streams.

○ Will the leaching of anions from a soil profile to groundwater be more severe at low pH or at high pH?

● Anion leaching will be most severe at high pH, when the soil particles possess the greatest negative charge.

Question 8.8

As a soil weathers and its minerals are transformed over millennia, how would you expect the soil's cation exchange capacity and anion exchange capacity to change?

8.8 Soil structure

Soil structure is the arrangement of sand, silt, clay and organic matter into larger units. Aggregates that form naturally are called **peds**, whereas those that form artificially, as during ploughing or digging, are known as **clods**. Peds, which may range from a millimetre to tens of centimetres in size, develop through soil-forming processes over decades and centuries as the soil matures. However, human interference can very quickly modify or destroy this structure. Soil structure greatly influences water infiltration, susceptibility to erosion and ease of root penetration and seedling emergence. For these reasons, much effort has been directed toward understanding the factors that promote and maintain good soil structure.

Formation of soil structure

Soil peds that one may collect from a field or a garden are comprised of a multitude of smaller aggregates. Examination of a single ped with a hand lens will reveal the many smaller constituents, which, in turn, are made up of still smaller units. Crushing a ped in your hand will similarly reveal this diverse array of ever-smaller aggregates. The interactions among the primary particles of sand, silt, clay and humus are classified as either biological or abiotic. Biological factors dominate in the larger aggregates and in sandy soils, while abiotic factors are most important in cementing the primary particles.

Abiotic factors

Aggregation begins with the attraction of clay particles for one another in a process known as **flocculation**. This attraction occurs because cations like calcium (Ca^{2+}) and magnesium (Mg^{2+}) act as bridges between the negatively charged clay particles. Large amounts of sodium (Na^+), however, have a repulsive effect, and can cause the clay to disperse instead of flocculate. It is for this reason that soils high in sodium often have very poor structure. Humus, with its high charge, serves to bind the individual clay particles with each other and also to silt particles, thereby increasing the aggregate size. Iron and aluminium oxides also act as cementing agents and may give rise to very stable aggregates, which become even more stable upon drying.

Shrinking and swelling, which are most pronounced in montmorillonite-type soils, also promote structure because the soil is forced to move along planes of weakness, and in this way they help to define the ped boundaries. These processes are most common in regions with many wet–dry or freeze–thaw cycles.

Biological factors

Bacteria, fungi, and plant root hairs all produce polysaccharides (long, chain-like molecules made up of sugar molecule subunits) that coat the surfaces of single mineral grains or clays. These intimate organo–mineral associations are ubiquitous in soil and help to bind **microaggregates** into larger units, visible with the naked eye, called **macroaggregates** (Figure 8.18). Microaggregates may also be bound physically by plant roots and fungal hyphae to form larger aggregates. In addition, earthworms and termites promote structure by creating channels as they burrow, and by modifying ingested soil as it passes through their gut.

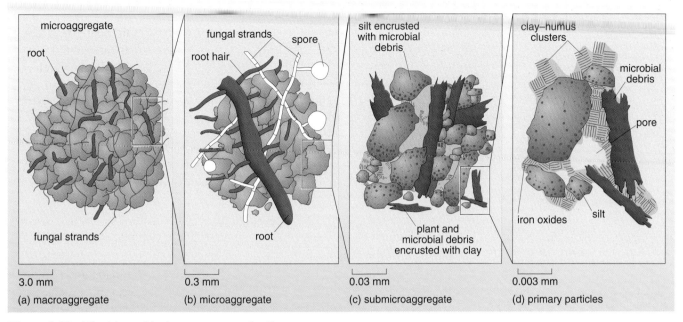

Figure 8.18 The hierarchical arrangement of soil aggregates.

○ Human interference can very quickly destroy soil structure. Why is it important to minimize this interference and preserve favourable soil structure?

● Good soil structure will enhance water infiltration, ease of seedling emergence and root penetration, and reduce susceptibility to erosion.

8.9 Soil solution

All soils, even those in the harshest sun-baked deserts, contain some water. Soil water is valuable other than just for sustenance of plants; it also plays a major role in the stability of the soil against erosion and the supply of oxygen to plants. In addition, the soil solution contains many dissolved chemicals that can provide vital nutrition to organisms in some cases, or be toxic in other cases. Organic matter increases the amount of available water both directly, due to its ability to absorb water and retain it, and also indirectly, through its beneficial effects on soil structure and pore space.

Water is essential for life. The growth and maintenance of plants and soil organisms, and the functioning of terrestrial ecosystems, are intimately tied to soil moisture. Nearly all the biological, chemical, and physical processes in a soil, including those responsible for soil formation, are mediated by water. However, despite the importance of water, many of the world's soils are unable to provide adequate moisture to maintain plant growth, even where the soil beneath a single square metre of the surface may contain a hundred kilograms of water. How can drought occur when so much water is present in a soil?

The answer is found by realizing that water in soil is quite different from water in a lake, a river, or a drinking glass. In the next section we shall explore some of the far-reaching consequences of soil water content for both the physical and the chemical behaviour of soil.

8.9.1 Soil water content

The wetness, or **water content** of a soil can be measured by simply drying the soil in an oven. A sample of field-moist soil is weighed and then placed in an oven at 105 °C, where it is dried for about 24 hours. At this temperature, just slightly above the boiling point of water, all the water is driven off but the organic matter, biota (albeit dry and dead) and other soil constituents are retained in the soil. After drying the soil is weighed again. The mass loss represents the amount of soil water.

○ How is this technique different from the % LOI technique used to measure soil organic matter content, which also involves mass loss after the soil is heated in an oven?

● The difference is that LOI is measured on soil that is burned, not just dried. It uses a much hotter oven (400 °C versus 105 °C); one that is hot enough to combust all the organic matter.

More 'high tech' methods of measuring soil water content, for example those exploiting the effect of water on electrical properties of different materials, have been developed and can be used as probes in the field, without the need to collect samples and bring them into a lab. These techniques can be used to provide fairly continuous 'real time' measurements of the water content of a soil as it changes due to rainfall, temperature and other factors.

8.9.2 Porosity and permeability

Soil is, by definition, made of unconsolidated material, and the spaces between the solid material in the soil are the **pores**. Soil water and soil air are held in the pores of a soil. Pores range in size from the tiny interlayer spaces of clay minerals to large cracks. Those smaller than about 0.08 mm are known as **micropores**, while those larger than 0.08 mm are called **macropores**. Generally, macropores dominate in sandy soils, while micropores are most abundant in fine-textured soils, especially those with little or no aggregation or other structure. Macropores can also be created by cracks opening in clayey soil that has wetted and dried, or by soil freezing and thawing, or by living organisms such as roots, earthworms, or termites.

The distinction between macro- and micropores is important. In most soils most of the time, micropores are filled with water. For this reason, fine-textured soils (those dominated by clays and silts) have a greater water-holding capacity than do sandy soils. However, soils dominated by micropores can be poorly drained and are vulnerable to flooding.

The rapid movement of water and air occurs mainly through the macropores. These larger pores may also be filled with roots and diverse tiny soil animals.

The **porosity** of the soil is the volume of the pores in relation to the total volume of the soil. It can be determined by taking an undisturbed core of soil, saturating the soil with water and weighing it, then drying the soil completely and re-weighing it. Porosity is determined as:

$$\text{porosity (\%)} = \frac{[\text{saturated mass} - \text{dry mass (converted to volume)}]}{\text{volume of core}} \times 100$$

The saturated mass minus the dry mass gives the mass of water that occupies the pores at saturation. Say this difference is 50 g, and the volume of the core is 100 cm^3. We know that the density of water is 1 g cm^{-3} at room temperature, and therefore 50 g of water occupies 50 cm^3 in total. Thus the porosity of this soil is 50 cm^3 of pores in 100 cm^3 of undisturbed soil, or 50%.

But simply knowing the porosity does not tell us much about the soil properties.

○ In addition to total pore space, what is important to know for estimating important properties of soils, such as ease of drainage?

● It is important to know the proportions of different sizes of pore, particularly the relative proportions of macropores to micropores.

Water is held most strongly in the smallest pores. In fact, despite its high porosity, much of the water in a clay-textured soil (that held in pores with sizes <0.0002 mm or <0.2 μm) may be held with such strength that it is not available to plants. Such a soil may have a high water content but poor water availability. On the other hand, free-draining sandy soils, with large pores, may have little water in those pores, most of them being filled with air. Most of the time it is the intermediate-sized pores (0.2–80 μm) that store water that is available for plant use.

So, if we know the porosity, and know the proportion of macropores to micropores in a soil, we can make some estimates about how well that soil will drain, how available water will be to plants and how well aerated the soil will be.

The **permeability** of a soil is the ease with which water passes through the soil. Permeability is related to porosity, pore size distribution and how well the pores are interconnected. Pores must be large enough for water flow to occur, so clay soils have high porosity but low permeability.

A soil is said to be **saturated** when all its pores are filled with water. Imagine that a saturated soil is placed in a flowerpot that has holes in the bottom to allow for drainage. Water will drain easily from the soil simply due to gravitational forces. When the last drop of water has fallen from the pot, the soil is said to be at **field capacity**. In this condition, the soil is holding as much water as it can against the force of gravity. The soil water content at field capacity differs greatly for different textures of soil. In general, the finer the texture of the soil, the more water it can hold at field capacity. Soils with high clay content can hold several times more water at field capacity than very sandy soils (Figure 8.19, red curve).

A plant growing in a soil at field capacity will initially be able to extract water easily to satisfy its needs. However, the plant will have progressively more difficulty in obtaining water as the soil dries. Eventually, even the most drought-adapted plants will be unable to extract sufficient moisture to survive. At this point the soil is said to be at its **permanent wilting point** (Figure 8.19, green curve). The amount of water held between field capacity and permanent wilting point, the **available water**, is the amount that is potentially available to plants:

available water (%) = soil water content at field capacity −
soil water content at permanent wilting point

Like field capacity, the soil water content at the permanent wilting point differs greatly for different textures of soil (Figure 8.19). A clay-textured soil may

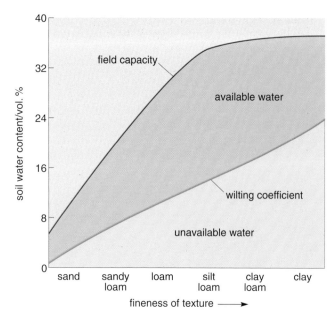

Figure 8.19   Generalized relationship between soil water content and soil texture.

contain large amounts of water at the permanent wilting point, but this water is held so tightly in very small pores that it is not available to plants.

In summer, the combination of high temperatures and high plant growth causes high rates of water loss from the soil. This causes the available water to decrease, and the soil experiences a **soil moisture deficit**. The soil moisture deficit is defined as the excess of evaporation (+ plant water loss to the atmosphere) over precipitation for a given time period (usually on a monthly basis). The soil moisture deficit is usually highest in mid-summer, and lowest in the winter.

Soil water content is useful to quantify changes in the hydrology of one soil over time, and how that may affect the growth of plants on that soil. For instance, if the soil water content at the wilting point is known for a particular soil, steps can be taken to ensure that over the summer the soil water remains well above this level. However, as Figure 8.19 shows, the soil water content at both field capacity and wilting point differs greatly for different soil textures. Therefore, to be of any real use for agriculture, the soil water content at field capacity and wilting point must be determined for each soil and each crop.

For determining how well plants will grow in different soils then, it is more useful to quantify the *strength* (or **water potential**) by which water is held in those soils. This gives a direct measurement of the actual availability of water to plants, and is a value that can be compared across a wide range of soil textures. This important measure of soil–water–plant interaction will be covered in detail later in the course.

Question 8.9

Pot A and pot B contain different soils. A bean plant is grown in each and the soils are gradually allowed to dry out until the bean plants are completely wilted. (a) How would you determine the amount of water that remained in each of the soils? (b) You determine that soil A has 5% soil water remaining, but soil B still has 17%, over three times as much water. Why do you think the bean plant in soil B has wilted? Using Figure 8.19, can you estimate the textures of the two soils?

8.9.3   Soil water movement

Infiltration is the process whereby water from rainfall, irrigation or snowmelt enters the soil pores and becomes soil water. The rate of infiltration often varies during rainfall or irrigation. If the soil is dry initially, the large pores will be empty, thereby allowing rapid infiltration. As the large pores fill with water over time, the infiltration rate decreases and surface puddling may become apparent. If the rainfall intensity is very high, the maximum potential infiltration rate ('infiltration capacity') of the soil may be exceeded and excess water will run off the surface, taking soil particles with it. This is one cause of soil erosion.

Once it has infiltrated, water moves through the soil by **percolation**. If the soil is wet, or the rainfall intensity is high, the water percolating through the soil will move rapidly through the larger pores. The rate of percolation is highly dependent on soil structure and texture, with the rate of water movement greatest in sandy soils, or very well structured soils. If the soil is dry or rainfall intensity is low, water will move more slowly than in wetter soils, and it will move through the finest pores. These differences are important because water moving more slowly through finer pores has more time to undergo chemical reactions with organic matter, clay particles, soil minerals and the soil biota.

The movement of water into and through a soil has implications for soil formation, plant growth, surface run-off and erosion, and also for the transport of nutrients and pollutants. There are three mechanisms of soil water movement: (i) saturated flow, (ii) unsaturated flow, and (iii) as water vapour.

Saturated flow occurs when all the pores are filled with water, such as after heavy rainfall or irrigation. The speed of water movement *(v)* down a slope through saturated soil between two points in a catchment that are a distance *l* metres apart is partly determined by the difference in height, *h*, between them (Figure 8.20). This difference in height is called the **hydraulic head** of water. This rate of flow can be described by **Darcy's law**:

$$v = K\,h/l \tag{8.14}$$

K is the **hydraulic conductivity** of the soil, and has the same units as speed (e.g. $cm\,d^{-1}$ or $m\,s^{-1}$). Hydraulic conductivity is the ability of a soil, specific for each soil, to conduct or transmit water. Hydraulic conductivity at saturation is mostly determined by the porosity of the soil and the sizes of the pores, including the extent to which cracks and fissures run vertically through the soil. In saturated soil the flow rate is highest in sandy soils, well-aggregated soils, and soils with many cracks and fissures. It is also affected by the amount and type of organic matter in the soil.

In most cases, however, the large pores are filled with air, and water must move by means of **unsaturated flow**. In these unsaturated soils, only the smallest pores may be transmitting water and therefore the flow rate of water is greatly reduced because the water must follow a tortuous path through the maze of capillary-sized micropores. Consequently, the flow rate of water is much slower in soils that are unsaturated.

Water may also move into or out of a soil as water vapour. This is usually of practical importance only in dry soils, such as those of desert regions, where the atmosphere at night may have a much higher humidity than during the day, and water vapour may move into soils at night and out of them during the day.

In many soils, water will move by all three mechanisms simultaneously.

Question 8.10

(a) Generally, water flows more quickly through a sandy soil than a clayey soil. Which of the variables in Equation 8.14 encompasses the effect of texture on flow rate? (b) Can you think of a situation where water will flow more rapidly at saturation through a clayey soil?

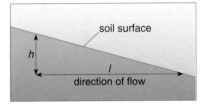

Figure 8.20 The flow of water through a saturated soil: *h* is the hydraulic head of water; *l* is the distance between two points on a slope.

8.9.4 Soil solution chemistry

Soil water is never pure water. As rainfall, water picks up numerous ions from the atmosphere and from dust. As this solution moves through the soil it exchanges ions and other soluble substances with clays (at anion and cation exchange sites), rocks (by weathering), roots and biota. Thus the composition of soil solution is extremely dynamic.

The major components found in the soil solution include:

Ions derived from atmospheric deposition. Rain and snow typically contain: basic cations from dust; anions (e.g. NO_3^- and SO_4^{2-}) from car exhaust and power stations; ammonium (NH_4^+) from agricultural emission of ammonia (NH_3); Na^+, Mg^{2+} and Cl^- from sea spray and volcanoes; as well as CO_2 from the atmosphere and H^+ from acids. Dust particles that have been filtered out by leaves (conifers are particularly effective at this), atmospheric chemicals adsorbed on leaves or bark, or aerosols absorbed into leaves can all be washed to the soil surface in rain or snowfall.

This rainfall, with its dilute chemicals, percolates through the soil. The cations are often vital nutrients for vegetation, and also play an important role in maintaining a high soil pH. The anions, particularly nitrate, are also plant nutrients. Sulfate is a plant nutrient but is usually not required in large amounts and the excess either washes through the soil solution or is adsorbed by hydrous oxides or organic matter. Chloride is barely used by most plants, and washes through the soil.

○ What ultimately happens to this Cl^-?
● Ultimately it returns to the sea.

Carbon dioxide from root respiration. As discussed previously, levels of dissolved CO_2 in soil solution can easily exceed atmospheric CO_2 concentrations by a factor of 10 or more, due to root and microbial respiration, especially in warm, wet climates where respiration is high. By creating high concentrations of CO_2 in the soil, living organisms strongly influence the geochemical process of rock weathering on land.

Dissolved nutrient ions from the decomposition of organic matter. Organic material is rich in carbon and nitrogen, and also contains phosphorus and other elements. These nutrients are released into the soil solution during decomposition, making them available for use by plants and microbes. Dissolved carbon, nitrogen and phosphorus compounds are often rapidly recycled back into living matter.

Dissolved inorganic ions from weathering. The basic cations Ca^{2+}, Mg^{2+}, K^+ and Na^+ are made available to the soil solution via cation exchange with clays and organic matter, but their primary source is usually rock weathering. Phosphorus is also supplied to the soil from the weathering of rocks, in particular those containing the mineral apatite. The concentration of dissolved silica in the soil solution is also regulated by the rates of chemical weathering of the silicate matrices in clays, and the minerals in silicate rocks.

The presence and amount in the soil solution of different chemical compounds such as the above can shed clues about the geology, vegetation and ecosystem processes in a catchment. For instance, a large amount of calcium with little soluble silica in the soil solution would indicate the presence of carbonate rock. High concentrations of ammonium or nitrate in soil solution, in areas not influenced by agriculture, may suggest that plant activity is low since these nitrogen-containing ions are not commonly in excess where plants are growing. Indeed, soil solution (and, therefore runoff stream water) NH_4^+ and NO_3^- concentrations typically peak in mid-winter, when plant activity levels are low.

8.10   Air

It may not be immediately obvious that soil contains air. In fact most soils contain about 25% air by volume. Whereas most soil water is held in the smallest pores, soil air is usually held in larger pores. Among other factors, such as climate and season, the degree of aeration of the soil is strongly influenced by its porosity, particle density and bulk density, as described below.

8.10.1   Soil density

Particle density is simply the mass per volume of the different particles in the soil. Most soil minerals have a density of 2.6 to 2.7 grams per cubic centimetre ($g\,cm^{-3}$), while a few minerals, such as hematite and goethite, have densities as high as $4\,g\,cm^{-3}$. Organic matter is much less dense, with an average density of $1\,g\,cm^{-3}$. The particle density of a soil is the weighted average of these individual densities, and is around $2.6\,g\,cm^{-3}$ for most soils.

Bulk density, on the other hand, is defined as the mass of a specific volume of dry soil. This volume includes pore space as well as solids. For this reason, bulk density will always be lower than particle density. For example, a soil with a particle density of $2.6\,g\,cm^{-3}$ and with 50% pore space will have a bulk density exactly one-half the particle density ($1.3\,g\,cm^{-3}$). Generally, sandy soils have higher bulk densities than do clayey soils. In other words, sandy soils have less pore space than clayey soils. This may at first seem difficult to believe, because sandy soils usually appear more porous than clayey soils. However, sandy soils have only a relatively few very large pores and almost no small pores, whereas clayey soils have many extremely small pores: the total pore space is greater in clayey soils. Organic matter decreases bulk density, mainly by increasing aggregation and promoting structure, i.e. 'opening up' soils. Channels such as earthworm burrows lower the bulk density, and can take soil air as deep as several metres.

Bulk density generally increases with soil depth because there is less organic matter with increasing depth and also because of compaction from the weight of the overlying soil. As a general rule, bulk density is highest in sandy subsoils with little organic matter ($1.6–1.8\,g\,cm^{-3}$), and lowest in organic matter-rich, well-aggregated, clayey surface soils ($1.0–1.2\,g\,cm^{-3}$).

Soil porosity and bulk density are influenced by many factors, such as soil texture, organic matter content and structure. Porosity and bulk density, in turn, affect water and air infiltration and ease of root penetration and seedling emergence.

Question 8.11

A surface soil has an average particle density of $2.4\,\mathrm{g\,cm^{-3}}$ with 45% pore space. What is the bulk density of this layer?

8.10.2   Soil aeration

Most plants are harmed if their roots are deprived of oxygen for longer than a few days, so maintaining adequately aerated soil is critical to many agricultural and forestry operations. The amount of oxygen in soil air and soil water that is available for aerobic respiration is also a fundamental property of soils. The amount of O_2 is controlled mainly by macropore space, soil water content and oxygen consumption by soil biota.

Because oxygen diffuses through water 10 000 times more slowly than it does through air, soils that are persistently waterlogged are poorly aerated. Oxygen levels decline rapidly in these situations, and most of the organisms living in these environments do not require oxygen for respiration: they are **anaerobic**. There are no multicellular anaerobic organisms. Plants that live naturally in waterlogged habitats (e.g. rice) have the ability to bring oxygen down to their roots. You will revisit this area in more detail later in the course.

Soil aeration has implications for a range of soil processes and properties, including organic matter decomposition, plant growth and soil colour; it also controls the form and concentration of many nutrients and pollutants. In poorly aerated soils the rate of microbial activity is greatly reduced, and consequently organic matter breaks down more slowly than it accumulates. It is for this reason that organic soils dominate in poorly drained marshy areas, where the anaerobic conditions also adversely affect the growth of most plants. However, some plants, such as rice, cranberries and reeds, have evolved ways to maintain normal growth even though their roots are submerged in water.

Many nutrients and pollutants change their form in response to a change in soil aeration (Table 8.6). In well-aerated soils, carbon is present mainly in the form of carbon dioxide and carbohydrates. Under waterlogged conditions, however, a significant portion of carbon may be present as methane, an important greenhouse gas. The oxidized forms of nitrogen and sulfur, NO_3^- and SO_4^{2-}, are readily absorbed by plants. However, under waterlogged conditions, these elements may be released to the atmosphere as the gases N_2, N_2O and H_2S. Highly insoluble oxides of Fe^{3+} and Mn^{4+} predominate under aerobic soil conditions. As a soil becomes more anaerobic, the solubility of these oxide minerals increases as Fe^{3+} and Mn^{4+} are reduced to Fe^{2+} and Mn^{2+}. The different forms can affect both the solubility, mobility and, in some cases, toxicity, of the elements or ions.

Table 8.6   Reduced and oxidized forms of important elements in soil.

Element	Aerated soil	Waterlogged soil
carbon	CO_2, carbohydrates	CH_4 (methane gas)
nitrogen	NO_3^-	N_2, N_2O, NH_4^+
sulphur	SO_4^{2-}	H_2S
iron	Fe^{3+} oxides	Fe^{2+}
manganese	Mn^{4+} oxides	Mn^{2+}

Question 8.12

Toxic ions such as selenium and arsenic can change their form, and thus their mobility and toxicity, if an aerated soil becomes waterlogged or vice-versa. As(III) is the dominant form of arsenic under anaerobic conditions and is much more toxic to humans than is As(V). Manganese oxides in soils and sediments can oxidize As(III) to As(V) and thus reduce the toxicity of this element.

If a wetland soil that was once flooded becomes well drained, which forms of iron, manganese or arsenic would predominate? What would happen to the toxicity of arsenic? Can you think of any other consequences of such drainage?

8.10.3 Soil air chemistry

Just as the chemical composition of the soil solution differs from that of rainfall, soil air is very different from atmospheric air. Plant respiration and decomposition releases carbon dioxide into the soil air and removes oxygen, thus soil air usually contains much higher levels of carbon dioxide and lower concentrations of oxygen than does the atmosphere. Because of this concentration gradient, oxygen tends to diffuse into soils and carbon dioxide tends to diffuse out. In addition, unless the soil is very dry, the soil air is saturated with water vapour. Other compounds in soil air can include methane and nitrous oxide, both from the activities of bacteria living under conditions of low oxygen.

8.11 Summary of Section 8

1   Soils are naturally occurring, unconsolidated materials consisting of mineral and organic components that are potentially capable of supporting plant growth. They are composed of inorganic matter, organic matter, water and air.

2   The mineral fraction of soils is made up of sands, silts and clays. Sand and silt contribute mainly to a soil's physical properties, whereas clays contribute both to the physical and chemical properties.

3   The organic material in soil is composed of matter that is or was living. Common living organisms in soil include bacteria, fungi, algae, protoctists (known also as protists or protozoa), earthworms, snails, millipedes, centipedes, mites, and springtails. The non-living organic matter is ultimately degraded to humus — dark-coloured, highly degraded organic compounds that resist further decay. Soil organic matter influences such fundamental soil properties as nutrient supply, ion exchange, structure, water-holding capacity and soil colour.

4   Isomorphous substitution results in most 2 : 1 aluminosilicate clays having a permanent negative charge. Aluminosilicate clays, hydrous oxide clays and organic matter can also possess pH-dependent charge due to the binding and release of H^+ in solution with edge hydroxyl groups. Unlike permanent charge, pH-dependent charge created is reversible, so if chemical conditions in the soil change, the charge may change as well. Hydrous oxide clays often possess positive charge, especially in moderately to strongly acid soils.

5   Cation exchange capacity (CEC) is the ability of a soil to adsorb and exchange cations with the surrounding soil solution. CEC is expressed in

units of moles of charge per mass of soil. The CEC of a soil depends primarily upon the amount of organic matter and the amount and type of clays in the soil. Soils with a high proportion of montmorillonite clays have a high CEC, whereas those with a high proportion of illite clays can trap and hold K^+ and NH_4^+ and release them slowly to the soil solution.

6   The capacity of a soil to retain anions is known as its anion exchange capacity (AEC). Soils with a low pH and a high proportion of hydrous oxide clays, such as highly weathered tropical or semi-tropical soils, can have substantial anion exchange capacity.

7   Base saturation is the ratio of the basic cations Ca^{2+}, Mg^{2+}, K^+ and Na^+ to the total CEC. In general, soils with a high base saturation have a higher pH and are more fertile than soils with a low base saturation. Base saturation can be increased by liming soils.

8   Soil structure is the arrangement of sand, silt, clay, and organic matter into larger units or aggregates. Aggregates are formed by negatively charged particles in clays or humus becoming connected by cation 'bridges'. Hydrated oxide clays, bacteria, fungi, and plant root hairs also enhance soil structure and may give rise to very stable aggregates. Good soil structure facilitates water infiltration, root penetration and seedling emergence, and also minimizes erosion. Poor cultivation practices can damage soil structure.

9   Water is held in pores that range in size from large cracks to the tiny interlayer spaces in clay minerals. A soil is saturated when all its pores are filled with water. At field capacity, the soil is holding as much water as it can against the force of gravity, and at wilting point the soil water is held too tightly in small pores for plants to extract. The amount of water held between field capacity and wilting point is the available water, the amount that is potentially available to plants.

10   The speed of water movement down a slope through saturated soil between two points in a catchment is determined by the hydraulic head, the hydraulic conductivity of the soil, and the distance between the two points, through Darcy's law.

11   The major components found in the soil solution include ions derived from atmospheric deposition, carbon dioxide from root respiration, dissolved organic nutrient ions and dissolved inorganic ions from weathering.

12   Soil air is held in larger pores. Chemical constituents of soil air include oxygen, carbon dioxide, methane and nitrous oxide. The level of aeration of a soil influences organic matter decomposition, soil colour, plant growth and the form and concentration of many nutrients and pollutants.

13   The distribution of micropores and macropores within a soil influences such fundamental soil properties as ease of drainage, aeration and root penetration. Generally, macropores dominate in sandy soils, while micropores are most abundant in fine-textured soils.

Soils and ecosystems

9.1 Soil development

On a geological time-scale, soils come and go in an eye blink. They are transient and dynamic. Many soils in the northern hemisphere are less than 10 000–15 000 years old, having been stripped away by glaciers before that. New soils can be (and are today) formed on the **till** or debris left from glaciers, on the ground left from volcanic eruptions, on wind-blown sand in dunes, on river deposits, on the craggy peaks of mountains, in cracks in rocks, as well as on spoil heaps from mines.

The factors controlling soil formation were identified principally by two soil scientists: V.V. Dukochaev in Russia in the late 19th century, and more recently in the mid 20th century by Hans Jenny in the United States. Extensive field observations led these two researchers to conclude that the formation of all soils could be attributed to five factors. These factors are parent material, climate, biota, topography and time.

9.1.1 Soil-forming factors

Parent materials are the geological or, rarely, organic precursors from which the soil forms. Examples of parent material are loose bedrock, glacial till, lake deposits, **alluvium** (river deposits) and **eolian** (wind) deposits. The parent material may profoundly affect the properties of the soil. For example, a soil formed from a clayey lake deposit would possess mainly small pores that prevent the rapid movement of water through the profile. Such a soil may be vulnerable to flooding. In contrast, soils developed from sandy eolian deposits would be very porous and allow easy movement of water. However, these latter soils may also be susceptible to wind erosion.

Climate refers to the temperature and precipitation regimes under which soil development occurs. Higher temperatures increase both the rate of chemical reactions and the amount of plant growth. Greater precipitation increases the rate of soil formation by increasing the amount of water that percolates through the unweathered parent material. Therefore soil forms much more quickly in the warm, humid tropics than in deserts or environments that are commonly cool or frozen.

Biota includes all living organisms, both plant and animal, associated with the developing soil. The biota contribute to soil formation mainly through the addition of organic material and the formation of soil structure. Also, vegetation stabilizes the soil surface and reduces the amount of natural soil erosion, thereby increasing the rate of soil development.

The organisms that first colonize new surfaces, the so-called 'pioneer organisms', are often able to get by on low levels of nutrients. Dissolved carbon dioxide from their respiration, and organic acids secreted as waste products, hasten the chemical weathering of rocks. In addition to this chemical weathering, roots can pry into cracks and break them apart: a form of physical weathering. In turn, water percolating through these fresh rock surfaces contains dissolved acids that also weather the rock.

Topography refers to the form of the Earth's surface and includes elevation, slope, and position in the landscape. Topography influences soil formation mainly through water redistribution. Low-lying areas receive water run-off from higher areas and are therefore able to support a more abundant and diverse vegetation. In addition, if the slopes are sufficiently steep, erosion of the surface soil layers from the upper slope positions may be considerable.

Time is the fifth soil-forming factor. Although organic material may accumulate to form a darkened surface layer within a matter of decades, most soil characteristics require hundreds or thousands of years to develop. For example, a soil developed from glacial deposits on the North American plains may be 15 000 years old, whereas a highly weathered soil in Brazil may be several million years old.

○ Under what conditions would you expect to find the most highly weathered soils?

● Soil weathering will be strongest in very old soils of the warm, humid tropics where leaching has been intense.

It is important to keep in mind that all five soil-forming factors are interdependent and occur simultaneously. For example, across a landscape, rainfall will accumulate in depressions and this increases leaching and gives rise to a more dense vegetation. The nature of the parent material itself may vary over the landscape; certain parent materials give rise to soils that are naturally more fertile than others.

9.1.2 Soil-forming processes

The four soil-forming processes are:

(1) *addition* of materials such as organic matter from leaves and roots, or atmospheric dust;

(2) *loss* of components, such as leaching of soluble salts or erosion of particulate matter from the soil surface;

(3) *transformation*, such as organic matter breakdown or mineral weathering; and

(4) *translocation* of material from one depth of the soil to another.

These four processes help us to understand how soils vary across landscapes and ecosystems. When describing and analysing a soil profile it is useful to ask yourself the following questions:

• What materials have been added to or lost from the soil?

• How have the organic and mineral components been transformed?

• To what extent have these materials been translocated through the profile?

Adding vegetation and other matter builds up the soil. More vegetation of course means more organic matter, but it also means a more diverse biological community, including organisms which feed on the living vegetation (consumers), and some which feed on the dead organic material (detritivores and decomposers). These organisms are rare until adequate vegetation is present; a lack of water or other necessary requirements, or the general unpalatability of the vegetation that is there will also affect their numbers. Thus, decomposition rates are initially fairly low, but as more and more organic matter is produced, more and more different organisms can make use of it, increasing rates of production and decomposition.

Some of the material that is added as eolian, alluvial or fluvial deposits is eventually eroded away by the forces of wind, water or gravity. In theory at least, the rate of gain of matter to the soil (biological production and material transport inward) will eventually match the rate of loss of matter to the soil (decomposition and material transport outward) and the soil neither accumulates nor is reduced in size. This final state, determined primarily by the climate, can be thought of as a quasi-steady state.

Steady state is an elusive (some may say illusory) quality. In reality, very few if any soils are in a steady state with respect to organic matter accumulation. Fluctuations in climate, changes in the vegetation or consumer community, natural events such as storms (which can blow down or uproot vegetation) and other environmental factors can all contribute to year-to-year net gains or losses in organic matter, even in a soil that has been established for hundreds or thousands of years. However, in the long run these fluctuations may tend to balance out. *Long-term changes* in factors such as climate or vegetation may, however, contribute to long-term changes in soil organic matter storage. There is some evidence that nitrogen deposition from acid rain, together with a warmer climate, is indeed changing the chemical quality of humus in some forest soils in North America and Europe.

9.2 The soil profile

Over time, and with the accumulation of organic material at the surface, leaching of water and weathering of minerals, a soil often develops a distinct vertical **profile** which can be seen by digging a shallow pit to expose a metre or more of the soil in a single face. Often when we dig such a soil pit, or see one in a road cut or quarry, we see horizontal layers that are different in colour or texture. These layers are called **horizons**, and they reflect the sum total of the processes of organic matter deposition and decomposition, water movement, rock weathering, and biological activity that have formed the soil thus far (Figure 9.1).

Soils formed in different climates or with different parent materials show different characteristic profiles. Often, particularly in soils developed under forest vegetation, there is a clearly defined layer at the top of the profile, which is composed of partly or fully decomposed organic matter. This layer is dark, sometimes nearly black, due to the presence of humus. It is known as the organic, or **O horizon**.

Below the O horizon are the mineral horizons. Unlike the O horizon, which is essentially entirely organic material, the mineral horizons are composed of both organic material and inorganic fragments of sand, silt and clay. Accumulation of organic matter in the upper part of the profile produces a dark-coloured **A horizon**. This A horizon receives water mainly from the O horizon, which is usually acidic due to organic acids and dissolved carbon dioxide from the decomposition and respiration of vegetation.

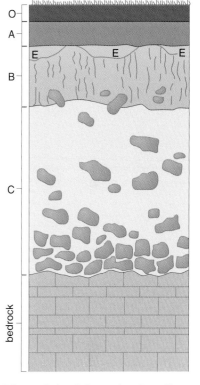

Figure 9.1 Schematic of a soil profile, showing O, A, E, B and C horizons.

○ What is the consequence of this?

● The acidic water causes more primary minerals in the A horizon to weather, as well as potentially further weathering the clay minerals present.

With the continued percolation of acidic water through the profile, more calcium, magnesium, sodium and potassium are leached out of primary minerals, forming clays. As leaching continues, the clays themselves (both aluminosilicate and hydrated oxide) and other fine particles may be transported from the upper part of the soil profile through soil pores to lower layers. This removal of clays and oxides from the upper soil layers often results in a fairly light coloured layer just below the A horizon. This is called the **E horizon**, or **eluvial** horizon (*eluvial*, washed from). The formation of an E horizon can take as little as decades or as long as hundreds to thousands of years.

As percolating water continues to pick up calcium and magnesium, some of the acidity is neutralized. The change in chemical conditions causes an increase in the soil pH. In addition, water is now passing into material that is less weathered and has much less biological activity (e.g. root channels, worm holes, mixing due to movements of organisms through the soil), and is consequently more dense. The flow rate of the water therefore usually slows down, often dramatically.

The slowing in the flow rate, increase in the soil water pH and other physical and chemical changes to the percolating water allow the clays to be eventually deposited. This results in a lower, clay-rich part known as a **B horizon** or **illuvial** horizon ('washed into'). Some of the illuvial horizon may be stained red from precipitated iron oxides.

Below the O, A, E and B horizons is unconsolidated rock, or parent material. This is known as the **C horizon**. Over long periods of time (tens to thousands of years, depending on the intensity of the weathering), as weathering continues, the A and B horizons may grow and become deeper, as more and more of the underlying rock is modified by the weathering process. Clearly, if weathering is more intense (e.g. in warm or wet climates) soils will be deeper and more developed.

○ Can you think of another factor that might increase the depth of weathered soil horizons?

● Enhanced acidification of soil water by acid rain is a candidate.

In fact, just such accelerated weathering of soil profiles has been observed in highly acid rain-impacted areas in Europe. But is acid rain the real culprit? We shall return to this question in Section 10.1.

In addition to the master horizon designations O, A, E, B and C, each horizon can be subdivided through further descriptions depending on its particular characteristics. For example, a B_s horizon indicates accumulation of organic matter and iron and aluminium oxides. A well-developed B_s horizon in a soil usually indicates very strong weathering, either natural or human-accelerated (e.g. by acid rain). There are also some soils in which one or more of the master horizons do not occur. For instance, the highly organic soils that develop on moors or in bogs can have O horizons that are metres in thickness, sitting directly on the parent material.

Examination of a soil profile can give us insight into the environmental conditions prevailing during the time the soil has been forming. Soils that formed long ago (e.g. during the last glacial period) and which are now buried, are particularly useful for reconstructing the climates of the past. These soils are known as **palaeosols** ('old soils'). There is evidence for fossil soils as old as 500 million years, although most of the palaeosols that have been studied are much younger. Examination of palaeosols can reveal much about past climates. In extreme cases, a palaeosol may possess the properties of a highly weathered tropical soil, even though the current climate of the region may be cold and dry.

A striking example of how palaeosols can be used to reconstruct past climates is an impressive sequence of atmospheric dust deposition followed by soil formation, spanning millions of years, found in Xian, China. Atmospheric dust is believed to be deposited during drier periods, when wind transport and deposition occur easily. Soil formation in these fresh deposits can begin in earnest only when the climate becomes warmer and wetter. A sequence of more than 30 palaeosols, each capped by an atmospheric dust layer, reveals an elaborate series of wet–dry climatic cycles that continued for at least 2.5 million years. The intensity and duration of each wet period can be inferred from profile depth and the degree of mineral weathering in the corresponding palaeosol.

Question 9.1

Give a qualitative description of the following soil profiles, including what you might expect for the relative depth of the organic (O) horizon, the soil texture and the degree of weathering and leaching. *Soil A*: 15 000 years old, derived from lake sediments, forming under grassland, temperate climate. *Soil B*: 1 million years old, derived from granite, forming under dense tropical forest vegetation, high rainfall and temperature.

9.3 Soil classification

9.3.1 International systems

There are two international hierarchical classification systems currently in place to classify the world's soils, and numerous national systems. Both international systems have evolved over the years as our understanding of soil formation and soil properties has improved. There are likely to be additional modifications to each system as our understanding is advanced further. One of the classification systems has been developed by the United Nations Educational, Scientific, and Cultural Organization (UNESCO) jointly with the Food and Agriculture Organization (FAO). This FAO–UNESCO classification scheme comprises 30 main groups, and includes an ever-increasing number of subdivisions at lower levels in the classification hierarchy. A second classification system has been developed by a team of international scientists at the US Department of Agriculture (USDA), and is currently in use to varying extents in more than 50 countries. This latter system is called Soil Taxonomy and is composed of six categories arranged hierarchically: order, suborder, great group, subgroup, family, and series.

Because the USDA system has fewer main groups than the UNESCO system we will present its classification system to broadly illustrate the diversity of the world's soils. In this system, all the world's soils can be assigned to one of 12 orders. The classification of a soil into an order is made on a variety of chemical and physical properties, including the presence or absence of certain 'diagnostic horizons'. Among the most important of these properties are mineralogy, texture, organic matter content, structure, colour, profile depth, moisture and temperature regimes, as well as a range of chemical criteria. These properties also help to define the diagnostic horizons.

For example, soils developed under grassland vegetation commonly have dark horizons that are rich in nutrients and organic matter. All soils with these distinct horizons are placed within the order mollisols. On the other hand, acidic soils formed under forest vegetation may have in their B horizon large amounts of humus and iron and aluminium oxides. Soils with this diagnostic humus- and oxide-rich B horizon would be classed as spodosols. Each of the other ten soil orders has its own diagnostic horizon and unique features. Table 9.1 gives a brief description of each of the 12 orders of Soil Taxonomy, some of which are illustrated in Figure 9.2.

Table 9.1   The 12 orders of Soil Taxonomy.

Order and name origin	General characteristics
alfisols — *alf*, nonsense symbol	moderately leached soils; formed under forests
andisols — *ando*, Jap., black soil	formed from volcanic ejecta parent materials
aridisols— *aridus*, L., dry	dry soils that occur in arid regions
entisols — *ent*, nonsense symbol (recent)	young soils; very little profile development
gelisols — *gelid*, Gk., cold	permanently frozen within 1 m of the surface
histosols — *histos*, Gk., tissue	organic soils
inceptisols — *inceptum*, L., beginning	show only the beginnings of profile development
mollisols — *mollis*, L., soft	grassland soils with dark, organic-rich surface horizons
oxisols — *oxide*, Fr., oxide	highly weathered soils; mainly found in the tropics
spodosols — *spodos*, Gk., wood ash	leached, acidic forest soils; occur in temperate regions
ultisols — *ultimus*, L., last	strongly leached and acidic; formed under wet tropical and subtropical forests
vertisols — *verto*, L., turn	self-churning soils rich in swelling clays

Jap, Japanese; Gk, Greek; L, Latin; Fr, French.

Figure 9.2 Some soils of the world: (a) histosol, developed under a peat bog; (b) gelisol, showing freeze–thaw perturbations , or 'cryoturbation'; (c) andisol, showing a new soil profile being formed on top of an old profile from a previous volcanic eruption; (d) spodosol (podzol in UK system); (e) oxisol; (f) inceptisol (rendzina in UK system).

9.3.2 Soils of the world

Approximately 13% of the Earth's ice-free land surface is covered by rock or shifting sand; these areas are considered not to have soil. Within the whole ice-free land surface, entisols are the most abundant soils, covering 16% of the area. They are found globally and occur in many climatic regions, but are most abundant in the Sahara desert, central Australia, and the upland regions of Iran and Pakistan (Figure 9.3). Aridisols are the next most abundant soils, occupying 12% of the Earth's ice-free land surface. The largest areas of aridisols occur in southern Australia, the Gobi Desert in China, and the deserts of western United States and southern Argentina.

Global Soil Regions

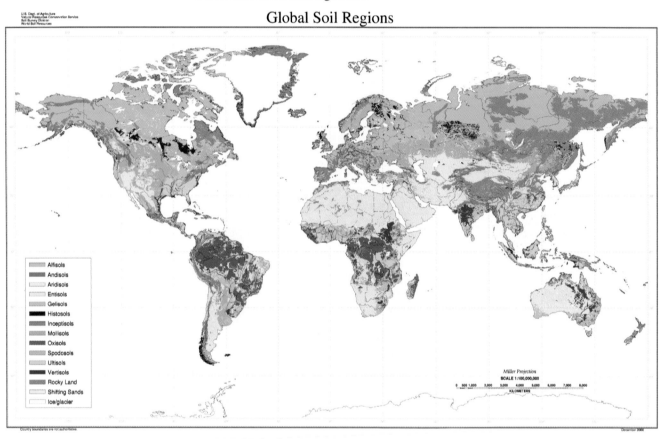

Figure 9.3 World map showing distribution of soil types.

Alfisols and inceptisols each occupy 10% of the global land area. Alfisols are found in cool, moist climates, with the largest area occurring in the Baltic States and western Russia. Inceptisols are found in most climatic regions and on all continents, where they are often the dominant soils of mountainous areas. Gelisols comprise 9% of the Earth's soils and occur mainly in the far north, although smaller areas may be found at higher elevations in less northerly latitudes. The productivity of these soils is limited not only by their permafrost, but also by the short growing season in the far north.

Oxisols (8%) generally occur in the tropics, most notably in Brazil and central Africa, where weathering conditions are most intense. Ultisols (8%) also occur in the warm, humid areas of the lower latitudes, often in association with oxisols. Mollisols are only slightly less abundant, comprising 7% of the land area. Due to their high fertility and suitability for agriculture, few mollisols remain uncultivated. They cover vast areas of the Ukraine, Russia, Kazakhstan, as well as the interior plains of North America.

Only 3% of all soils are classed as spodosols, yet these soils support many of the important coniferous forests in northern Europe and north-eastern North America. By virtue of their low fertility, however, few spodosols are used for intensive agriculture. Although vertisols are also limited in their extent, comprising only 2% of the world's soils, important areas occur in Australia, India, and the Sudan. Those in India are especially significant, as they are used to grow crops that feed one of the world's most populous regions.

Histosols (1%) occur mainly in the wet, cool regions of northern Canada, Siberia and Finland. Despite their low abundance, these soils form an integral part of many environmentally important wetland ecosystems. Although andisols are the least abundant of all soils, occupying less than 1% of the Earth's ice-free surface, significant areas occur in Japan, Chile, and Mexico. Their low abundance does not, however, diminish their importance, as they are used successfully for intensive agriculture in many populous regions of the Pacific rim.

9.3.3 National systems: UK

A number of countries have developed soil classification schemes that pre-date the development of an international system, and continue in many cases to be used in those countries because they are well-suited to the particular types of soil found there. For instance, the Canadian soil classification system places a great emphasis on finer classifications of boreal, peaty and tundra soils than do the international systems.

The 1990 classification of British soils is based on six major soil groups, which are further divided into subgroups. The groups are lithomorphic soils, brown soils, podzols, gley soils, peat soils and man-made soils.

Lithomorphic soils are soils which are underlain by bedrock or broken rock at shallow depth, and where soil development is restricted. There are five groups of lithomorphic soils, including **rendzinas** (soils developed over chalk) and **rankers** (soils developed over siliceous rocks, often in upland areas). Lithomorphic soils also include coastal sands, rockslides, and recent alluvial deposits. Cultivation of lithomorphic soils is severely restricted by shallowness and stoniness, erosion or flooding risk, or poor water holding capacity.

Brown soils are fairly well-drained soils with distinct brownish-reddish brown subsurface horizons, usually denoting the formation of iron oxides. This group includes those soils traditionally classed as brown earths, as well as well-developed calcareous soils. They are a common soil type in the UK, occurring mainly in humid lowland areas.

Podzols are generally coarse-textured and highly leached. They possess distinct horizons: generally a bleached grey upper layer where the clay content has been reduced by water leaching and, below this, an illuvial B horizon where clays, organic matter and aluminium and iron compounds accumulate. Podzols occur widely in cooler wetter areas, especially in Scotland, where leaching rates are high. They are often highly acidic and their use in cultivation is restricted, although they are widely used for forestry.

Gley soils are characterized by a greyish or mottled horizon near the surface, indicating waterlogged conditions and the reduction of iron oxides. Gley soils often form in wet cool areas or where soil permeability is restricted. Together with brown soils they are the most widely distributed soils in the UK.

Peat soils possess a thick organic layer (at least 30 cm deep if directly over bedrock, and 40 cm deep if over mineral soil). Organic matter accumulates in peat soils due to the restriction of decomposition under cool, wet conditions. Peat soils cover about 10–15% of Scotland and Ireland.

Man-made soils are those soils that have been extensively re-worked by human modification such as extensive ploughing, large addition of manure, major disturbance or removal and replacement.

Question 9.2

The soil profile we described in Dunsford Wood in Block 1 showed a dark upper horizon, a striking greyish layer below that, and a fairly abrupt change to a yellow–brown layer below the grey. The soil is strongly leached, with clear profile development and low soil pH. Into which great group in the USDA system should it be placed? Into which group in the British classification system?

9.4 Summary of Section 9

1   The formation of all soils can be attributed to five factors: parent material, climate, biota, topography, and time. These five factors interact through four soil forming processes to produce the soil profile.

2   The four soil-forming processes are: (1) additions of materials such as organic matter from leaves and roots, or atmospheric dust; (2) losses of components, such as leaching of soluble salts or erosion of particulate matter from the soil surface; (3) transformations, such as organic matter breakdown or mineral weathering; and (4) translocation of material from one depth of the soil to another.

3   Soils often develop a distinct vertical profile due to the long-term accumulation of organic material at the surface, leaching of water and weathering of minerals. The horizons within a soil profile, which may be different colours or textures, reflect the sum total of the processes of organic matter deposition and decomposition, water movement, rock weathering, and biological activity that have formed the soil thus far.

4   In the US Department of Agriculture soil classification system, all of the world's soils are assigned to one of 12 orders. The classification of a soil into an order is based on a variety of chemical and physical properties, including the presence or absence of certain diagnostic horizons. Among the most important of these properties are mineralogy, texture, organic matter content, structure, colour, profile depth, moisture and temperature regimes, as well as a range of chemical criteria.

5   National soil classification schemes are often used in addition to international systems because they are well-suited to the particular types of soils found in a particular country.

Soil processes and properties in the environment

10

Having described how the environment forms the soil we will now discuss a few major soil processes and properties, which in turn form, influence and interact with the environment around it. This 'environment' could be many things, for instance a catchment (including plants, animals and the cycling of matter and energy in the catchment), a river system, even an ecosystem many hundreds or thousands of kilometres away from the soil itself.

Nearly all soil organisms are profoundly influenced by the chemical and physical properties of the soils they are found in. For some organisms this is more obvious than for others. Earthworms, for example, are generally intolerant of soil pH levels lower than about 4.5 and so acid soils are usually devoid of these creatures. Many plants have fairly strict requirements for or intolerances of levels of water or chemical nutrients, which is why willow trees are not found in deserts and *Sphagnum* moss is rare in calcareous soils. Other organisms may be tolerant of wider ranges in soil chemical and physical properties but their distributions are centred on a narrow range of soil conditions.

Humans are by no means exempt from the influences of physical and chemical soil characteristics. Crops such as maize and wheat require fertile, well-drained soils with a high base saturation and neutral to slightly alkaline pH. On the other hand, rice, the staple food of most of the world's population, requires poorly-drained, slightly acid soils but with high levels of nitrogen and phosphorus.

Where ideal soil conditions do not exist, or are in some way deficient, humans have often modified the conditions, perhaps *amending* soils by adding lime (to raise the pH), organic material (to add nitrogen and increase clumping or aggregation to improve drainage) or mineral fertilizers (to add nutrients such as nitrogen, potassium, or phosphorus to the soil solution), or by ploughing, tilling and draining to improve aeration, or by creating dams and flooding soils to reduce drainage and encourage waterlogged conditions (e.g. for rice).

Enhanced soil fertilization due to greatly increased levels of industrial nitrogen fixation in the last 50 years and the widespread use of irrigation are the principal soil amendments that have led to increased food production. But the amendments cost money, and there are some regions, such as desert or arctic areas, or poorer regions of the planet where such amendments are too costly.

Soil properties are affected by the organisms that live in and on the soil. By simply existing (growing, reproducing, metabolizing), microbes, plants and animals can alter the physics and chemistry of the soil, in some cases irreversibly. By altering the soil they are in turn affecting essential components of ecosystems.

In this section we will consider examples of major soil properties and processes, and how they influence the environment: soil acidification, nutrient release and retention, erosion and salinity.

10.1 Soil acidification

Soils can receive relatively large loads of acid over time. Some are the result of natural processes such as decomposition (which releases high concentrations of dissolved CO_2) or plant growth (organic acids are leached from roots and hydrogen ions are actively pumped into the soil solution, as we shall see below). In addition, human-caused factors such as acid deposition contribute to the load. In the 1980s, as widespread forest decline began to be documented in Europe, the German soil scientist Professor Bernard Ulrich used the concept of buffering systems to describe the chemical changes in forest soils as they are progressively acidified. Box 10.1 gives more information about the important concept of buffers; Box 10.2 describes the measurement of soil pH, which is critical to our concept of progressive soil acidification.

Box 10.1 Buffering reactions

A solution is said to be **buffered** if its pH remains constant even though small amounts of acid or base are added to the solution. The classic example of buffering occurs when a weak acid and one of its salts are present together in solution. Let us take for example acetic acid and sodium acetate in solution together. If the salt is completely ionized, the species in solution are sodium (Na^+) ions, acetate (AcO^-) ions, H^+ ions and undissociated acetic acid (HOAc). The relationship between the acetate ions and undissociated acid is shown by the equilibrium equation; the Na^+ ions do not interfere with the equilibrium:

$$HOAc\,(aq) \rightleftharpoons H^+(aq) + AcO^-\,(aq)$$

The equilibrium is governed by the equation below, where K is the equilibrium constant, which stays the same however the concentrations of HOAc, H^+, and AcO^- in the solution might change:

$$K = \frac{[H^+][AcO^-]}{[HOAc]}$$

When the sodium acetate is first added to an aqueous solution of acetic acid, this introduces a high concentration of acetate ions, and so more HOAc is formed by the acetate ions 'mopping up' many of the H^+ ions. A solution is left therefore with few H^+ ions and a great excess of acetate ions. Now if a little acid is added, there are plenty of acetate ions to neutralize

it, and because they are in great excess, this hardly alters the concentration of acetate at all. Similarly, the amount of new HOAc that is formed is small, and the concentration of HOAc remains close to the starting value. As a result, the H^+ concentration, and hence the pH, also remains very close to its original value. Of course, if a large amount of acid is added, much of the acetate is used up, and its concentration falls. As a result, more HOAc ionizes generating more H^+ ions, and the pH drops.

If a small amount of base is added, the system operates in reverse. As H^+ is neutralised, some of the reservoir of acetic acid ionizes to generate more H^+. Once again the changes in concentration of the acetate ion (increase) and the acetic acid (decrease) are both small in comparison with the concentration of each species originally present. The equilibrium is therefore **buffered** by the AcO^- ions, which act as a resource to remove added acid from solution, or restore it if the pH increases.

In the environment, several different buffering mechanisms are encountered. Each operates over a different pH range, but their role is always to provide a continuous and reversible means of removing added acid from solution, and preventing it from lowering the pH of water in contact with the soil. However, like the acetate ions, their buffering capacity can be exceeded if the concentration of acid becomes too great.

Box 10.2   Soil pH

The pH of the 'soil' may either be considered the approximate pH of the soil solution (which is often our best estimate of the hydrogen ion concentration to which roots and soil biota are exposed), or the potential pH of the soil if all the H^+ adsorbed by soil were in the soil solution. For the former, a water–soil slurry (typically 2.5 parts water to 1 part soil sample) is made up, and the pH of the solution is measured after a given amount of time. To estimate the potential pH of the soil it is mixed with a solution of a salt like $CaCl_2$. The adsorbed H^+ on the soil surface exchanges with the Ca^{2+} and is released into solution, where the pH is measured.

○   Would the potential pH of the soil (measured in $CaCl_2$) be usually lower or higher than the actual pH of the soil?

●   Potential pH is usually lower (i.e. higher $[H^+]$) than pH in water, because the potential pH includes the hydrogen ions that are currently adsorbed onto the soil surface.

When hydrogen ions from either natural processes or acid rain are added to a soil dominated by carbonate rock, the ions are neutralized by reaction with the calcium carbonate:

$$CaCO_3 + H^+ \rightarrow Ca^{2+} + HCO_3^- \tag{10.1}$$

These weathering reactions are rapid, and as long as there is sufficient carbonate in contact with the acid, the soil is effectively buffered against acidification and remains at a pH of 8–6. Such soils are in the **carbonate buffer range**. Soils developed on calcareous bedrocks often show little effects from acid deposition because of the high acid neutralization capacity of calcium carbonate.

In catchments dominated by aluminosilicate rocks, some H^+ is neutralized by weathering the silicate minerals in these rocks to form clays, e.g.:

$$2KAlSi_3O_8 + 2H^+ + 9H_2O(l) \rightarrow 2K^+ + 4Si(OH)_4 + Al_2Si_2O_5(OH)_4 \tag{10.2}$$

orthoclase   kaolinite

The protons consumed in silicate buffering are converted to the very weak soluble silica or silicic acid, which is ultimately transformed to SiO_2 and H_2O in rivers or the sea. The cations liberated through silicate weathering can be bound as exchangeable cations in clay minerals or humus in the soil.

Weathering of silicate minerals is not simple dissolution, but also involves reactions on internal mineral surfaces. The rate of acid neutralizing depends upon the structure of the silicate mineral as well as on the proportion of basic cations in the mineral and the concentration of acid in the soil solution. Overall, however, the amount of acid neutralized by reaction with silicate minerals is much less than that neutralized by reactions with carbonate minerals. Weathering of primary silicates to form clays takes place under all pH conditions, but is the major buffer acting in soils between pH 6 and 5. Soils dominated by silicate rocks and with soil pH values in this range are in the **silicate buffer range**.

As long as the rate of input of acid does not exceed the rate at which that acid can be neutralized by reactions with either carbonate or silicate minerals, the soil will remain between a pH of 8 and 6 (if carbonate bedrock), or 6 and 5 (if silicate bedrock). These soils are conducive to biological activity; organic material is rapidly broken down, and earthworms and other burrowing organisms mix the soil throughout the rooting zone, resulting in good aeration and structure.

If the rate of H^+ input exceeds the rate at which that acid can be neutralized by weathering, however, basic cations such as Ca^{2+} and Mg^{2+} will be removed from pH-dependent exchange sites, particularly of humus, and exchanged for the hydrogen ions in solution:

$$\text{clay–humus—}Ca^{2+} + 2H^+ = \text{clay–humus—}2H^+ + Ca^{2+} \qquad (10.3)$$

This **exchangeable cation buffer range** can occur over the whole pH range of the soil, but as pH begins to decline below 5, aluminium, manganese and some heavy metals in clay lattices or minerals begin to become soluble, hydrolysing and forming different charged species (e.g. $Al(OH)_2{}^+$, $Al(OH)^{2+}$, Al^{3+}, collectively considered as Al^{n+}) depending on the pH. These ions can also become adsorbed onto the cation exchange sites.

As long as the base saturation is above around 10–15%, acid loads to soils can be buffered by cation exchange reactions. However, the soil is effectively 'drawing on its reserve' of accumulated adsorbed base cations from silicate weathering. Unless the rate of acid input declines enough to return the soil to the silicate buffer range, the supply of exchangeable base cations will gradually be depleted, as weathering is not fast enough to replenish the cation exchange sites with basic cations.

Many soils in forests in central Europe are in this state. Professor Ulrich estimated that it would take about 50 years to completely leach the exchangeable calcium from soil exchange sites in a soils containing 10% clay to a depth of 50 cm, given acid loads typical of central Europe in the 1980s. These acid loads have since declined, but in many cases are still well above the level that can be buffered by silicate weathering. Therefore many soils are still depleting their reserves of base saturation.

As base saturation declines and more and more exchange sites become occupied by aluminium, the danger of aluminium toxicity increases. Dissolved aluminium impairs the functioning of many important soil biota, such as earthworms, and dissolved aluminium in soil solution is toxic to plant roots. Many bacteria are also sensitive to dissolved aluminium in the soil solution, reducing the efficiency of organic matter decomposition. The decline in earthworm activity reduces the turnover and aeration of the soil, and structure becomes more compact, reducing water percolation. In addition, dissolved aluminium leaching into stream water or lakes is toxic to fish. Many soils in central European forests show characteristics of early stages of aluminium toxicity.

If the base saturation declines below around 10–15%, and acid inputs continue, most of the exchangeable ions on cation exchange sites are acidic cations (H^+, Al^{n+}, Mn^{2+}, Fe^{3+}). These ions exchange with the soil solution instead of basic cations. Soil solution pH declines further until, at pH 4.2, further declines in pH are buffered by the dissolution of aluminium *itself* from clay minerals. This can be summarized as:

$$Al(OH)_3 + 3H^+ \longrightarrow Al^{3+} + 3H_2O$$

The soil is now in the **aluminium buffer range**, and aluminium toxicity is a major problem. The rate of dissolution, and therefore the rate of buffering of excess hydrogen ions, depends upon the type of clay: hydrated oxide clays dissolve faster than the aluminium bound in silicate lattices. In extreme cases, the rate of input of hydrogen ions exceeds the rate of buffering by dissolving aluminosilicate and hydrous oxide clays, and soil pH declines further until it is buffered by the dissolution of iron oxides, at around pH 3.2: the **iron buffer range**.

The pioneering work of Professor Ulrich and his students in explaining the process of soil acidification through acid rain is reflected today in current European legislation. The molar ratio of Ca^{2+} to Al^{n+} in the soil solution is the major criterion for setting pollution abatement targets for forests across Europe. A molar Ca : Al ratio above 1 in soil solution, and exchangeable Ca saturation above 5% is considered the minimum level for forests to place them in the silicate/cation exchange buffer zones and reduce the likelihood of aluminium toxicity. This criterion has resulted in the widespread liming of many areas of central Europe and southern Scandinavia (Figure 10.1).

Figure 10.1 Liming by helicopter.

Soils can, and do, acidify without acid rain, however. In fact there is some criticism that some areas of Scandinavia may currently be limed that have always been naturally acidic. Many of these areas are developed on highly organic soils. But even in low-organic soils acidification is a natural part of the development of a soil. This is due in part to leaching of basic cations from the CO_2-enriched acid soil solution percolating through the soil. However, it is also due to the physiological requirements of plants for basic cations.

Earlier we discussed how vegetation, particularly tree roots, removes cations from the soil solution. These cations come from cation exchange sites on organic matter and clays and, ultimately, from mineral weathering. Like all organisms, trees need nutrients such as calcium, magnesium, potassium and nitrogen to grow. They acquire these nutrients from the soil through their roots. Because most of these nutrients occur as positively charged ions (Ca^{2+}, Mg^{2+}, K^+, NH_4^+), we might expect that the action of nutrient uptake would lead to an excess of positive charge inside the root. This cannot occur — these tissues must remain electrically neutral.

Figure 10.2 Exchange between H^+ and basic cations mediated by plant roots.

Plants maintain an internal charge balance whilst taking up cations by releasing hydrogen ions from the root into the soil solution. These hydrogen ions in turn interact with the soil, releasing more basic cations into the soil solution, and lowering the base saturation. (Figure 10.2).

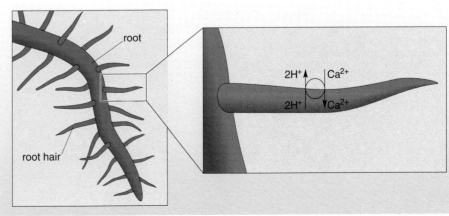

If the vegetation is growing rapidly, over time enough hydrogen can be released into the soil solution, and basic cations released from soil exchange sites, that the soil base saturation declines below the level at which soil pH is buffered by silicate weathering, and the soil goes into the cation exchange or (if base saturation declines to less than ca 10–15%), the aluminium buffer range. As with any other form of soil acidification, this can trigger a rapid decline in soil pH, the release of Al^{3+} to the soil solution, and damage to vegetation and downstream biota.

The remarkable thing about such a scenario is that soil acidification can, in theory at least, be triggered by growing vegetation alone. The ambitious programme of reforestation by conifers introduced in many parts of Europe in the latter half of the 20th century could provide just such a trigger. This rather remarkable hypothesis provoked a spirited response when it was first introduced in the early 1980s. Most scientists agree that the overall pattern of soil acidification seen across Europe is primarily due to acid rain. However, rapidly growing plantations of trees *do* acidify soils. The relative importance of this phenomenon and its relation to modern forestry practices is still not clear.

Why then don't trees acidify soils all the time? The answer, in part, is that they do. However, the basic cations that are removed and stored by trees are returned to the soil when those trees shed their leaves and die (this, incidentally, is one of the reasons that conifers acidify soils more intensively than broadleaf trees). It takes the continuous removal of vegetation to eventually deplete the soil of basic cations and trigger soil acidification. Human removal of trees, viewed in this way, is exporting basic cations from the ecosystem.

To test for yourself whether trees really do accumulate basic cations, try Activity 10.1.

Activity 10.1 (optional) Do trees accumulate basic cations?

For this you will need some soil pH paper or a pH tester kit from your local garden centre, and some ash from a barbecue or wood fire. Try to get pH paper or a kit with a fairly wide range in pH: 5–9 should do. Take a tablespoon full of ash and put it in a small container, then add 2 tablespoons of water from the tap, or follow the instructions in the kit. Mix it thoroughly and let it sit for 5 minutes, then measure the pH. Is the slurry acid or alkaline?

Allow about 20 minutes for this activity.

10.2 Soil fertility

10.2.1 Nutrient availability

Essential elements

Eighteen elements are essential for plant growth (Table 10.1). All these elements must be available in sufficient quantities if normal plant growth and reproduction are to occur. Nine of the essential elements are required by plants in relatively large amounts and are called **macronutrients**. The remaining nine elements are needed in much smaller quantities and are therefore known as **micronutrients**. Several other elements may also be taken up by plants, but they are believed to be not essential for growth.

Plants obtain carbon, hydrogen and oxygen from air and water; the other elements must be acquired from soil minerals and organic matter. Deficiencies of micronutrients are rare, as most soils are able to supply them in the small amounts required for normal plant growth. The much greater demand for macronutrients, however, often results in deficiencies of these elements. Much of soil fertility is therefore concerned with the macronutrients that are derived from the soil solids: nitrogen, phosphorus, and potassium.

○ What do you think would happen to a plant if the soil is unable to provide sufficient quantities of only *one* essential element?

● Growth will be restricted, despite the abundance of the other nutrients. In extreme cases the plant may fail to reproduce, or it may even die.

Nutrient pools

Plant roots cannot obtain nutrients by ingesting whole soil particles. Rather, the nutrient elements within these particles must first be dissolved in the soil solution, from where they can then be absorbed by plant roots and microbes. Nutrient elements in the soil solution are therefore said to be in the **plant-available** pool. The amount of nutrients in the soil solution is usually less than 1% of the total nutrient content of a soil. However, nutrients within the soil solution are subject to considerable changes in concentration over short periods of time, due to uptake by plants or losses by leaching.

○ How can the type of soil clay affect the nutrient retention?

● Large amounts of cations can be fixed in the interlayer spaces of clay minerals such as illite and slowly released to the soil solution. Some cations may be so strongly fixed that they are not exchangeable. Clays such as kaolinite have low cation exchange capacities and will not retain many nutrient ions.

During cation exchange, various cations in solution exchange with nutrient cations (e.g. potassium) held at mineral and organic surfaces, thereby replenishing the soil solution with nutrients. Nutrients held at surfaces in such a loose manner are said to be exchangeable and, because they can move readily between the surface and solution, comprise the **labile pool** of nutrients. Those

Table 10.1 Elements essential for plant growth.

Macronutrients	Micronutrients
carbon (C)	iron (Fe)
hydrogen (H)	manganese (Mn)
oxygen (O)	boron (B)
nitrogen (N)	zinc (Zn)
phosphorus (P)	copper (Cu)
potassium (K)	chlorine (Cl)
sulphur (S)	molybdenum (Mo)
calcium (Ca)	cobalt (Co)
magnesium (Mg)	nickel (Ni)

nutrients held within very soluble minerals and rapidly decomposable organic matter are also considered part of the labile pool. The exchangeable nutrients within the labile pool are considered to be plant-available because they can move rapidly into soil solution.

Approximately 90% of a soil's nutrients are unavailable to plants because they are locked up in the soil minerals (e.g. feldspar, mica), in non-exchangeable form in clays, or in resistant organic matter. These nutrients comprise the **stable nutrient pool** and can be released only over long periods of time, through mineral weathering or the breakdown of stable organic matter. The stable nutrient pool can be considered a long-term reservoir of essential elements.

○ Which type of soil, sandy or clay-textured, would have the greater amount of plant-available nutrients?

● A clay-textured soil will have a much larger pool of plant-available nutrients. This is due mainly to the greater cation exchange capacities of clay soils.

Question 10.1

The availability of any nutrient can be estimated by knowing its distribution among the three nutrient pools. Describe in qualitative terms the distribution of potassium in a sandy soil among the three nutrient pools described.

10.2.2   Nutrient deficiency and toxicity

If all nutrients, except one, are present in amounts sufficient to ensure optimal plant growth, a plant's development can be no greater than is allowed by that one limiting nutrient. A nutrient that limits plant production in this way is therefore called the **limiting nutrient**. However, a plant's growth may be limited not by nutrients, but rather by insufficient water, sunlight, or heat — a more general term is then a **limiting factor**. Increasing the amount of a factor that is not limiting will do little to increase plant growth. For example, if the growth of a rose bush is limited by an insufficient supply of the element boron, adding nitrogen, water, or any other non-limiting factor will do little to improve the plant's health or growth rate.

As the concentration of a limiting nutrient increases, plant growth will increase correspondingly. Once the nutrient has reached a concentration that is optimal for plant growth, further additions of this nutrient do not increase plant production. If available nutrient concentrations increase beyond the sufficiency range, the plant may take up excess nutrient and in some cases symptoms of toxicity can appear. Toxicity can result in reduced growth and even plant death in extreme cases.

Macronutrient toxicity rarely occurs, even at very high concentrations. Because macronutrients are much more abundant and available than micronutrients, plants have evolved ways of coping with, and indeed using, higher concentrations of these elements. However, during evolution, plants have been exposed to only low levels of the micronutrients and most plants have not evolved biological pathways to deal with high levels; consequently, high amounts of these elements may be toxic.

Because rocks differ in their chemical composition, the proportions of the different mineral elements in the soils derived from them also vary. Calcareous rocks such as chalk and limestone are largely composed of calcium carbonate, so soils derived from them can be deficient in many of the essential mineral elements. Soils derived from rocks that are chemically similar to granites are rich in the lighter elements such as K and Ca, but contain relatively lower quantities of the heavier micronutrients such as Mn, Fe and Mg. On the other hand, soils derived from rocks chemically resembling basalts contain a greater proportion of the heavy elements and as such are described as 'base-rich'. Both granites and basalts are classified as siliceous, because they have a 'skeleton' composed of silicates (SiO_4^{1-}) to which the positively charged cations are attached. In addition, the granites contain a lot of aluminium (the aluminosilicates), which can cause problems of aluminium toxicity in acid conditions.

Plant roots pump out hydrogen ions and remove nutrient ions such as calcium, magnesium, potassium, nitrogen and phosphate from the soil, often leaving the soil solution deficient in some of these nutrients, particularly nitrogen (found as either the ammonium ion, NH_4^+ or the nitrate ion, NO_3^-) and phosphorus (found as the phosphate ion, PO_4^{3-}). Basic cations are supplied to soils from weathering rocks in the area as well as from dust in rain and snowfall. The supplies of calcium, magnesium, sodium and, to a lesser extent, potassium from these sources, are generally sufficient to supply the needs of vegetation.

Not so with nitrogen and phosphorus. The supply of nitrogen to organisms has, until very recently, been limited by the rate at which atmospheric N_2 is converted to biologically-available NH_4^+ and NO_3^- by certain microbes. Humans have since doubled this global rate through developing an industrial process that converts atmospheric nitrogen to nitrate. Most of this available nitrogen is added to agricultural land as fertilizer. Some of this fertilizer finds its way into rivers and lakes through drainage from agricultural land, and some ammonia (NH_3) is volatilized from the land and is dissolved into rain to form ammonium ions, NH_4^+. This NH_4^+, together with NO_3^- from fossil fuel burning, has increased the nitrogen content of rain and snow falling on even 'pristine' areas. Cultivating organic soils and burning forest also release bio-available nitrogen into the environment. Despite the increase in nitrogen in deposition, it still remains the major limiting nutrient for terrestrial vegetation, due to high biological demands for the element.

Phosphorus is supplied in even lower amounts than nitrogen to most terrestrial ecosystems, although it is required in smaller amounts by organisms. The main source of phosphorus is from weathering rocks, and there is only one mineral, apatite, where it is present in any substantial quantities. Apatite commonly occurs in igneous rocks, but never as a major mineral constituent. Some phosphorus also reaches terrestrial catchments as dust in rain and snowfall. Some land areas that receive elevated nitrogen deposition have become phosphate-limited, or limited by magnesium, potassium or trace elements.

After nitrogen and phosphorus, potassium is the nutrient most likely to limit plant growth. Potassium is an essential plant nutrient because it is, among other things, a key component of many important enzymes. It is also essential for optimal efficiency of water use, conferring to the plant considerable drought tolerance. Soil potassium originates almost entirely from the inorganic components, principally micas and feldspars.

10.3 Erosion

10.3.1 Overall effects

Soil erosion is the removal of soil by the action of wind or water. Erosion of land surfaces is a natural process that has been occurring for millennia. The rate of this erosion is, however, generally very slow in the absence of human activity. For example, the erosion rate of most vegetated, undisturbed soils is slower than the rate of their formation, which is about 1 cm every 100–400 years. The mere presence of soils is proof that they form faster than they erode.

Disruption of vegetation or the soil surface by humans or animals can greatly increase the rate of erosion. This accelerated erosion may be more than 100 times as rapid as that which occurs naturally. The cutting of forests on hillsides, especially in areas of high rainfall, may cause rapid loss of soil through water erosion. Similarly, overgrazing or excessive tillage in arid and semiarid regions can greatly increase soil losses through wind erosion. The devastation caused to many soils on the North American plains during the 1930s 'Dustbowl' is a testament to the destructive forces of wind combined with human misuse of soil.

The effects of soil erosion may be confined to the soil itself, or they may appear off-site, far from the place of erosion through, for example, silting of rivers or estuaries. Regardless of where the damage occurs, the far-reaching consequences of soil erosion result in significant social, economic, and environmental costs.

On-site losses

The soil surface horizon, which is rich in organic matter and nutrients, is constantly exposed to the erosive forces of wind and water. Loss of soil through erosion reduces soil quality significantly. Erosion is destructive not only because it may remove vast amounts of soil, but also because it preferentially removes the most valuable components: organic material and the fine mineral particles. These losses reduce cation exchange and water holding capacities, and also decrease biological activity.

In cases of severe erosion, the entire surface horizon may be removed, thereby exposing the underlying B or C horizon. In such extreme instances the concomitant deterioration of soil structure greatly reduces water infiltration rate. Consequently surface runoff is increased and erosion is accelerated further.

○ What are some of the consequences if the surface horizon of a soil has been lost through erosion?

● In addition to increased susceptibility to further erosion, much of the organic matter and available nutrients will be lost, thereby reducing soil productivity and biodiversity significantly.

Distant effects of soil erosion

Many deleterious effects of soil erosion appear downstream or downwind from the site of erosion. Sediments eroded from soils may enter streams and lakes where they increase turbidity and disrupt aquatic ecosystems. These sediments may also collect in reservoirs, resulting in reduced water storage capacity. The cost of recovering these reservoirs through dredging is often many times higher than the cost of implementing measures that would control soil erosion and prevent this costly sediment accumulation. Sediments from eroded arable soils frequently contain high levels of nitrogen, phosphorus, and also pesticides, which may contaminate water used for recreation or drinking.

Wind erosion of soils may also have far-reaching consequences. The clay-sized particles within dust clouds present a human health hazard, as they irritate the lining of lungs, causing inflammation and other damage. Moreover, dust storms resulting from soil erosion during dry, windy weather may create hazardous driving conditions on motorways.

In light of the many direct and indirect consequences of soil erosion, protecting this fragile resource must be a priority. Fortunately, the environmental, social, and economic costs of erosion can be controlled by implementing sound soil management practices. The effective implementation of these practices requires that one first have an understanding of the causes and mechanisms of soil erosion.

10.3.2 Water erosion

Large raindrops fall to the earth's surface at a speed of about 30 km an hour. Despite the small mass of a single raindrop, this high speed gives the drop considerable energy, which is dissipated upon impact with the soil surface. Such violent collisions detach soil particles, particularly fine sands and silts, displacing them up to 2 m from the point of impact. Sustained impact by raindrops may destroy aggregates, especially those that have been weakened by prolonged wetting. When the disaggregated material dries, a hard crust may form at the soil surface. These crusts often inhibit seedling emergence and restrict water infiltration, thereby encouraging runoff during subsequent rainfall.

If a rainstorm is of sufficient duration and intensity, the soil's infiltration capacity will be exceeded and water will begin to flow over the surface. The particles detached during raindrop impact will then be transported downslope by the flowing water. If the water flows smoothly, then the detached particles will be removed from the soil uniformly. If, however, the water begins to flow preferentially in lower places and small depressions, erosion will occur by **channelized flow**. Water flowing in this manner moves faster and becomes turbulent, thus scouring the soil surface and accelerating erosion. Channelized flow occurs initially in small channels. As erosion proceeds, the channels enlarge until they form gullies, which allow increasingly faster and more turbulent flow, leading ultimately to **gully erosion**. Although gullies are the most visible form of water erosion and often scar the landscape severely, most soil is moved by the less obvious, slower movement of water passing uniformly over the surface and in small channels.

Control of water erosion

Water erosion can be reduced by controlling soil detachment and transport. Detachment is controlled most effectively by maintaining vegetative cover. Undisturbed grasslands and forests provide the greatest protection. For soils that have been disturbed, such as intensively farmed cropland, a layer of crop residue can be left after harvesting. This surface mulch of organic material serves to disperse the energy of raindrop impact.

Soil transport, which is most problematic in hilly areas, is controlled by reducing the flow of water downslope. Terraces may be constructed to reduce slope gradient and thus minimize water flow and soil transport (Figure 10.3). However, due to the high cost of construction and maintenance, terracing is recommended only for intensively farmed areas where arable land is in short supply. Contour cultivation, which involves tilling and planting at right angles to the natural slopes of the field, can effectively reduce soil transport on moderately hilly land.

Figure 10.3 Terraced fields in Asia.

10.3.3   Wind erosion

Wind erosion, like water erosion, is a global problem. Every continent has soils that have been degraded by wind erosion. Although wind erosion occurs principally in arid and semiarid regions, soils of humid climates that experience occasional drying are also vulnerable to the erosive forces of wind. The peat soils of the East Anglian fens are an example of this.

Wind erosion becomes significant only when ground-level wind speed exceeds about $25\,km\,h^{-1}$. As air moves rapidly over the soil surface, small grains become detached from aggregates and clods. Although the rapidly moving air is itself an erosive agent, the airborne particles are a far more destructive force, dislodging grains as they collide with the soil aggregates.

The way in which the detached soil particles are transported by wind depends mainly on their size. Most of the grains move by **saltation**, a mechanism that involves short bounces across the soil surface. Particles moving by this method rarely rise more than 30 cm above the ground. Particles with a fine sand size (0.05–0.5 mm) are transported particularly easily by saltation. For this reason, soils with fine sandy textures are among the most vulnerable to wind erosion.

The smallest particles detached during erosion, principally silts and clays, may be carried in suspension by wind to great heights and transported thousands of kilometres from their origin. This latter method of particle transport gives rise to dust storms that frequently originate over dry, unprotected soil during windy weather conditions. Despite the dark and ominous appearance of many dust clouds, particles transported in this way generally account for less than 15% of the total amount moved by wind.

Larger grains, up to about 1 mm in size, may move simply by rolling along the soil surface in a process known as **soil creep**. This is the principal means by which many sand dunes move (Figure 10.4). In extreme cases of erosion, wind removes all the finer particles, leaving only a very coarse soil surface known as **desert pavement**.

○   Which types of soil are the most vulnerable to wind erosion?

●   Soils with fine sandy textures, poor structure, no vegetation, and few plant residues at the surface.

Control of wind erosion

Wind erosion can be controlled most effectively by increasing soil moisture and reducing soil surface wind speed. Moisture greatly increases soil particle cohesiveness, and therefore also increases the wind speed required to detach individual grains. Where irrigation is available, it is often advisable to moisten the soil surface if dry, windy weather is anticipated. However, few people have access to irrigation waters and such quick preventative measures.

Ground-level wind speed can be reduced by maintaining well-anchored crop residue, such as stubble from a previous crop. Tillage practices that increase surface roughness, and in this way lower wind speed and trap soil, have also been used with success. Wind speed can also be reduced by planting trees. These windbreaks have the additional advantage of providing shelter for farm buildings and a habitat for wildlife.

Figure 10.4 Migrating sand dunes.

Question 10.2

Describe the ways in which one may minimize (a) wind erosion; and (b) water erosion.

10.4 Soil salinity

10.4.1 Overall effects

Common salt (sodium chloride, NaCl) is an essential part of animal diets and is used as a flavour-enhancing addition to many foods, yet an excess of this substance can kill plants and lay waste to vast stretches of the world's most fertile land. The ancient Romans were well aware of salt's destructive influence on plant growth and used this knowledge to strategic advantage. Following the defeat of Carthage more than 2000 years ago, the Romans spread salt over the land to ensure that this once great society could not re-establish itself. Even today, many of the world's soils are sufficiently degraded by salts that they are unable to support the growth of plants that provide food and fibre for humans, or habitats for wildlife.

Salts that are more soluble than gypsum ($CaSO_4$) are the most harmful to soil quality. These so-called **soluble salts** are composed mainly of calcium (Ca^{2+}), magnesium (Mg^{2+}), sodium (Na^+), chloride (Cl^-), and sulphate (SO_4^{2-}). Magnesium sulphate ($MgSO_4$) and NaCl are among the most common soluble salts. These salts originate from mineral weathering and from ancient deposits laid down aeons ago. They are also added to the soil through rainfall or irrigation. Soils containing an excess of soluble salts are said to be **saline**. These salt-affected soils occur mainly in arid and semiarid environments, where evaporation exceeds precipitation.

Movement of soluble salts

Saline soils occur most frequently in low lying areas, where soluble salts accumulate in what are known as saline seeps. These features form when salt-rich groundwater flows above an impermeable layer, toward a lower position in the landscape. From here, the salty water can move to the soil surface where it is evaporated or transpired through plants, thus leaving the salts behind (Figure 10.5). During the early stages of salinization, the salts may be visible as white surface crusts near the periphery of the seep area. As salt accumulation increases, the entire surface of the seep area may be covered with a white salt crust that is visible even from aircraft passing overhead.

Figure 10.5   Salt crusts.

○ Can you think of other examples in which soluble salts have collected at the soil surface?

● Salts can be seen to collect, for example, around the edges of a drying up pond or lake in the summer.

Extreme examples of this are seen when water from lakes is drained for human use, such as Mono Lake in California or the Aral Sea in Kazakstan. Here the water level has dropped so much that the soils are saline and the remaining water is too salty for most uses. Deeply furrowed soils in saline areas also often show a salt crust on their ridges. Salt-rich water in these soils moves upward via capillary action to the ridge crests. The water then evaporates at the ridge surface, leaving a salt crust behind.

Of all the ions released by soluble salts, sodium (Na^+) is the most problematic. Sodium disperses clay and organic matter, thereby degrading soil structure and reducing macropore space. Consequently, soils high in sodium are poorly aerated and have restricted permeability to water.

The electrical conductivity (EC) of water increases as salts dissolve and release their electrically charged ions. An indirect measure of soluble salt content in soils can therefore be obtained by measuring the EC of the soil water.

Effects of salinity on plants

Plants vary in their sensitivity to salty soils. Those with the lowest salt tolerance include tomatoes, onions and lettuce. At the other extreme are the **halophytes**, which occur most frequently in saltmarshes, beaches, and other saline environments. One of the most noteworthy halophytes is the saltwort (*Salicornia europea*), a highly salt-tolerant plant that can tolerate inundation by seawater (e.g. Flow Point in the Teign Valley).

The deleterious effects of salts on plants are due to both the loss of water from the plant, and to toxic levels of sodium and chloride. Fruit crops and woody ornamentals are especially sensitive to high levels of these elements. It is common practice in many cities of temperate regions to use salts during winter months as de-icing agents on roads and pedestrian walkways. Heavy applications of these salts frequently damage nearby vegetation. With sufficient precipitation, however, these salts are leached out of the soil and the vegetation can recover. However, some road verges are now so saline that they are being colonized by species usually restricted to salt marshes.

○ In terms of soils, which would be a more desirable salt to use as a de-icing agent: KCl or NaCl?

● Because of the undesirable effects of sodium on soil properties, KCl is a more suitable salt.

10.4.2 Reclamation of salt-affected soils

The reclamation of salt-affected soils can be difficult and lengthy. Despite the challenges inherent in such a task, one can improve the chances for successful remediation. The first step is to establish good internal drainage to facilitate the leaching of salts. Many salt-affected soils currently have adequate drainage in place, others require the implementation of artificial drainage systems, or the addition of deep-rooted, salt-tolerant plants to lower the water table. Second, excess sodium in sodic and saline–sodic soils is replaced by calcium, which improves soil structure, facilitates leaching of the sodium (by cation exchange), and is an essential plant nutrient. This is achieved most economically by the addition of gypsum. And third, the salts must be leached from the soil. This can be accomplished by the liberal application of good quality irrigation water. Where irrigation water is not available, natural precipitation must be relied upon to leach the salts.

Question 10.3

Many soils that contain excess quantities of sodium have poor structure and are frequently saline. Attempts to remediate these soils often involve mechanically incorporating gypsum ($CaSO_4$) to considerable depths. How can such a technique improve base saturation and soil structure without exacerbating the problems of salinity?

10.5   Summary of Section 10

1   Soils can resist changes in pH by chemical reactions between hydrogen ions in the soil solution and certain key chemical constituents in the soil. Soils pass through different 'buffer zones' as they are progressively acidified — the buffer zones correspond to the dominant soil component that removes hydrogen ions from solution at the pH of the soil. As pH decreases, soils go through the carbonate buffer zone (if there is carbonate bedrock), silicate buffer zone (if silicate bedrock), aluminium buffer zone and iron buffer zone. The exchangeable cation buffer zone occurs over all normal soil pH ranges. Soils that have reached the aluminium buffer zone release Al^{3+} to the soil solution, which can damage vegetation and downstream biota.

2   The eighteen elements essential for plant growth are divided into nine macronutrients (carbon, hydrogen, oxygen, nitrogen, phosphorus, potassium, sulphur, calcium and magnesium) and nine micronutrients (iron, manganese, boron, zinc, copper, chlorine, molybdenum, cobalt and nickel).

3   A plant's development can be no greater than is allowed by the limiting nutrient (usually nitrogen or phosphorous) or limiting factor (often water, sometimes sunlight or heat).

4   Erosion is the removal of soil by the action of wind or water. Disruption of vegetation or the soil surface by humans or animals can greatly increase the rate of erosion. Erosion is destructive not only because it may remove vast amounts of soil, but also because it preferentially removes the most valuable components: organic material and the fine mineral particles. These losses reduce cation exchange and water holding capacities, and also decrease biological activity.

5   Soils containing an excess of soluble salt are said to be saline. Salts originate from mineral weathering and from ancient deposits laid down aeons ago. They are also added to the soil through rainfall or irrigation. Once in a seep area, salty water moves to the soil surface where it is evaporated or transpired through plants, thus leaving the salts behind.

Learning outcomes for Sections 7–10

After working through these sections you should be able to:

1   Describe the three components that comprise a soil texture, and use a soil texture diagram to describe the texture of a soil given the proportions of two of its three major textural components. (*Question 8.1; Activity 8.1*)

2   Outline the overall physical and chemical properties that sand, silt and clay contribute to a soil. (*Question 8.2*)

3   Explain the physical and chemical properties that soil organic matter imparts to a soil.

4   Describe the effects of cultivation, climate and vegetation on the organic matter content of soils. (*Question 8.3*)

5   Explain how isomorphous substitution occurs in clay minerals and the type of charge that it generates. (*Question 8.4*)

6   Explain how pH-dependent charge is generated in clay minerals, humus and hydrous oxide clays. (*Question 8.8*)

7   Define cation exchange capacity and describe the factors that contribute to cation exchange capacity in a soil. (*Questions 8.4 and 8.5*)

8   Name the factors that affect the strength of adsorption of different cations to soil exchange sites. (*Question 8.5*)

9   Describe the difference between exchangeable and non-exchangeable cations in a soil. (*Question 8.4*)

10 Calculate cation exchange capacity given the molar composition of a CEC percolate solution. (*Question 8.6*)

11 Given a full soil chemical analysis including LOI, pH, vegetation type, and exchangeable ion concentrations, calculate cation exchange capacity and base saturation, and use this information to hypothesize about the soil's fertility and resistance to acidification. (*Question 8.7; Activity 10.1*)

12 Define anion exchange capacity and describe the type of soils that have the highest anion exchange capacity. (*Question 8.8*)

13 Describe the relationship between soil water content at field capacity and soil texture, and explain the reasons for this relationship. (*Question 8.9*)

14 Describe the variables in Darcy's law as they pertain to calculating the speed of water as it moves down a slope through saturated soil. (*Question 8.10*)

15 Outline the major chemical components one may expect to find in (a) the air and (b) the solution of a soil.

16 Explain the difference between particle density and bulk density and, given one of these and the soil porosity, calculate the other. (*Question 8.11*)

17 Describe the consequences of waterlogging soil in terms of soil gases and soil solution chemistry. (*Question 8.12*)

18 Sketch a general soil profile showing O, A, E and B horizons and explain how these horizons form. (*Questions 9.1 and 9.2*)

19 Given a soil profile, speculate on the environment under which that soil formed and, given information on an environment, make an educated guess about what the soil profile may look like. (*Question 9.1*)

20 Briefly outline the chemical and physical properties used to classify soils, and give a few examples of soil orders in either the USDA system or the British system. (*Question 9.2*)

21 Briefly outline five buffer ranges that may be found in soils and describe how growing vegetation can acidify soil. (*Activity 10.1*)

22 Describe the types of nutrient pools in soils, the factors that influence the availability of nutrients, and the concept of limiting factors. (*Question 10.1*)

23 Outline some of the major consequences of soil loss through erosion, and the factors that make a soil more susceptible to erosion. (*Question 10.2*)

24 Describe the process by which soils become saline and how they may be remediated. (*Question 10.3*)

Comments on activities

Activity 2.1

Perhaps the first thing to notice is that the peridotite and basalt are dark in colour and the andesite and granite are much paler. The hand specimen of granite shows clearly that the rock is made up from several distinct types of crystal, to judge from their different colours and appearances. We might reasonably assume that the different crystals, which interlock with one another, are of different minerals. The peridotite specimen also shows different types of crystal, but their dark colours make individual crystals less easy to pick out. The presence of different minerals is less obvious in the hand specimens of the other two rocks, because their individual crystals are much smaller, but under the microscope all four rocks have a similar interlocking crystal structure (Figure 2.8).

The two rocks that formed well below the Earth's surface, granite and peridotite, contain large crystals, easily visible to the naked eye. This is a result of a long, slow cooling process. Basalt and andesite cooled quickly as their magma poured out of volcanic vents or flowed along fissures close to the surface of the Earth. There was therefore little time for the mineral crystals to grow before they were locked into the crystal matrix.

Activity 2.2

Both the hand specimens have a granular texture, like fragments stuck together, rather than the crystalline texture of the igneous rocks. The view through the microscope confirms that these rocks are constructed of unconnected fragments rather than interlocking crystals. The view of the polished face of the limestone shows that each separate rounded grain is surrounded by a material that fills in the spaces between the grains. The microscope view of the sandstone shows this too, although in this rock the grains seem to be touching each other.

Activity 2.3

Whereas the crystals of the minerals that comprise the igneous rocks appear to be jumbled together in a random way, the crystals of these two metamorphic rocks show definite alignment. This is particularly obvious in the hand specimen of the phyllite, a schist, but it can also be seen, in a rather coarser form, in the gneiss. The thin sections of both rocks confirm the 'stretched out' appearance of the crystals.

Most metamorphic rocks demonstrate this type of appearance, but interestingly it is not shown by either marble or quartzite (Activity 2.4), even though these are also metamorphic rocks.

0.5 mm

(a)　　　　(b)

Figure 2.8 Appearance of crystalline (igneous) rock specimen under the microscope: (a) actual and (b) corresponding sketch.

Activity 2.4

At a first glance these two rocks look very much alike, and both lack the variety of colour and texture shown by the previous specimens. Instead they have a granular-like character, reminiscent of the sedimentary rocks. They both appear to be crystalline, but it is impossible to distinguish individual crystals in the hand specimens. However, under the microscope, the interlocking crystal structure can be seen, even though only one type of mineral is present.

The gneiss and the phyllite are composed of several different mineral types, as is seen in the different colours and forms of the crystals, but we said that the quartzite and marble each consisted of a single mineral. An explanation of the uniform appearance of these two rocks must lie in this fact, since in order to demonstrate alignment, new minerals must be formed.

Activity 3.1

Gabbro appears to be a mixture of light and dark crystals, which are clearly visible in the hand specimen. The thin section shows that many of these crystals are larger than 1 mm long, so this must have cooled slowly.

The hand specimen of the diorite has a similar speckled appearance to the granite, and it is also pale in colour overall. Under the microscope several of the crystals are seen to exceed 1 mm in length, and so it can be grouped with the intrusive rocks. Again it must have cooled slowly.

The rhyolite is the most difficult rock to classify in the hand specimen, as the most obvious feature is the presence of a few large crystals. However, under the microscope it is clear that a high proportion of the field of view is taken up by a matrix of tiny crystals.

As the bulk of the rock is fine-grained, this indicates that it cooled rapidly, at or near to the surface of the Earth. The presence of larger crystals suggests that some degree of fractional crystallization had begun to take place before the magma poured out of a volcanic vent. The solution to this puzzle is that the larger crystals would be carried up to the surface in the magma, and become embedded in the fine-grained matrix of this *extrusive* rock.

Each of these three rocks is to be found in the Earth's crust, and so we are now able to group the crustal rocks as: extrusive (basalt, andesite and rhyolite) and intrusive (gabbro, diorite and granite).

Activity 3.2

Your completed table should look like Table 3.8.

Table 3.8   Mineral composition of igneous rocks (completed).

Rock type	Extrusive or intrusive?	Feldspar	Mica	Pyroxene/ amphibole	Olivine	Quartz
basalt	extrusive	✔ (plagioclase)		✔ (pyroxene)	✔	
gabbro	intrusive	✔ (plagioclase)		✔ (pyroxene)	✔	
andesite	extrusive	✔ (plagioclase)		✔ (pyroxene)		
diorite	intrusive	✔ (plagioclase)	✔ (biotite)	✔ (amphibole)		✔
rhyolite	extrusive	✔ (potassium feldspar/ orthoclase)				✔
granite	intrusive	✔ (potassium feldspar)	✔ (muscovite/ biotite)			✔

Activity 3.3

Your completed table should look like Table 3.9.

Table 3.9   Characteristics of selected minerals (complete).

Mineral	Metal ions and silica	Hardness	Colour
olivine	Mg, Fe, SiO_2	6–7	grass green
pyroxene	Mg, Fe, Ca, $2SiO_2$	5–6	almost black
amphibole	Mg, Fe, Ca, Al, SiO_2	5–6	blue–grey to black
feldspar (plagioclase)	Ca, Na, Al, SiO_2	6	white
feldspar (orthoclase)	K, Al, SiO_2	6	pink
mica (muscovite)	K, Al, SiO_2	2–4.5	white
mica (biotite)	K, Mg, Fe, Al, SiO_2	2.5–3	dark brown
quartz	SiO_2	7	colourless/grey

Activity 4.1

Table 4.2   Completed table of minerals in metamorphic rocks.

Rock type	No.	Mica	Feldspar	Quartz	Other minerals
gneiss	4	✔	✔	✔	
schist	26	✔		✔	garnet
phyllite	11	✔		✔	chlorite
slate	2	✔		✔	clays

Activity 5.2

Table 5.5   Characteristics of silicate-based sedimentary rocks.

Rock type	No.	Contents	Cement	Roundness	Sorting	Bedding/ lamination
conglomerate	29	pebbles, sand (quartz?)	silica/quartz (probably)	rounded	very poor	none seen
micaceous sandstone	20	quartz, little mica	calcite	sub-angular	well sorted	yes bedding
coarse sandstone	10	quartz, some feldspar, iron oxide	silica/quartz	rounded to sub-rounded	well sorted	none seen
mudstone (shale)	15	quartz, clay	not identified	difficult to tell	very well sorted	yes lamination

Table 5.6 Characteristics of carbonate sedimentary rocks.

Rock type	No.	Muddy matrix	Crystalline matrix	Major fossil	Ooids
Limestone 1	21	matrix, and inside shells	inside shells only	shell debris	no
Limestone 2	22	no	cement	none	yes
Limestone 3	24	no	cement	crinoids	no

Limestone 1 has a muddy matrix, and the other major component is shell debris, which is biologically generated (by the shellfish!). Limestone 1 is therefore a 'shelly limestone'. In both Limestone 2 (the oolitic limestone) and Limestone 3, the matrix, or cement, is crystalline calcite. The 'popular' name for Limestone 3 would be 'crinoidal limestone'.

Activity 8.1

There is no 'answer' to this activity — the answers will be as many as the number of people trying the activity and the number of soils they look at. However, we can make some general comments that may or may not be true for your soil.

Most soils in residential areas, gardens or farms are highly organic as they have been amended with compounds such as compost or topsoil, so they will be dark. Clumps and aggregates in the soil generally indicate good drainage, and are often found in fertile soils. The fertility of many soils used by people also means that you may well find animals such as earthworms or pillbugs in your trowelful or spadeful of soil. Many such soils have little difference in colour between upper parts and lower parts in a spadeful of soil because they have been dug and turned over repeatedly. In contrast, some forest soils, or less disturbed soils you may see along a road cut, may show a pronounced darkening of the uppermost part of the soil because of contact with organic material from above. Roots may also be more concentrated in the darker, more organic-rich parts of the soil. The presence of rocks, gravel, and sand (which make a soil feel gritty if it is rubbed between the fingers) tells something about the geology in the area you live. For example, soils from coastal areas may feel gritty, indicating sandiness.

Activity 10.1

You should find that the pH of the ash is well above 7. It is alkaline, suggesting that the original tree from which the charcoal came had accumulated basic cations at the expense of acidity pumped into the soil.

Answers to questions

Question 1.1

The statement reminds us that the three most abundant elements in the Earth's crust are oxygen (45.2%) (which combines with silicon in the silicate minerals, and aluminium and silicon in the aluminosilicate minerals), silicon (27.4%), and aluminium (8.0%). Figures from Table 1.1.

Question 2.1

Igneous rocks are formed by the cooling and crystallization of magma. They consist of interlocking crystals of several different minerals. Small crystals indicate rapid cooling at the Earth's surface, and large crystals slower cooling at depth.

The fragmentary (or *sedimentary*) rocks form from fragments of more ancient rocks, or the remains of ancient life forms, which accumulate as sediments, and are then compressed and cemented together in the lithification process.

Metamorphic rocks are former igneous or sedimentary rocks, which have been transformed by heat and pressure, often as a result of deep burial. They too contain interlocking mineral crystals, often distributed in layers.

Question 2.2

Specimen (a) is a sedimentary rock. The granular texture and the glassy grains (probably quartz) suggest that it is a poorly cemented sandstone. If the rounded grains had been opaque, it could have been an oolitic limestone, another sedimentary rock. Unfamiliar terms will be explained later in the text.

Specimen (b), Larvikite, is an igneous rock. The large interlocking crystals, and the irregularly dispersed dark mineral indicate that it has crystallized slowly from cooling magma deep within the Earth.

Specimen (c) is a metamorphic rock. The sheen on the surface is due to elongated crystals of the platy mineral mica, and the large dark red crystals are the metamorphic mineral garnet. It is therefore described as a garnet mica schist.

Question 3.1

(a) The silicate portion of this mineral has the formula Si_2O_6. Reference to Table 3.7 shows that this is in the pyroxene group of minerals, and Table 3.5 will show that they have structures consisting of single chains, where silicate tetrahedra are joined to each other at two oxygen atoms.

(b) The presence of magnesium and iron in the structure indicates that this is a mafic mineral.

(c) The part of the formula $[Mg,Fe^{2+}]$ shows that either magnesium or iron (II) can contribute two of the positive charges needed to balance the negative charges on the silicate chain. In different specimens of augite, the atom ratio of magnesium to iron could in theory lie anywhere between $1:99$ and $99:1$; in practice less extreme values are found.

(d) As we saw in the answer to (c), the proportions of magnesium to iron can vary. However the formula of the mineral shows that for every calcium atom in the structure, there is *either* one magnesium *or* one iron atom present.

(e) The formula of augite is $Ca(Mg,Fe^{2+})Si_2O_6$. Six oxygen atoms have 12 negative charges, and these are balanced by 2 positive charges from calcium (Ca^{2+}), 8 positive charges from the two silicon atoms (Si^{4+}), and therefore 2 positive charges are needed from (Mg,Fe^{2+}). As magnesium *must* have 2 positive charges (Mg^{2+}), and either magnesium or iron must occupy a place in the crustal structure, only Fe^{2+} would have the correct number of charges to achieve neutrality.

Question 3.2

Norite is almost exactly on the boundary of mafic and intermediate rocks, but probably has enough calcium, iron and magnesium to be classified as mafic. Syenite, from Table 3.3, has a silicon and aluminium composition quite close to andesite or diorite, although it has quite a lot less iron, magnesium and calcium. It would be placed in the intermediate group of rocks.

Question 5.1

Table 5.7 A completed version of Table 5.4, Properties of rocks and minerals.

	Rock or mineral	Type	Texture/occurrence	Appearance	Other properties
1	shale	2. sedimentary rock	6. fine-grained, thin layers	7. dark colour	3. soft, crumbles easily
2	conglomerate	7. sedimentary rock	5. pebbles of various sizes and rock types	1. various coloured components	4. fragments set in finer-grained matrix, like concrete
3	aluminium-rich mica (muscovite)	4. felsic mineral	1. component of granites	6. shiny silvery surface	2. soft, splits easily into thin sheets
4	slate	1. metamorphic rock	7. may retain signs of the original layered structure	4. dark grey	7. hard, formed from mudstones by heat and pressure
5	shelly limestone	6. carbonate rock	2. broken fragments of fossil shells	3. pale grey, rough surface	1. cemented by calcite
6	kaolinite	3. clay mineral	3. very fine particles <0.004 mm	5. tiny white plates under electron microscope	5. layered structure with tetrahedral and octahedral sheets in each layer
7	amphibole	5. mafic mineral	4. present in diorite and andesite	2. dark green to black, shiny	6. include OH^- or F^- ions within their structure

Question 5.2

The 1 : 1 and 2 : 1 refer to the sheets of silicate tetrahedra and aluminate octahedra that bond together to form the separate layers in an aluminosilicate clay. In a 1 : 1 clay mineral, like kaolinite, one sheet of each type forms the separate layers. The layers are held together by hydrogen bonds. A 2 : 1 clay, like montmorillonite, has layers composed of a sheet of aluminate octahedra sandwiched between two sheets of silicate tetrahedra. This more complex structure is also subject to some substitution of both silicon and aluminium atoms by aluminium, iron and magnesium. This generates negative charges on the outside of the layers, which attract positively charged metallic ions, which hold the layers together by electrostatic attraction.

Question 6.1

Water evaporating from a solution of Ca^{2+}(aq) + $2HCO_3^-$(aq) will increase the concentration of those ions, driving Equation 6.3 to the left.

Question 6.2

Pyroxene will be the most readily weathered, because mafic minerals are more susceptible to chemical weathering than felsic minerals. The conditions at the Earth's surface are very different from those under which they were formed. The sodium-rich feldspar and the mica (especially the magnesium and iron-rich mica) will be the next most readily weathered, with the quartz being the most resistant to weathering.

The melting temperatures of these mineral are in inverse order to their readiness to undergo weathering, with quartz having the lowest melting temperature, pyroxene the highest, and the mica and feldspar melting temperatures falling in between.

The first stages of weathering will be physical and bioweathering. Weaknesses in the rocks, at cracks and joints, will be exploited by atmospheric and climatic forces; e.g. frost shattering, and by the growth of plants, to break rock into smaller fragments. Aqueous acid is formed as rainwater dissolves carbon dioxide from the air, according to Equation 6.1.

Plant growth and decay will also generate acidic solutions, many times more concentrated than rainwater. These acids can enter the minerals at weaknesses along the cleavage planes, and interact with the oxygen atoms of the silicate tetrahedra.

Sodium feldspar, with formula $NaAlSi_3O_8$, will weather to give a clay mineral (kaolinite or illite), 'soluble silica' — $Si(OH)_4$, and sodium ions in solution. Pyroxene will also yield soluble silica together with magnesium (Mg^{2+}) and iron (II) ions (Fe^{2+}) in solution, although the latter will rapidly be oxidized to Fe^{3+} ions. Micas, which already have clay mineral-like structures, will give all these products too. Quartz will hardly weather at all, and be left as sand grains.

Question 6.3

Hydrated oxide clays are formed under conditions of intense weathering, in which the first breakdown products are further transformed, resulting in this new set of minerals. The intensity of weathering is greater in tropical climates. The higher temperature increases the rate of chemical reactions, the higher rainfall leads to more rapid leaching of the ions liberated in the weathering processes, and the luxuriant and more rapid growth generates soil water with a high CO_2 concentration and low pH. Waterlogged ground also provides a less oxygen-rich environment, allowing the more soluble iron (II) ions to migrate to lower levels in the soil where they are oxidized to insoluble iron (III). Under these warm, wet, acid conditions, silicate structures break down further, and much more silicon is lost as well as the ions that weather out easily, like Na^+, K^+, Ca^{2+}, and Mg^{2+}. For example, the aluminosilicate clay kaolinite is further weathered to the hydrated aluminium oxide clay, gibbsite:

$$Al_2Si_2O_5(OH)_4(s) + 5H_2O(l) \rightarrow 2Al(OH)_3(s) + 2Si(OH)_4(aq)$$

Only the ions of resistant elements like iron, aluminium and manganese survive. When they are hydrated, they generate more acid, thus accelerating the weathering process further, as in the oxidative hydration of iron (II) to goethite:

$$4Fe^{2+}(aq) + O_2(aq) + 6H_2O(l) \rightarrow 4FeO(OH)(s) + 8H^+(aq)$$

Question 8.1

(i) a clay

(ii) a silt loam

Question 8.2

(a) The floodplain soil has the highest clay content and the lowest sand content, therefore, from Table 8.3, it would be expected to have the highest water-holding capacity and fertility, but also the greatest vulnerability to compaction. The pine–fir woods, grassland, pasture land and river bank have the highest sand content, and, among these, the pasture land and pine–fir wood have the lowest clay content, suggesting these two soils would have the highest drainage capacity and be the most vulnerable to erosion.

(b) Basing these conclusions on texture alone ignores other contributing factors such as the following:

- The water-holding capacity, the drainage capacity and the vulnerability to erosion are all affected strongly by the slope of the site, the existing vegetation, and the management (e.g. how the soil is tilled or ploughed if agricultural).

- Fertility can be affected by the bedrock or parent material, the climate, the existing vegetation and the management (e.g. addition of fertilizers or lime).

- Vulnerability to compaction can be affected by management.

- The data shown are only for one depth in the soil. Other depths could have different properties.

You may have thought of other factors as well.

Question 8.3

The order is peat >> grassland > forest. Peatlands are generally found in climates that are cool and moist, and these often lead to poor drainage conditions, ideal for the accumulation of organic matter. Grasses have extensive non-woody rooting systems that add a large amount of organic matter directly into the soil. Trees tend to have larger, more woody roots that do not form the dense network that grass roots do. Forests contribute organic matter mainly through above-ground litter.

Question 8.4

(a) As a 1 : 1 clay, kaolinite would have no isomorphous substitution. If crystals are large and regularly ordered they would have a low surface area, and few exposed hydroxyl groups. All of these factors would make the CEC of kaolinite relatively low.

(b) CEC is not the amount of *negative* charge, but the capacity of a clay mineral to exchange *positive* ions. This depends not only on the negative charges, but also on the amount of surface area exposed to soil solution. Clays with more weakly held layers, such as montmorillonite, have a higher proportion of exchangeable cations than clays such as illite, despite the former having a lower permanent negative charge. Cations can become trapped in the tightly held layers of illite clays and be essentially non-exchangeable.

Question 8.5

With its single charge and large size (radius of 0.167 nm), Cs^+ would behave most similarly to potassium, K^+ (single charge, ionic radius of 0.133 nm). Since it is a cation, radioactive caesium would persist for the longest time in soils with a high cation exchange capacity, such as highly organic soils or those rich in 2 : 1 clays such as montmorillonite and illite. Since it is similar to potassium, you may hypothesize that caesium may become trapped in soils high in illite clays, just like potassium does. This is indeed the case, and high-illite soils may retain, and slowly release, caesium over long time periods.

Fairly high levels of radioactive caesium persists today in areas with highly organic soil that received large amounts of the fallout from Chernobyl, such as in some parts of Lapland in Scandinavia, where incidences of contaminated reindeer meat (from deer that have grazed on grasses growing in radioactive soil) still occur.

Question 8.6

First we need to determine the mmol of charge per litre (for aluminium this is already given). To do this we multiply the concentration of each ion by its charge:

Table 8.7

	mmol l^{-1}	Charge	mmol charge l^{-1}
Ca^{2+}	12.5	2	25
Mg^{2+}	1.4	2	2.8
K^+	0.105	1	0.105
Na^+	0.24	1	0.24
H^+	6.5	1	6.5
Al^{4+}	33.8	varies	33.8

The total amount of charge per litre is then the sum of all of the individual charges: 68.4 mmol charge l^{-1}.

A 100 ml (one-tenth of a litre) solution with 68.4 mmol charge per litre contains 68.4/10 mmol charge = 6.84 mmol charge. There is thus 6.84 mmol exchangeable charge in 10 g of soil.

We have the answer for 10 g, but need to express it per kg. Since 100×10 g = 1000 g, or 1 kg, we must multiply our answer by 100:

6.84 mmol charge in 10 g soil $\times$ 100 = 684 mmol charge in 1 kg soil.

Our final step is to convert mmol charge (10^{-3} moles) into cmol charge (10^{-2} moles):

684 mmol charge = 684×10^{-3} moles charge = 68.4×10^{-2} moles charge, or 68.4 cmol charge.

The cation exchange capacity of the soil is thus:

68.4 cmol charge/kg soil.

Question 8.7

The CEC of the soil is calculated by multiplying each exchangeable cation value by its charge, and adding these values together. In this case you are already given the values for each ion in units of moles of charge, so you only need to add those together. The base saturation is calculated by determining the proportion of the CEC for each soil that is from the basic cations Ca^{2+}, Mg^{2+}, Na^+ and K^+. CEC and BS (%) are shown in bold in the completed table below.

The Trusham 1 and 2 soils both have base saturation above 50% and the Swanaford soil is quite close to that value . These would be the most desirable from an agricultural perspective. The high cation exchange capacities correspond to the high LOI, suggesting relatively high levels of organic matter even below the organic surface layer. The Swanaford soil probably has a high base saturation because it receives nutrients, including basic cations, from the river when it overflows its banks. The Trusham soils have similar CEC because they are the same soil series. However Trusham 2, the farm soil, has a much higher exchangeable calcium and base saturation, suggesting it has been limed. Growing forests have low levels of organic matter, remove large amounts of cations, and are well leached, thus the Yarner Wood and Haldon Forest have the lowest CEC and base saturation, and would be the poorest for agriculture.

Question 8.8

As a soil's mineralogy changes slowly over time, aluminosilicate clays will be replaced by various oxides of iron and aluminium. Because the layer silicates possess a net negative charge, they will have a very low anion exchange capacity. However, iron and aluminium oxides within an acidic environment will possess a net positive charge and they will, therefore, have considerable anion exchange capacity. Consequently, as a soil's mineralogy becomes dominated by oxide minerals, the CEC decreases and the AEC increases.

Question 8.9

(a) The water content of the soil is measured by weighing the soil, then drying it in an oven at 105 °C for about 24 hours. After drying, the soil is weighed again. The mass loss represents the amount of soil water.

(b) Soil B is likely to be of much finer texture than soil A. Plants are unable to extract water held in the smallest pores in such a soil. Using Figure 8.19 as a guide would suggest that soil A is a sandy loam and soil B is a clay loam.

Table 8.8 Chemical data for some soil samples in the Teign Valley, Devon, UK. (All exchangeable ions are given in mmol charge per 100 g soil.)

Soil series	Swanaford	Yarner Wood	Haldon Forest	Trusham 1	Trusham 2
Depth (cm)	8–23	0–28	0–13	0–28	0–23
Surface vegetation and environment	ash/willow, floodplain of River Teign	oak/bilberry	pine plantation	semi-natural meadow	agricultural pasture
LOI	8.1	3.5	3.4	7.0	5.2
pH (H_2O)	5.7	4.5	4.3	5.9	6.2
Exchangeable Ca^{2+} ($mmol_c$/100g soil)	12.0	1.40	1.20	25.0	32.8
Exchangeable Mg^{2+}	2.0	0.20	0.20	2.8	1.4
Exchangeable K^+	0.11	0.04	0.03	0.11	0.1
Exchangeable Na^+	0.20	0.05	0.05	0.24	0.13
Exchangeable H^+	2.0	0.51	1.1	0.7	0.03
Exchangeable Al^{3+}	16.2	7.5	9.1	16.4	1.02
CEC ($mmol_c$/100g soil)	**32.5**	**9.7**	**11.7**	**45.3**	**35.5**
BS (%)	**44**	**17**	**13**	**62**	**97**

Question 8.10

(a) The effect of soil texture on rates of saturated flow is incorporated in K, the hydraulic conductivity of the soil.

(b) After a long, dry period, clay textured soils will be heavily cracked and so will conduct water, and any dissolved nutrients and pollutants, very quickly. By Darcy's law, water may also flow through a clayey soil more rapidly than a sandy soil if the slope was much steeper (h was higher and/or l was shorter), although the differences in slope would have to be extreme.

Question 8.11

If pore space comprises 45% of the soil, the remaining 55% will consist of solid particles, with an average density of 2.4 g cm^{-3}. Therefore, 2.4 g cm^{-3} × 55% = 1.32 g cm^{-3}.

Question 8.12

The oxic conditions of a well-drained soil would favour the precipitation of various iron oxides (e.g. goethite) and manganese oxides, and would favour the transformation of As(III) to the less toxic As(V). There are many other consequences of drainage, including rapid decomposition of organic material leading to the release of CO_2, which adds to the problem of greenhouse warming. Drying the wetland soil can result in significant soil losses, land subsidence, erosion and loss of habitat. (You may have come up with others!)

Question 9.1

Soil A: This moderately weathered soil will have a relatively thick organic matter-rich surface horizon due to the extensive rooting systems of the grasses. Because the soil has developed from lake sediments the texture is likely to be silty to clayey. Under a temperate climate with only moderate precipitation, there is not likely to be significant leaching, although the B horizon may be well developed.

Soil B: This profile will be highly weathered. The soil will be deep, possibly several metres in thickness, although the O horizon may be thin due to a less dense rooting mass and rapid recycling of organic nutrients. The soil will be dominated by red and yellow oxide minerals. Highly weathered soils derived from granites commonly contain much sand-sized quartz grains that have survived the intense weathering.

Question 9.2

In the USDA system the Dunsford Wood soil best fits in the order spodosol — a leached, acidic forest soil in temperate regions. In the UK system, the presence of a bleached grey upper layer, evidence of leaching and low pH identify it as a podzol.

Question 10.1

Virtually all the potassium (>90%) will be unavailable to plants, as it is held in the stable nutrient pool as components of soil minerals, principally micas and feldspars. Because sandy soils generally have low cation exchange capacity, the amount of potassium within the labile pool, and thus plant-available, will be very low. The fraction of potassium in the soil solution will be much less than 1%, although the size of this pool may change rapidly in response to leaching or uptake by plants.

Question 10.2

Wind erosion may be controlled by increasing soil surface moisture through irrigation, or by reducing ground level wind velocity through the implementation of windbreaks. Water erosion can be controlled by maintaining soil vegetation and a layer of crop residue at the soil surface. In particularly hilly areas, especially in regions where land is in short supply, terraces may be built to minimize the flow of water downslope.

Question 10.3

The calcium from the gypsum will replace sodium on the exchange sites, thus enhancing particle aggregation and the formation of desirable structure. However, gypsum is sufficiently insoluble that it does not contribute significantly to soil salinity.

Appendix *Digital Kit*

Rock Kit

Code no.	Rock type
1	andesite
2	slate
3	basalt
4	gneiss
5	quartzite
6	porphyritic rhyolite
7	sandstone
8	chalk (micrite)
9	diorite
10	sandstone (eolian)
11	phyllite
12	sandstone (greywacke)
13	granite
14	dolerite
15	mudstone (shale)
16	amphibolite
17	marble
18	sandstone (arkose)
19	gabbro
20	sandstone (micaceous)
21	shelly limestone (biomicrite)
22	oolitic limestone (oosparite)
23	peridotite
24	crinoidal limestone (biosparite)
25	ignimbrite
26	garnet mica schist
27	mudstone
28	vesicular basalt
29	conglomerate
30	biotite gneiss

Mineral Kit

Code no.	Mineral name
I	pyrite
II	muscovite mica
III	quartz (a broken fragment from a large crystal)
IV	calcite rhomb
V	gypsum
VI	galena
VII	hexagonal crystal of quartz
VIII	twinned crystal of orthoclase
IX	orthoclase in Shap granite
X	garnet
XI	hematite
XII	Iceland spar crystal
XIII	amphibole
XIV	andalusite
XV	kyanite
XVI	biotite mica
XVII	olivine
XVIII	chert (flint)
XIX	talc
XX	zircon

Acknowledgements for Part 2 *Earth*

Figures 2.1, 2.8, 4.1, 4.3a, 4.5, 5.4a-c, 5.13: Andy Tindle/Open University; *Figures 2.2, 6.8a:* Linda McArdell, OUGS; *Figure 2.3:* Adapted from *Understanding Earth* by Frank Press and Raymond Siever © 1998, 1994 by W. H. Freeman & Company, used with permission; *Figure 2.4:* Kevin Church/Open University; *Figures 2.5, 6.3b, 6.4b, 6.10a, 6.14, 6.15:* Polly Rhodes, OUGS; *Figure 2.6:* Based on the 1:50,000 map with permission of Ordnance Survey on behalf of The Controller of Her Majesty's Stationery Office, © Crown copyright, The Open University, ED 100020607 and the British Geological Survey map, IPR/20-33 British Geological Survey © NERC. All rights reserved; *Figure 2.7:* Based on the 1:50,000 map with permission of Ordnance Survey on behalf of The Controller of Her Majesty's Stationery Office, © Crown copyright and The Environment Agency; *Figure 4.4:* Welsh Slate Museum, Llanberis; *Figure 5.6:* Naomi Williams/Open University; *Figure 5.8c:* Adapted from Waistnidge, D. (1998) *Encounter: Chemistry at work*, Chemistry Review **7**(4) p.25, Philip Allan Updates; *Figures 5.11, 6.17:* Adapted from Table 1 of Fieldes M. and Swindale L.D. (1954) *Chemical Weathering of silicates in soil formation*, New Zealand Journal of Science and Technology, vol. **36** (Sect B) SIR Publishing; *Figure 5.12:* Jackie Lees, Department of Geological Sciences, University College, London; *Figure 5.14a:* Clif Jordan, Carbonate Rocks; *Figures 6.1a, 6.1b and 6.9a:* Evelyn Brown/Open University; *Figure 6.2:* Stuart Bennett/Open University; *Figures 6.3a, 6.4a:* Geoslides; *Figure 6.5a* Duncan Woodcock, OUGS; *Figures 6.5b, 6.19:* Michael Gagan/Open University; *Figure 6.6:* Paul Everett; *Figure 6.7:* Simone Pitman/Open University; *Figure 6.8b:* Ross Wilson, c/o James Howden; *Figure 6.9b:* Carole Arnold; *Figure 6.10b:* Angela Coe/ Open University; *Figure 6.13:* P. R. Deakin and Treak Cliff Cavern, Derbyshire; *Figure 6.18:* Eric Condliffe, Electron Optics Laboratory, School of Earth Sciences, Leeds University; *Figure 6.20:* Strakhov, N.M. (1967) *Principles of Lithogenesis*, vol. **1** Oliver & Boyd; *Figure 6.21:* Hsu, P.H. (1989) *Aluminium oxides and oxyhydroxides*. In 'Minerals in Soil Environments' (J.B Dixon and S.B. Weed, eds.), SSSA Book Series **1**, pp.331–378, Soil Science Society of America, Madison, WI. *Figures 8.4a, 8.12, 8.14:* Wild, A. (1993) *Soils and the Environment – An Introduction*, Cambridge University Press; *Figure 8.6a:* Irvine Cushing/Oxford Scientific Films; *Figure 8.6b:* David Thompson/Oxford Scientific Films; *Figures 8.7 and 8.17:* Schlesinger, W. H. (1997) *Biogeochemistry – An Analysis of Global Change*, Academic Press Limited; *Figure 8.8:* Stevenson, F. J. (1994) *Humus Chemistry: Genesis, composition, reactions* (2nd edn), © 1994 John Wiley & Sons Ltd; *Figures 8.9, 8.18 and 8.19:* Brady, N. C. (1999) *The nature and properties of soils*, (12th edn), Prentice Hall; *Figure 8.16:* Buildings Research Establishment; *Figure 9.1:* Gerrard, J. (2000) *Fundamentals of Soils – Routledge Fundamentals of Physical Geography Series*, Routledge; *Figures 9.2a–e:* Deckers, J. A., Nachtergaele F. O. and Spaargaren O. C. (1998*) World Reference Base for Soil Resources: Introduction*, Acco, Belgium; *Figure 9.2f:* Trudgill, S. (1989) *Soil Types – A Field Identification Guide*, Field Studies Council; *Figure 9.3:* US Department of Agriculture; *Figure 10.1:* Environmental Images; *Figure 10.3:* F. Botts, Food and Agricultural Organisation, UN; *Figure 10.4:* Dermot Tatlow/Panos Pictures; *Figure 10.5:* Mark Edwards/Still Pictures.

Index

Note: Entries in **bold** are key terms. Page numbers referring to information that is given only in a figure or caption are printed in *italics*.